BARRON'S
The Leader in Test Preparation

STUDENTS' #1 CHOICE

AIMS–Math

ARIZONA'S INSTRUMENT TO MEASURE STANDARDS HIGH SCHOOL EXIT EXAM

Ed Anderson, M.A.
Chairperson, Mathematics Department
Marcos de Niza High School
Tempe, Arizona

President, Arizona Association of Teachers of Mathematics

Judy Reihard, M.A.
Mathematics Teacher
Marcos de Niza High School
Tempe, Arizona

Consultant, Arizona Department of Education

BARRON'S

All inquiries should be addressed to:
Barron's Educational Series, Inc.
250 Wireless Boulevard
Hauppauge, New York 11788
http://www.barronseduc.com

Library of Congress Control Number: 2006034726

ISBN-13: 978-0-7641-3568-2
ISBN-10: 7641-3568-6

Library of Congress Cataloging-in-Publication Data

Anderson, Ed, 1951–
    AIMS math Arizona instrument to measure standards high school exit exam / Ed Anderson, Judy
Reihard.
        p. cm.
    Includes index.
    ISBN-13: 978-0-7641-3568-2
    ISBN-10: 0-7641-3568-6
    1. Mathematics—Study and teaching (Secondary)—United States. 2. Mathematics—Study and
teaching (Secondary)—Standards—United States. 3. Mathematical ability—Testing. 4. Education—
Standards—Arizona. 5. Educational tests and measurements—Arizona. I. Reihard, Judy. II. Title.

QA13.5.A7.A64 2007
510.76—dc22

                                                                                2006034726

10%
POST-CONSUMER
WASTE
Paper contains a minimum
of 10% post-consumer
waste (PCW). Paper used
in this book was derived
from certified, sustainable
forestlands.

Contents

Introduction

Understanding mathematics and being skillful in using mathematics is very important for your future success. Mathematics is found in almost all jobs and occupations. The skills that you are developing as you use this book will be useful in your continued education and future career or occupation. The work you do in preparing for this test will not only help you achieve your goal of passing the AIMS test but also benefit you as you use mathematics throughout your future.

The Arizona Instrument to Measure Standards (AIMS) is a test that measures a student's knowledge related to the reading, writing, and mathematics standards as developed by the State of Arizona. The intent of this publication is to assist students in preparing for the mathematics portion of the AIMS test.

Each year during the spring semester the high school AIMS test is administered to students in the 10th grade. The test is administered again to juniors and seniors not meeting the Standards as required by the state for graduation in the fall and spring of each year.

Arizona reports the results on the AIMS test in four different categories: Exceeds the Standard, Meets the Standard, Approaches the Standard, and Falls Far Below the Standard. If you score between 100% and 88% on these practice tests, the likelihood of you exceeding the standard is great. A score of 87% to 60% on the practice tests in this book would equate with your meeting the standard. Approaching the standard would probably result if your score were between 59% and 48%, and a score between 47% and 0% probably means you would be in the "falls far below" category.

At the present time the Arizona universities have a scholarship program that includes exceeding the standard on the AIMS test as one of the qualifying requirements. Also, at the present time, a student must at least meet the standard in all three testing areas—writing, reading, and mathematics—to graduate from high school.

Sophomores taking the test for the first time as well as juniors and seniors who are retesting can use this book. A student could also use the book earlier than the 10th grade to begin preparing for the test. The book could be used as an introduction to high school mathematics.

The mathematics standard by Strand and Concept follow. The performance objectives covered in each strand and concept of the math standard can be found in the appendix.

2005 Arizona Mathematics Standard

Strand 1: Number Sense and Operations

    Concept 1: Number Sense

    Concept 2: Numerical Operations

    Concept 3: Estimation

Strand 2: Data Analysis, Probability, and Discrete Mathematics

    Concept 1: Data Analysis (Statistics)

    Concept 2: Probability

    Concept 3: Discrete Mathematics—Systematic Listing and Counting

Strand 3: Patterns, Algebra, and Functions

    Concept 1: Patterns

    Concept 2: Functions and Relationship

    Concept 3: Algebraic Representations

    Concept 4: Analysis of Change

Strand 4: Geometry and Measurement

    Concept 1: Geometric Properties

    Concept 2: Transformation of Shapes

    Concept 3: Coordinate Geometry

    Concept 4: Measurement

Strand 5: Structure and Logic

    Concept 1: Algorithms and Algorithmic Thinking

    Concept 2: Logic

This book begins with information about the test, followed by a diagnostic test. The diagnostic test is designed to help you find out what your strengths and weaknesses are on the material covered on the AIMS test. Take the diagnostic test as if you are in an actual testing situation. Do the best that you can without guessing. When you finish, check your answers for correctness, making note of the questions you missed. A diagnostic chart will direct you to the chapter(s) with which you seem to have difficulty.

Strands 1–5 contain instruction and explanation related to each of the strands. These sections highlight the important concepts and vocabulary in each strand. The material from each section is independent of the other sections; however, it is recommended that you master the strands in the order in which they are presented. The book contains sample problems related to each strand. You are encouraged to attempt each problem and check the accompanying solution. As in most math classes, the key to success is active participation. You need to work the problems on paper—don't just look at them and expect to have success. The manner in which you use the instruction/explanation sections will depend on how you learn best. You might want to read the instructions and then do the sample problems. Or you might want to do the sample problems first, see how you do, and then come back to the instruction.

At the end of the book, you will find two complete practice tests designed to help you prepare for the actual test. Once again, the best way to prepare for the test is to pretend that you are in an actual testing situation, taking the test as if it was the real thing. The actual AIMS test is administered in two consecutive sessions with a small break between. Work the test completely. Use the formula reference sheet to help you take the test. There are no time limits on the AIMS test, so don't worry about the amount of time you take—don't rush; take your time. The AIMS test does not permit the use of a calculator, so you are strongly urged to do all the problems in this book without one. Each of the problems in the practice tests is referenced with the instructional chapters. If you miss a problem or are not really sure about the solution, turn to the appropriate section to restudy the material. This also would be a good time to ask for assistance and/or additional practice problems from your mathematics teacher or a qualified tutor.

The format of the AIMS test is multiple-choice, four choices per question. Since you will be able to use the reference sheet on the test, it is advisable that you become familiar with it in your preparation. There is no penalty for guessing, so on the actual AIMS test you are encouraged to answer all of the questions. Eliminating some of the choices that appear to be obviously incorrect may help you determine your best answer.

The appendix contains a glossary of mathematical terms, a formula reference sheet, and a complete listing of the Arizona Math Standard, including all of the performance objectives, to assist you in working the problems. The appendix also contains the blueprint of the proportion of questions on the test covering each strand and concept. This blueprint is reviewed and published each year by the Arizona Department of Education.

Best wishes as you prepare for the math AIMS test. Your efforts preparing for this test will help you improve your score and give you a feeling of a job well done. You can do mathematics—go for it!

Diagnostic Test
Answer Sheet
Fill in the bubble completely. Erase carefully if an answer is changed.

1. A B C D
2. A B C D
3. A B C D
4. A B C D
5. A B C D
6. A B C D
7. A B C D
8. A B C D
9. A B C D
10. A B C D
11. A B C D
12. A B C D
13. A B C D
14. A B C D
15. A B C D
16. A B C D
17. A B C D

18. A B C D
19. A B C D
20. A B C D
21. A B C D
22. A B C D
23. A B C D
24. A B C D
25. A B C D
26. A B C D
27. A B C D
28. A B C D
29. A B C D
30. A B C D
31. A B C D
32. A B C D
33. A B C D
34. A B C D

35. A B C D
36. A B C D
37. A B C D
38. A B C D
39. A B C D
40. A B C D
41. A B C D
42. A B C D
43. A B C D
44. A B C D
45. A B C D
46. A B C D
47. A B C D
48. A B C D
49. A B C D
50. A B C D

Cut along dotted line.

Diagnostic Test
Answer Sheet
Fill in the bubble completely. Erase carefully if an answer is changed.

51. A B C D
52. A B C D
53. A B C D
54. A B C D
55. A B C D
56. A B C D
57. A B C D
58. A B C D
59. A B C D
60. A B C D
61. A B C D
62. A B C D
63. A B C D
64. A B C D
65. A B C D
66. A B C D
67. A B C D

68. A B C D
69. A B C D
70. A B C D
71. A B C D
72. A B C D
73. A B C D
74. A B C D
75. A B C D
76. A B C D
77. A B C D
78. A B C D
79. A B C D
80. A B C D
81. A B C D
82. A B C D
83. A B C D
84. A B C D

85. A B C D
86. A B C D
87. A B C D
88. A B C D
89. A B C D
90. A B C D
91. A B C D
92. A B C D
93. A B C D
94. A B C D
95. A B C D
96. A B C D
97. A B C D
98. A B C D
99. A B C D
100. A B C D

Diagnostic Test

Directions: Choose the best answer for each of the following.

1. The set of real numbers shown below is a subset of which of the following sets?

$$\left\{-7, -2\frac{2}{5}, 0.\overline{6}, 5, \sqrt{36}\right\}$$

A. The set of natural numbers
B. The set of rational numbers
C. The set of integers
D. The set of irrational numbers

2. What is the value of the following expression?

$$(3 \times 10^{-4})(5 \times 10^{7})$$

A. 1.5×10^{3}
B. 1.5×10^{-3}
C. 1.5×10^{4}
D. 1.5×10^{7}

3. Which of the following graphs shows a negative correlation between the two variables?

A.
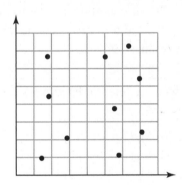

B.
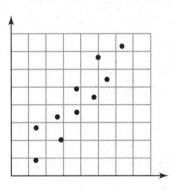

C.
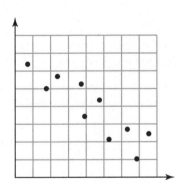

D.
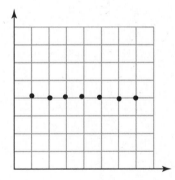

GO ON ➡

4. In the diagram below, which of the following pairs of angles are adjacent angles?

A. $\angle 1$ and $\angle 3$
B. $\angle 1$ and $\angle 4$
C. $\angle 2$ and $\angle 4$
D. $\angle 2$ and $\angle 3$

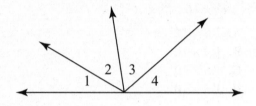

5. The 16-member a cappella choir, which contains 8 male voices and 8 female voices, is going on tour. In how many possible ways can the director assign the 8 females to 2 rooms in groups of 4 females each.

A. 1,820
B. 1,680
C. 70
D. 14

6. In how many different ways can the six letters in the word "pencil" be arranged?

A. 12
B. 36
C. 120
D. 720

7. Which of the following stories matches the graph below?

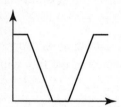

A. Mary is traveling to the store. The graph show her car's speed as it approaches an intersection with a stop sign and then resumes her travel to the store.
B. The graph shows the distance traveled from Bob's house to the store.
C. The graph shows the speed of Jennifer going down a hill and up the hill on her bicycle.
D. The air conditioner is set to turn on at 76°, and turn off when the temperature reaches 79°.

8. Which are three ordered pairs for the equation $y = 5 - x$?

A. (0, 5), (1, 6), (2, 7)
B. (0, 5), (2, 3), (−1, 6)
C. (1, 6), (2, 3), (8, 3)
D. (2, 7), (0, 5), (11, 6)

9. Which of the equations below can be written using this table of values?

X	Y
$\frac{2}{5}$	2
2	10
7	35
$\frac{3}{10}$	$\frac{3}{2}$

A. $y = x + 8$
B. $y = x - 8$
C. $y = \dfrac{x}{5}$
D. $y = 5x$

GO ON ➡

10. Which is the solution to the following inequality?

$$-5x + 3 < -7$$

A. $x < 2$
B. $x > 2$
C. $x > 0$
D. $x < 1$

---

11. What is the value of c?

$$\begin{bmatrix} 4 & -2 \\ -1 & 7 \end{bmatrix} + \begin{bmatrix} 1 & 0 \\ 5 & 3 \end{bmatrix} = \begin{bmatrix} a & b \\ c & d \end{bmatrix}$$

A. 5
B. 4
C. 0
D. -2

---

12. Which equation can be used to find the value of x in the right triangle shown?

A. $\cos 70° = \dfrac{5}{x}$

B. $\cos 70° = \dfrac{x}{5}$

C. $\sin 70° = \dfrac{5}{x}$

D. $\sin 70° = \dfrac{x}{5}$

---

13. Which of the following quadrilaterals does not have two pairs of parallel sides?

A. Parallelogram
B. Trapezoid
C. Rhombus
D. Rectangle

---

14. Ski lift cables are strung to the top of a 1,500-foot mountain as shown in the diagram. The angle of elevation of the cables is 30°. How long are the cables?

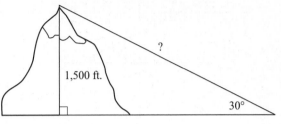

A. $1,500\sqrt{3} \approx 2,598$ ft

B. 3,000 ft

C. $1,500\sqrt{2} \approx 2,121$ ft

D. 1,500 ft

---

15. The length of $\overline{AB}$ is $\sqrt{13}$. What is the length of the image of $\overline{AB}$ when it is rotated about the origin 180°?

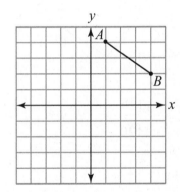

A. 4
B. 3
C. $\sqrt{13}$
D. $\sqrt{15}$

GO ON ➡

16. Which ordered pair is the solution to the system of equations represented by line *a* and line *b*?

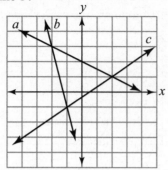

A. (−1, −1)
B. (0, 2)
C. (−2, 3)
D. (2, 1)

---

17. Given the right triangular prism whose bases are equilateral triangles as shown below, what is the total surface area of the triangular prism?

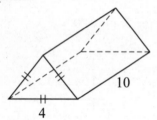

A. $8\sqrt{3} + 120$
B. $4\sqrt{3} + 120$
C. 120
D. $40\sqrt{3}$

---

18. What is the length of $\overset{\frown}{CD}$ if $\overline{CQ} \perp \overline{QD}$ and $CQ = 8$?

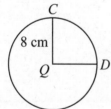

A. 2π
B. 4π
C. 16π
D. 64π

19. Which of the following procedures is a valid procedure for solving the inequality below?

$$-8x + 2(3x+4) \geq 10$$

A. $-8x + 2(3x+4) \geq 10$
$$-8x + 6x + 8 \geq 10$$
$$-2x + 8 \geq 10$$
$$-2x \geq 2$$
$$x \leq -1$$

B. $-8x + 2(3x+4) \geq 10$
$$-6x + 3x + 8 \geq 10$$
$$-3x - 8 \geq 10$$
$$-3x \geq 18$$
$$x \leq -6$$

C. $-8x + 2(3x+4) \geq 10$
$$-8x + 6x + 8 \geq 10$$
$$-2x + 8 \geq 10$$
$$-2x \geq 2$$
$$x \geq -1$$

D. $-8x + 2(3x+4) \geq 10$
$$-8x + 6x - 8 \geq 10$$
$$-2x - 8 \geq 10$$
$$-2x \geq 18$$
$$x \leq -9$$

---

20. What is the inverse of the following conditional statement?

If all the sides of a polygon are congruent, then the polygon is equilateral.

A. If all the sides of a polygon are not congruent, then the polygon is not equilateral.
B. If all the sides of a polygon are congruent, then the polygon is not equilateral.
C. If a polygon is not equilateral, then all of its sides are not congruent.
D. If a polygon is equilateral, then all of its sides are congruent.

GO ON ➡

21. Which of the following addition properties justifies the statement below?

$$3 + 0 = 3$$

- **A.** Commutative
- **B.** Identity
- **C.** Inverse
- **D.** Closure

22. Which of the values below is closest to the value of the expression below?

$$\sqrt{31}$$

- **A.** 3.1
- **B.** 3.9
- **C.** 4.4
- **D.** 5.6

23. In a drawer there are 10 forks, 15 spoons, and 12 knives. What is the probability of getting a spoon on the first random pick?

- **A.** $\dfrac{27}{37}$
- **B.** $\dfrac{10}{37}$
- **C.** $\dfrac{15}{37}$
- **D.** $\dfrac{12}{37}$

24. A mouse runs through a house. There are two paths by which it can go from the playroom to the living room, two paths by which it can go from the living room to the kitchen, and three holes by which it can leave the kitchen to get outside. By how many different paths might it make its way from the playroom to the outside?

- **A.** 7
- **B.** 12
- **C.** 72
- **D.** 144

25. If the following sequence continues, what is the eighth term in the sequence?

$$5, 6, 9, 14, 21, 30, \ldots$$

- **A.** 31
- **B.** 39
- **C.** 51
- **D.** 54

26. Which of the following represents a relation that is a function?

A. 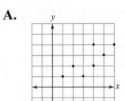 **B.**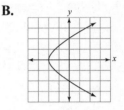

- **C.** {(3, 2), (2, 3), (0, 2), (4, 5)}
- **D.** {(3, 2), (2, 3), (2, 0), (4, 5)}

27. Evaluate the expression $-12(2x - 1) - x^2$ if $x = -3$.

- **A.** 75
- **B.** 38
- **C.** –45
- **D.** –57

28. Which number phrase below is represented by the variable expression below?

$$3(x - 2)$$

- **A.** Three less than two times a number x
- **B.** Three times the difference of a number x and two
- **C.** The product of three and two more than a number x
- **D.** Twice the sum of a number x and three

GO ON ➡

29. Given the proportion below, which of the following equations could be used to complete the solution to find x?

$$\frac{x}{6} = \frac{9}{20}$$

A. $20x = 54$
B. $6x = 180$
C. $9x = 120$
D. $54 = \dfrac{x}{20}$

30. Given $x \geq 0$ and $y \geq 0$, what is the value of the expression below?

$$\sqrt{36x^2} \cdot \sqrt{16y^{16}}$$

A. $24xy^4$
B. $48xy^4$
C. $24xy^8$
D. $48xy^4$

31. What are the values of x that make the following equation true?

$$x^2 - x - 20 = 0$$

A. $x = -6$ or $x = 4$
B. $x = -12$ or $x = 2$
C. $x = -2$ or $x = -8$
D. $x = 5$ or $x = -4$

32. Which of these is equivalent to the equation below?

$$I = PRT$$

A. $P = \dfrac{IT}{R}$
B. $T = IPR$
C. $R = \dfrac{I}{PT}$
D. $I = \dfrac{P}{RT}$

33. In the diagram below, $\overline{AD} \cong \overline{CB}$ and $\overline{DC} \cong \overline{BA}$. Which of the methods below would you use to most directly prove the triangles congruent?

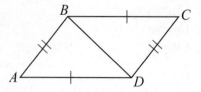

A. AAS
B. SSS
C. SAS
D. ASA

34. Which of the following equations is represented by the graph shown?

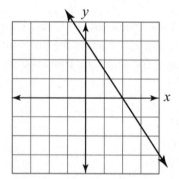

A. $y = \dfrac{2}{3}x$
B. $y = \dfrac{3}{2}x + 3$
C. $y = \dfrac{-3}{2}x + 2$
D. $y = \dfrac{-3}{2}x + 3$

35. What is the midpoint of the line segment joining points $(-3, -5)$ and $(5, -1)$?

A. $(1, -3)$
B. $(-8, -3)$
C. $(-1, -3)$
D. $(4, -2)$

GO ON ➡

36. What is the volume of the cube with an edge of 4 in., as shown below?

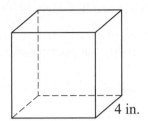

4 in.

A. 24 in.³
B. 64 in.³
C. 96 in.³
D. 256 in.³

37. Given: Regular square pyramid *ABCD*, with slant height $\overline{GH}$ and altitude (height) $\overline{GF}$ drawn. *AB* = 6 cm and the volume of the pyramid is 144 cm³. What is the length of altitude (height) $\overline{GF}$?

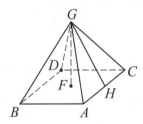

A. 18 cm
B. 12 cm
C. 6 cm
D. 4 cm

38. Which of the following could be the next line in the process of solving this equation for *x*?

$$4(x + 2) = 4 - (x + 1)$$

A. $4x + 8 = 4 - x + 1$
B. $4x + 2 = 4 - x - 1$
C. $4x + 8 = 4 - x - 1$
D. $4x + 6 = 4 - x + 1$

39. Which of the following is the value of this expression?

$$7 + 3^2 \cdot 2$$

A. 200
B. 126
C. 32
D. 25

40. Mary and her friends ate in a restaurant, and the bill for their dinner was $36.96 (including the tax). They want to give their waiter a tip that is 20% of the total. Which would be closest to that 20% tip?

A. $3.75
B. $4.50
C. $5.50
D. $7.50

41. In which of the following situations would it be more appropriate to use a census method to count as opposed to a sampling method?

A. How many people in the Phoenix metropolitan area listen to the KNIX radio station at least 3 hours during a Friday night to Sunday night time span?
B. How many trout fish are contained in Woods Canyon Lake?
C. How many swimming pools are in the state of Arizona?
D. How many books are in your library that were written by Charles Dickens?

GO ON ➡

42. Five distinct points are shown on the number line below.

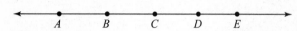

Which expression will determine how many distinct line segments have two of these points as endpoints?

A. $\dfrac{5!}{3!2!}$

B. $\dfrac{5!}{3!}$

C. $\dfrac{2!}{5!}$

D. $\dfrac{5!}{2!}$

43. What are the first four terms in the sequence that is defined below?

$$a_1 = 1$$

$$a_{n+1} = \left(a_n\right)^2 + 3$$

A. 1, 4, 19, 364
B. 1, 4, 8, 16
C. 1, 4, 16, 25
D. 1, 4, 7, 10

44. What is the domain of the function and graph shown?

$$y = \sqrt{x+1}$$

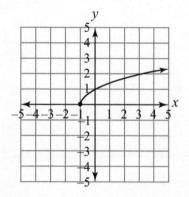

A. $x \geq 1$
B. $y \geq 0$
C. $x \geq 2$
D. $x \geq -1$

45. Which best describes the relationship between these lines?

$$y = 2x - 1 \text{ and } 3y = 6x - 3$$

A. Coincident
B. Intersecting but not perpendicular
C. Parallel
D. Perpendicular

46. Charlie is going to participate in a bowling league. He pays $25 when he signs up for the league. He pays $12 for each night he bowls in the league. Which of the following would represent his total cost (C) for bowling n nights?

A. $C = 25n + 12$
B. $C = 12n - 25$
C. $C = 12n + 25$
D. $C = 25n - 12$

47. What is the equation of the line passing through $(5, -3)$ and $(0, 4)$?

A. $y = -\dfrac{5}{7}x + 4$

B. $y = \dfrac{7}{5}x + 4$

C. $y = \dfrac{x}{5} + 4$

D. $y = -\dfrac{7}{5}x + 4$

48. What is the solution to the following equation?

$$\sqrt{4x} + 5 = 9$$

A. 49
B. 4
C. 1
D. $\dfrac{1}{4}$

GO ON ➡

49. Which of the following is a net of a square pyramid?

A.

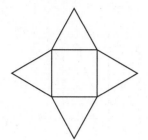

B.

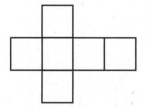

C.

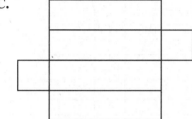

D.

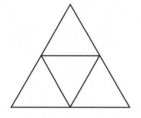

50. Which similarity theorem can be used to prove $\triangle BAC \sim \triangle EAD$?

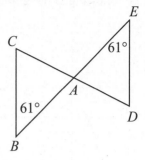

A. SSS
B. AA
C. SAS
D. AAS

Now would be a good time for you to take a break. When you actually take the AIMS test at your school, you will be given a break at this time.

Directions: Choose the best answer for each of the following.

51. Which of the following is the graph of $y = x^2 - 1$?

A.

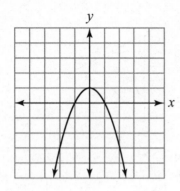

B.

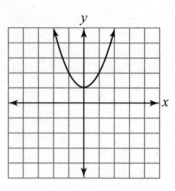

C.

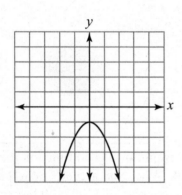

D.

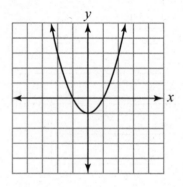

52. The graph of $y = 2x + 1$ is shown below.

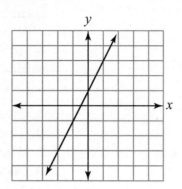

Which of the following would be the graph if the 2 were changed to a −3?

A.

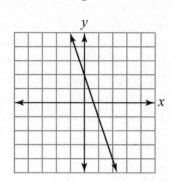

B.

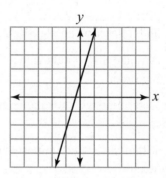

C.

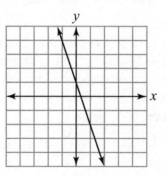

D.

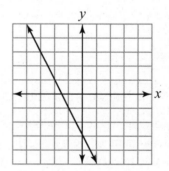

GO ON ➡

53. What is the value of *x* in the diagram?

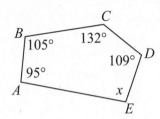

A. 60°
B. 99°
C. 180°
D. 279°

54. Steps 1 and 2 describe an algorithm.

Step 1: Arrange the 19 data values in descending order.

Step 2: Determine which is the tenth value in the list.

The algorithm above determines which of the following?

A. Mean
B. Median
C. Mode
D. Range

55. Which of the following is a finite set?

A. {three-digit whole numbers}
B. {rational numbers less than 5}
C. {positive integers}
D. {irrational numbers greater than 0}

56. What is the value of the expression below?

$$6+\left|-10+8\right|+\left|6-11\right|$$

A. −1
B. 13
C. 29
D. 41

57. Michael charges $11.75 each time he mows his neighbor's lawn. Which of the following is the most reasonable estimate of the number of times Michael will have to mow his neighbor's lawn in order to earn $240.00 to buy a tux for the prom?

A. 2 times
B. 10 times
C. 15 times
D. 20 times

58. The drama class is looking for suggestions for the spring musical. They are going to conduct a survey of the people who might participate in the musical to help them decide which of five possible musicals they will choose. Which of the following is the best group to survey?

A. The school secretaries
B. The school choir
C. The school soccer team
D. The school math club

59. In order to compare the expected probability of getting a 6 when rolling a die to the experimental probability, Josie rolled a die 18 times. The results of the experiment are in the table below.

Roll #	1	2	3	4	5	6	7	8	9	10	11	12	13	14	15	16	17	18
The number that came up on the die	2	2	6	3	5	4	1	4	3	6	6	5	2	1	6	1	2	4

How do the results of her experiment compare with the expected probability of rolling a 6 with a die?

A. Her results match the expected probability exactly.
B. Her results represent a probability less than the expected probability.
C. Her results represent a probability greater than the expected probability.
D. It is impossible to compare her results with the expected probability.

GO ON ➡

60. Study the diagram below.

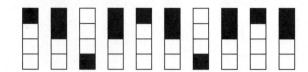

If the above pattern continues, which of the following would be next in the pattern?

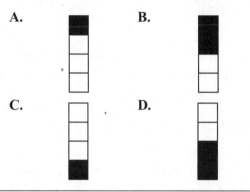

A.   **B.**

C.   **D.**

61. A bottle rocket shot off from ground level travels in a parabolic arc. A representation of the path of that arc has been graphed on the coordinate plane below. Notice the rocket leaves the ground at the point (0, 0).

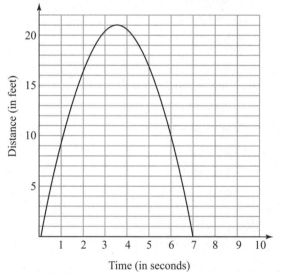

What is the maximum height that the rocket will reach?

A. 0 feet

B. 3.5 feet

C. 7 feet

D. 21 feet

62. Which pair of expressions is **not** equivalent?

A. $4 \cdot 3^2$ and $60 - 24$

B. $3x + 4y$ and $3y$ when $x = 2$ and $y = 3$

C. $4(a^2 - 1)$ and $4(a - 1)$ when $a = 0$

D. $5x^3$ and $3x + 2$ when $x = 1$

63. A campsite charges $14 a day and a one time $8 fee for parking. The table below shows the charges for campsite usage and parking per day.

# of days in campsite	1	2	3	4	5	6
Cost for camping	$22	$36	$50	$64	$78	$92

Which of the following equations represents the relationship between the cost (*C*) for camping and the number of days camping (*n*)?

A. $C = 14n + 8$

B. $C = 8n + 14$

C. $C = 14n$

D. $C = 8n$

64. What is the value of x in the solution for the system of equations below?

$$2x - 3y = -1$$
$$3x + y = -7$$

A. -1

B. 2

C. -2

D. 1

65. What is the slope of the line containing the points (6, –1) and (–3, 5)?

A. -2

B. $\dfrac{4}{9}$

C. $-\dfrac{2}{3}$

D. $-\dfrac{9}{4}$

GO ON ➡

66. Given $\overrightarrow{AB}$ and $\overrightarrow{CD}$ intersecting at P with the angles as shown. What is the value of x?

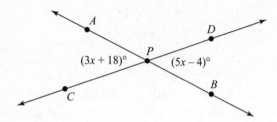

A. 11

B. 7

C. $\dfrac{11}{4}$

D. $\dfrac{7}{4}$

67. Jamie is constructing a garden that is in the shape of an isosceles triangle as shown below. She knows the measure of the angle at vertex A, but does not know the measure of the angles at vertices B and C.

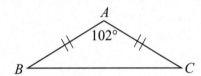

Which of the following sets of angle measures would you choose for $\angle B$ and $\angle C$?

A. $\angle B = 78°$ and $\angle C = 78°$

B. $\angle B = 30°$ and $\angle C = 60°$

C. $\angle B = 39°$ and $\angle C = 39°$

D. $\angle B = 45°$ and $\angle C = 45°$

68. What is the image of $A(3, -1)$ if it is moved 3 units to the right and 2 units down?

A. $A'(3, -2)$

B. $A'(5, -4)$

C. $A'(6, 1)$

D. $A'(6, -3)$

69. What is the distance between the points $(-5, 3)$ and $(-2, 9)$?

A. $\sqrt{40} = 2\sqrt{10}$

B. $\sqrt{42}$

C. $\sqrt{45} = 3\sqrt{5}$

D. $\sqrt{34}$

70. Which of the following procedures correctly simplifies the numerical expression below?

$$7(4^2 - 5) + 2(7 + 4)$$

A. $7(4^2 - 5) + 2(7 + 4)$
   $7(-1)^2 + 2(11)$
   $7(1) + 2(11)$
   $7 + 22$
   29

B. $7(4^2 - 5) + 2(7 + 4)$
   $7(8 - 5) + 2(11)$
   $7(3) + 2(11)$
   $21 + 2(11)$
   $21 + 22$
   43

C. $7(4^2 - 5) + 2(7 + 4)$
   $7(16 - 5) + 2(11)$
   $7(11) + 2(11)$
   $77 + 2(11)$
   $79(11)$
   869

D. $7(4^2 - 5) + 2(7 + 4)$
   $7(16 - 5) + 2(11)$
   $7(11) + 2(11)$
   $77 + 22$
   99

GO ON ➡

71. Which of the following sets is an infinite set?

    A. {the natural numbers}
    B. {−8, −9, −10, −11}
    C. {1, 2, 3, . . . , 14, 15, 16}
    D. {whole numbers between 0 and 10}

72. The marketing team at Sports USA is creating a design for the label on their new sports drink. They want the label of the can to appeal to athletes. Which of the following questions would **not** be appropriate for gathering information to help them choose the most appealing label?

    A. What do you think the name of the new Sports USA drink should be?
    B. What should the logo be on the new Sports USA drink?
    C. Should the new Sports USA drink highlight a particular sport?
    D. By what method should the new Sports USA drink be shipped to the retail outlets?

73. What is the probability that a randomly selected point lies in region B?

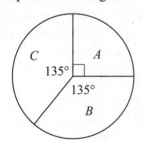

    A. $\dfrac{1}{6}$

    B. $\dfrac{1}{3}$

    C. $\dfrac{3}{8}$

    D. $\dfrac{2}{3}$

74. The spinner shown is spun once. What is the probability it will land on a prime number?

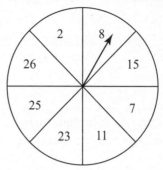

    A. $\dfrac{3}{8}$

    B. $\dfrac{1}{2}$

    C. $\dfrac{1}{4}$

    D. $\dfrac{5}{8}$

75. If Tim's grandfather gives him one dollar on Sunday and tells him he will double the gift each day for a week, what amount of money will his grandfather give him on Saturday?

    A. $2.00
    B. $32.00
    C. $64.00
    D. $128.00

76. Which of the following is equivalent to this expression?

    $(4a^2b^2)(5a^2b^4)$

    A. $20a^2b^3$
    B. $20a^3b^5$
    C. $20a^4b^6$
    D. $20a^4b^8$

77. For which equation is 4 a solution?

    A. $\dfrac{x}{3} = 12$
    B. $3x + 1 = 4x$
    C. $4(x + 2) = 8$
    D. $8x = 7x + 4$

GO ON ➡

78. Given $y \geq 0$, which of the following expressions is equivalent to $\sqrt{12} \cdot \sqrt{3y^2}$?

A. $36y^2$
B. $6y$
C. $6y^2$
D. $36y$

---

79. $\triangle XYZ$ is an isosceles right triangle. Which of the statements below is **not** always true for $\triangle XYZ$?

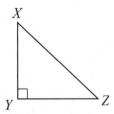

A. The sum of the angles is 180°.
B. The triangle has three equal sides.
C. The triangle has two equal angles.
D. The sum of the squares of the legs of the triangle is equal to the square of the hypotenuse.

---

80. In circle C, AB is tangent at B, and AD is tangent at D. What is the length of $\overline{AB}$?

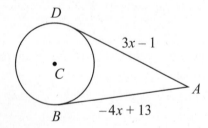

A. 2
B. 5
C. 12
D. 61

---

81. Which inequality has the solution shown in the graph?

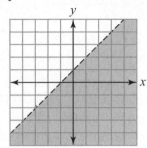

A. $y < x + 1$
B. $y > x + 1$
C. $y < -x + 1$
D. $y > -x + 1$

---

82. Using the similar quadrilaterals $ABCD$ and $WXYZ$, what is the value of y?

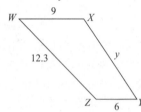

A. 1.5
B. 8
C. 15
D. 12

---

83. The Excel High School basketball team scored a total of 88 points in a game. Carlos, a member of the team, scored 24 of the 88 points. Which of the following operations would find what percent of the total Carlos scored?

A. $(24 \div 88) \cdot 100$
B. $(88 \div 24) \cdot 100$
C. $(88 \cdot 100) \div 24$
D. $(24 \cdot 88) \div 100$

---

84. Which of the following could be a row in a stem-and-leaf plot for the set of data below?

24, 28, 47, 48, 48, 52, 54, 58

A. $5 \mid 24$
B. $4 \mid 788$
C. $4 \mid 78$
D. $2 \mid 8$

GO ON ➡

85. The chart below shows a portion of a federal income tax table.

If taxable income is:		And you are:			
At least	But less than	Single	Married filing jointly	Married filing separately	Head of household
$16,800	$16,850	$2,551	$2,311	$2,810	$2,313
$16,850	$16,900	$2,601	$2,409	$2,865	$2,447
$16,900	$16,950	$2,635	$2,451	$2,911	$2,458
$16,950	$17,000	$2,657	$2,516	$2,947	$2,575

What is your tax if you are single and have a taxable income of $16,892?

A. $2,601
B. $2,635
C. $2,657
D. $2,810

86. The graph below shows the changes in a bank balance over time. If the bank balance is zero at A, describe what happens during the time period from D to E.

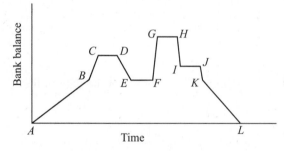

A. Withdrawals were made from the account during that time.
B. Deposits were made to the account during that time.
C. No deposits or withdrawals were made during that time.
D. Interest was added to the account.

87. What is the y-intercept of the equation below?

$$3y - 2x = 12$$

A. $\left(0, \dfrac{1}{4}\right)$
B. $(0, -6)$
C. $(0, 4)$
D. $\left(0, \dfrac{1}{6}\right)$

88. What is the measure of $\overset{\frown}{ABC}$?

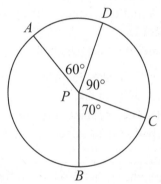

A. 90°
B. 110°
C. 140°
D. 210°

89. Given the point $A(-3, 2)$, which of the following is the point A'' if you first reflect A over the y-axis and then reflect A' over the x-axis?

A. $(-3, -2)$
B. $(3, -2)$
C. $(3, 2)$
D. $(-2, 3)$

GO ON ➡

90. In the figure, all the angles are right angles. Find the area.

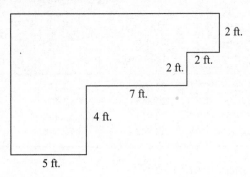

A. 1,120 sq. ft.
B. 112 sq. ft.
C. 72 sq. ft.
D. 44 sq. ft.

91. Assuming the following statements are true, which of the following conclusions can you reach?

If you do not take Algebra, then you will not take Geometry.
Kelly does not take Algebra.

A. Kelly will take Geometry.
B. Kelly will not take Geometry.
C. Kelly does not take Algebra.
D. Kelly does take Algebra.

92. Mary purchased her prom dress for $154.49. She gave the clerk two $100 bills. Which of the following represents the change she would receive using the least number of bills and coins?

A. 1 penny, 2 quarters, 1 $5 bill, 2 $20 bills
B. 1 penny, 2 quarters, 1 $10 bill, 2 $20 bills
C. 1 penny, 1 dime, 2 quarters, 1 $5 bill, 2 $20 bills
D. 1 penny, 1 nickel, 1 dime, 1 quarter, 1 $10 bill, 2 $20 bills

93. The National Basketball Association has prepared a box-and-whisker plot that displays the height in inches of all of the current professional basketball players. Which of the following is a measure of central tendency that would most easily be determined using this graph?

A. Median
B. Mode
C. Mean
D. Range

94. Given the box-and-whisker plot below from Ms. Johnson's 9th grade classes.

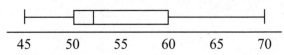

About what percent of her 9th grade students are less than 60 inches tall?

A. 25%
B. 50%
C. 75%
D. 100%

95. Which equation represents the data in the table?

x	y
2	0
0	4
−2	8

A. $x + y = 5$
B. $x - y = 5$
C. $y = 3 - 2x$
D. $y = 4 - 2x$

GO ON ➡

96. Which of the following numbers would be a counterexample for the following statement?

If $a < b$, then $a^2 < b^2$.

A. $a = 3$ and $b = 7$
B. $a = \sqrt{9}$ and $b = \sqrt{13}$
C. $a = -7$ and $b = -3$
D. $a = 1.2$ and $b = 3.5$

---

97. The sequence below uses the rule, $A_n = |n - 2|$, beginning with $a_1 = 1$.

$\{1,0,1,2,3, \ldots\}$

If the sequence continues and $A_n = 13$, what is the value of n?

A. $n = 12$
B. $n = 13$
C. $n = 14$
D. $n = 15$

---

98. Using the same relationship between x and y as is displayed in the table below, what is the value of y when x is 8?

x	y
1	3
2	5
3	7
4	9
5	11

A. 13
B. 15
C. 17
D. 19

99. For which of the following scatterplots would it be appropriate to draw a line of best fit?

A.

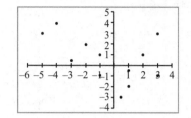

B.

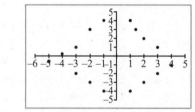

C.

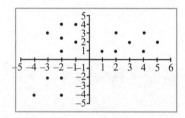

D.

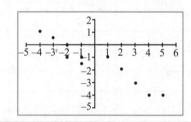

100. Which of the following experiments describes an independent event?

A. There are 20 socks in your drawer, and 12 of them are white. You grab a sock without looking. Then you grab a second sock without putting the first one back.

B. You keep your math, English, science, and world history homework assignments in four folders. You randomly choose one folder, finish your assignment in that class, and then choose a new folder.

C. From a standard deck of cards, Jamie draws a club, sets it aside, and then draws a diamond.

D. Monica tosses a fair coin two consecutive times, and it lands on heads both times.

Solutions: Diagnostic Test

Answer Key

1.	B	14.	B	27.	A	40.	D	53.	B	65.	C	77.	D	89.	B
2.	C	15.	C	28.	B	41.	D	54.	B	66.	A	78.	B	90.	C
3.	C	16.	C	29.	A	42.	A	55.	A	67.	C	79.	B	91.	B
4.	D	17.	A	30.	C	43.	A	56.	B	68.	D	80.	B	92.	A
5.	C	18.	B	31.	D	44.	D	57.	D	69.	C	81.	A	93.	A
6.	D	19.	A	32.	C	45.	A	58.	B	70.	D	82.	D	94.	C
7.	A	20.	A	33.	B	46.	C	59.	C	71.	A	83.	A	95.	D
8.	B	21.	B	34.	D	47.	D	60.	C	72.	D	84.	B	96.	C
9.	D	22.	D	35.	A	48.	B	61.	D	73.	C	85.	A	97.	D
10.	B	23.	C	36.	B	49.	A	62.	B	74.	B	86.	A	98.	C
11.	B	24.	B	37.	B	50.	B	63.	A	75.	C	87.	C	99.	D
12.	A	25.	D	38.	C	51.	D	64.	C	76.	C	88.	D	100.	D
13.	B	26.	C	39.	D	52.	C								

Diagnostic Chart

Question #	Correct	Incorrect	Strand to study	Page number(s) to review	Question #	Correct	Incorrect	Strand to study	Page number(s) to review
1			1	39	23			2	86
2			1	50	24			2	91
3			2	77	25			3	108
4			1	371	26			3	113
5			2	93	27			3	127
6			2	92	28			1	46
7			3	124	29			3	138
8			3	133	30			3	132
9			3	133	31			3	143
10			3	137	32			3	142
11			3	150	33			4	197
12			3	147	34			4	218
13			4	184	35			4	222
14			4	178	36			4	212
15			4	232	37			4	214
16			4	221	38			5	263
17			4	212	39			1	49
18			4	209	40			1	55
19			5	263	41			2	75
20			5	254	42			2	93
21			1	41	43			3	108
22			1	56	44			3	112

Diagnostic Chart

Question #	Correct	Incorrect	Strand to study	Page number(s) to review
45			3	123
46			3	134
47			3	135
48			3	143
49			4	187
50			4	197
51			4	224
52			3	116
53			4	195
54			5 (2)	265 (72)
55			1	44
56			1	48
57			1	55
58			2	64
59			2	79
60			3	108
61			3	125
62			3	127
63			3	134
64			3	139
65			3	117
66			4	190
67			4	175
68			4	231
69			4	222
70			5 (1)	263 (49)
71			1	44
72			2	64

Question #	Correct	Incorrect	Strand to study	Page number(s) to review
73			2	89
74			2	83
75			3	108
76			3	129
77			3	137
78			3	132
79			4	174
80			4	204
81			4	219
82			4	200
83			1	46
84			2	65
85			2	66
86			3	124
87			3	120
88			4	209
89			4	232
90			4	210
91			5	252
92			1	45
93			2	69
94			2	69
95			3	116
96			5	258
97			3	108
98			3	116
99			2	76
100			2	87

# You Got Correct	
Grading Scale for AIMS	
Exceeds	100–88
Meets	87–60
Approaches	59–48
Falls Far Below	47–0

Your Classification

Answers Explained

1. **B** Choice B is correct because all of the numbers in the set are rational numbers. Choice A is incorrect because 5 is the only number that is a natural number; choice C is incorrect because the set contains fractions and decimal numbers, which are not integers; choice D is incorrect because there are no irrational numbers in the set.

2. **C** Choice C is correct because to find the product of two scientific numbers you multiply the two numbers that are between 1 and 10, and you multiply the powers of ten; $3 \times 5 \times 10^{-4} \times 10^7 = 15 \times 10^3$. Written in scientific notation, this is 1.5×10^4, which is answer C. A, B, D are incorrect because the exponents of ten are wrong.

3. **C** Choice C is the correct choice because as you read the graph from left to right the flow of the points is going down which indicates a negative correlation. Choice A is incorrect because the points are all over the graph. Choice B is incorrect because it indicates a positive correlation. Choice D is incorrect because it indicates a constant correlation; y does not increase or decrease as x increases.

4. **D** Choice D is correct because adjacent angles are angles that share a common side (the sides of angles are rays). The only pair of angles that share a common side is $\angle 2$ and $\angle 3$. Choices A, B, and C are angles that do not share a common side.

5. **C** The assignment of the eight females to the rooms in groups of four is a combination problem. The order in which the females are assigned in each room does not matter. You are finding the number of combinations of 8 things taken 4 at a time. Choice C is the correct answer because the rule for finding the number of groups (combinations) of 8 things taken 4 at a time has been followed:

$$_8C_4 = \frac{8!}{(8-4)! \cdot 4!} = \frac{8!}{4! \cdot 4!} = \frac{8 \cdot 7 \cdot \cancel{6} \cdot 5 \cdot \cancel{4 \cdot 3 \cdot 2 \cdot 1}}{\cancel{4} \cdot \cancel{3} \cdot 2 \cdot 1 \cdot \cancel{4 \cdot 3 \cdot 2 \cdot 1}} = \frac{280}{4} = 70$$

Choices A, B, and D either do not apply the combinations formula at all, or they apply it incorrectly.

6. **D** This is a permutation problem because you are asked to find the number of different arrangements of the letters—this means the number of different orders in which all six of the letters can be written. Choice D is correct because the rule for permutations,

$$_nP_r = \frac{n!}{(n-r)!}$$

has been applied correctly:

$$_6P_6 = \frac{6!}{(6-6)!} = \frac{6!}{0!} = 6! = 6 \cdot 5 \cdot 4 \cdot 3 \cdot 2 \cdot 1 = 720$$

This problem contains a special case of the factorial notation, that is 0!; 0! = 1. Another way to work this problem is by using the counting principle. Think of the six positions of the letters of the word "pencil" as slots to fill: __•__•__•__•__•__. You have six choices for the first slot, five for the next slot, then four, three, two, and for the last slot, only one: $6 \cdot 5 \cdot 4 \cdot 3 \cdot 2 \cdot 1 = 720$. Choices A, B, and C either do not use the permutation rule or do not use it correctly.

7. **A** Choice A is correct because, as Mary slows down, her rate of travel will decrease, she will not be moving at all while she is stopped at the stop sign, then her rate of travel will increase, and then level off at a constant speed. B is incorrect because Bob's distance from home will increase, not decrease, as he travels to the store. C is incorrect because, as Jennifer goes down the hill, her speed will increase, not decrease. As she goes up the hill, her speed will probably decrease, not increase. D is incorrect—the graph is actually upside down for choice D.

8. **B** A good way to test for the correct answer in this question is to substitute the values of x and y in the ordered pairs into the equation to see if they satisfy the equation. The first ordered pairs in choices A and B would satisfy the equation. You know right away that choices C and D are incorrect because the first ordered pairs do not satisfy the equation. By testing the second ordered pairs in choices A and B you find that the second ordered pair in A does not satisfy the equation. Choice B is the correct answer because all of the ordered pairs in choice B satisfy the equation.

9. **D** The first and third ordered pairs, $\left(\dfrac{2}{5}, 2\right)$ and (7, 35), will both satisfy choice D. The second ordered pair, (2, 10), will satisfy equations in choices A and D. The fourth ordered pair satisfies the equation in choice D. The only equation that satisfies all four of the points is in choice D. The slope of the line defined by the second and third points, (2, 10) and (7, 35), is 5. That is the slope of the equation in choice D. Choice D is the correct answer.

10. **B** Solving an inequality is very much like solving an equation. The first step is shown below:

$$-5x + 3 < -7$$
$$-5x < -10$$

However, the next steps requires you to divide both sides by negative five. Dividing (or multiplying) both sides of an inequality by a negative number reverses the sense of the inequality. Therefore,

$$-5x + 3 < -7$$
$$-5x < -10$$
$$x > 2$$

Therefore, choice B is the correct choice.

11. **B** To add two matrices you add the elements that are in the same row and column. We want to find the value of c, which is in the second row, first column. The numbers in the matrices you are adding that are also in the second row, first column are negative one and five. Therefore, the problem is to add –1 + 5, which is 4. Therefore choice B is the correct choice.

12. **A** To solve for x, you must use one of the trigonometric functions. The definitions of the functions can be found on the reference sheet. In the triangle, x is the hypotenuse of the right triangle and 5 is the leg adjacent to the 70° angle. The function that contains these terms is the cosine. Therefore, choices C and D would be incorrect. The cosine is defined as the ratio of the leg adjacent to the angle over the hypotenuse. Choice B is incorrect because the ratio is backwards. Choice A is the correct choice.

13. **B** Choice B is the correct answer because a trapezoid has exactly one pair of parallel sides. Each of the other quadrilaterals are parallelograms, and parallelograms have two pair of parallel sides.

14. **B** The triangle shown in the problem is a 30-60-90 triangle. The relationship between the sides of this triangle and the angles is given on the reference sheet in picture format.

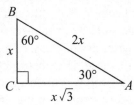

The height of the mountain, 1500 ft, is x. You are asked to find the length of the cable, which is $2x$ or 3000 ft. The correct choice is B. Be sure you take advantage of the reference sheet provided. It can be a great asset.

15. **C** Rotation does not change the length of an image. The length of $\overline{AB}$ and $\overline{A'B'}$ will be exactly the same. Choice C is the correct choice. Choices A, B, and D are incorrect because the lengths are different than the length of $\overline{AB}$.

16. **C** To solve a system of equations using a graph requires that you determine the ordered pair of the point where the lines intersect. Lines a and b intersect at the point $(-2, 3)$. Choice C is the correct choice. Choices A, B, and D are not the correct ordered pairs of the intersection of the lines.

17. **A** The total surface area of any prism can be found by finding the sum of the two bases and the lateral surface area. The bases of the prism are equilateral triangles. To find the area of any triangle you use the formula from the reference sheet, $A = \frac{1}{2}bh$. The base of the triangle is 4. Using the 30°-60°-90° triangle relationship, the height of the triangle is $2\sqrt{3}$. The area of the triangle can be found by substituting the values for b and h into the formula.

$$A = \frac{1}{2} \cdot 4 \cdot 2\sqrt{3}$$
$$= 4\sqrt{3}$$

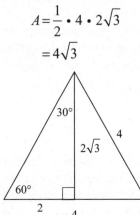

The area of two bases will be $8\sqrt{3}$.

The lateral area can be found by looking at the lateral surface as three rectangles whose length and width are 10 by 4. The area of one rectangle is 40 square units. The three rectangles would have a total area of 120 square units.

Adding the two bases to the lateral area results in a total surface area of $8\sqrt{3} + 120$. The correct choice is A. Choice B is incorrect because the answer does not have the area of the base twice. Choice C is incorrect because it only includes the lateral area. Choice D is actually the volume of the prism.

18. **B** Arc length can be found by using the formula on the reference sheet,

$$\text{Length of arc} = \frac{\text{m of arc}}{360°}(2\pi r). \text{ If } \overline{CQ} \perp \overline{QD}, \; m\overset{\frown}{CD} = 90°.$$

$$\text{Length of } \overset{\frown}{CD} = \frac{90°}{360°}(2\pi \cdot 8)$$

$$= \frac{1}{4}(16\pi)$$

$$= 4\pi$$

Choice B is the correct answer.

19. **A** Choice B is incorrect because the multiplication $2(3x + 4)$ must be done before any addition. Choice C is incorrect because in the last step, when dividing by -2, the sense of the inequality must be reversed. Choice D is incorrect because $2(3x + 4) = 6x + 8$, not $6x - 8$. Choice A is the correct answer because all of the rules for order of operations and solving inequalities have been followed.

20. **A** The inverse of an implication is created by negating both the if and the then clauses in the implication. The inverse of the given statement, "If all the sides of a polygon are congruent, then the polygon is equilateral" is written by placing the word *not* into the sentence in the if and then clauses: "If all the sides of a polygon are *not* congruent, then the polygon is *not* equilateral." Choice A is the correct choice.

21. **B** Choice B is the correct answer because zero is the identity element for addition. Choices A, C, and D are incorrect because they identify different properties.

22. **D** Choice D is correct because $\sqrt{31}$ is between $\sqrt{25}$ and $\sqrt{36}$, which are 5 and 6. Therefore, $\sqrt{31}$ is between 5 and 6. Choices A, B, and C are not within the correct range of values.

23. **C** Choice C is correct. To find probability you must first find the number of possible picks. That would be the total of the number of forks, spoons, and knives, which is 37. That number becomes the denominator of your fractional answer. The numerator is the number of items you want; you want fifteen spoons. Choices A, B, and D are incorrect because they have the incorrect numerator in each of the fractions.

24. **B** Using the counting principle rule, you can find the total number of outcomes by multiplying the number of choices in each group. Since there are two paths from the playroom to the living room, two paths from the living room to the kitchen, and three holes to get outside, you would multiply $2 \cdot 2 \cdot 3 = 12$. The other three choices do not follow the counting principle correctly.

25. **D** The difference between the consecutive terms in the sequence forms a pattern: $6 - 5 = 1$, $9 - 6 = 3$, $14 - 9 = 5$, $21 - 14 = 7$, $30 - 21 = 9$. The pattern is 1, 3, 5, 7, 9; these are consecutive odd numbers. If we continue the pattern of odd numbers, they would be 1, 3, 5, 7, 9, 11, 13, ... By adding 11, the seventh term in the given pattern would be 41, and by adding 13, the eighth term in the given pattern would be 54. Choices A, B, and C do not represent a possible eighth term in the indicated pattern.

26. **C** Choice C is the correct choice—every x-coordinate is mapped to one and only one y-coordinate. Choice A is incorrect because the points (4, 2) and (4, 4) map the x value of 4 into two different y values. Choice B is incorrect because the graph fails the vertical line test—a vertical line drawn on the coordinate plane will intersect the curve in more than one place. Choice D is incorrect because the points (2, 3) and (2, 0) map the x-value of 2 into two different y-values.

27. **A** The phrase "evaluate the algebraic expression" means you are to substitute the given value for the variable and determine the numerical value of the expression. Substituting the negative three for x results in this numerical expression, $-12(2[-3] - 1) - (-3)^2$. Using the rules for order of operations would result in the expression, $-12(-6 - 1) - 9$, and then the expression $-12(-7) - 9$, which gives the results, $84 - 9 = 75$. This is choice A, which is the correct answer. Each of the other choices is incorrect because of incorrect usage of the rules for order of operations.

28. **B** Changing an algebraic expression into a word expression is similar to being a court stenographer and writing the words as the algebra symbols happen. $3(x - 2)$ means 3 times the expression $(x - 2)$, and $(x - 2)$ means the difference between x and 2. Therefore, choice B is correct because it says "three times the difference of a number x and two." Choices A, C, and D are incorrect because the algebraic expressions and the word phrases do not say the same thing.

29. **A** Proportions are solved by using the means-extremes product property. This property states that the product of the means of a proportion equals the product of the extremes. In the given proportion, the product of the means is 54 and the product of the extremes is $20x$. Therefore, choice A, $20x = 54$, is correct.

30. **C** To simplify this expression, you must simplify each of the square root expressions and then multiply.

$$\sqrt{36x^2} \cdot \sqrt{16y^{16}} = 6x \cdot 4y^8 = 24xy^8$$

Remember, when you multiply exponential expressions, you add the exponents.

$$\sqrt{36x^2} = 6x \text{ because } 6x \cdot 6x = 36x^2 \text{ and}$$

$$\sqrt{16y^{16}} = 4y^8 \text{ because } 4y^8 \cdot 4y^8 = 16y^{16}$$

Therefore, the correct answer is choice C.

31. **D** The equation is a quadratic equation and can be solved using the quadratic formula or by factoring.

From the reference sheet, the quadratic formula is	If you solve by factoring,
$$x = \frac{-b \pm \sqrt{b^2 - 4ac}}{2a}.$$ The values of a, b, and c are 1, -1, -20. Therefore, $$x = \frac{-(-1) \pm \sqrt{(-1)^2 - 4(1)(-20)}}{2(1)}.$$ Simplifying, $x = \dfrac{1 \pm \sqrt{81}}{2}$ and $x = 5$ or $x = -4$.	$x^2 - x - 20 = 0$ $(x-5)(x+4) = 0$ $x = 5$ or $x = -4$

Choice D is the correct choice.

32. **C** To solve the equation for any one of the variables, $I, P, R, T,$ you must isolate that variable. To isolate $P, R,$ or $T,$ you would divide by the other two variables. Choice B is incorrect because it is solved for T but does not indicate division. Choice D is incorrect because I cannot be equal to what is in the given equation and what is in choice D. The correct answer is either A or C. But to solve for $P,$ as shown in choice A, you would have to divide I by both T and $R.$ Choice A is incorrect. Choice C is the correct choice.

33. **B** The tick marks on the diagram indicate that there are two pairs of congruent sides given. That eliminates choices A and D. To get the triangles congruent, you must find another pair of congruent sides or a pair of included congruent angles. Because $\overline{BD}$ is a common, shared side it is congruent to itself. Choice B is the correct choice because you have three pair of congruent sides.

34. **D** The graph of the line has a y-intercept of $(0, 3)$. Choices A and C are incorrect because they do not have y-intercepts of 3. The graph shows a line with a negative slope. Choice B is incorrect because its slope is a positive number. Choice C is the correct choice. Check its slope, down 3 and 2 to the right.

35. **A** The formula for finding the midpoint between two given points, (x_1, y_1) and (x_2, y_2), is given on the reference sheet as $\left(\dfrac{x_2 + x_1}{2}, \dfrac{y_2 + y_1}{2} \right)$. Notice the formula states that you find the sum of the x-coordinates and divide by two, and you find the sum of the y-coordinates and divide by two.

$$\left(\frac{5 + (-3)}{2}, \frac{-1 + (-5)}{2} \right) = \left(\frac{2}{2}, \frac{-6}{2} \right) = (1, -3)$$

Choice A is the correct choice.

36. **B** A cube is a rectangular prism and the formula for finding its volume ($V = Bh$) can be found on the reference sheet. The capital B stands for area of the base. Because the edges of a cube are all the same the length and width of the base are each four. The area of the base is 16 sq. in. Multiplying the area of the base times the height ($16 \cdot 4$) results in a volume of 64 cu. in. Cubic units are sometimes written using an exponent of three—cu. in. is the same as in.[3]. The correct choice is B.

37. **B** The formula for finding the volume of a pyramid is given on the reference sheet as $V = \dfrac{1}{3} Bh$. V represents the volume of the pyramid, B is the area of the base, and h is the height (altitude) of the pyramid. The base of the pyramid is a square, and its area can be found by squaring the measure of the side of the base.

$$V = \frac{1}{3} Bh$$

$$144 = \frac{1}{3} (6)^2 \cdot h$$

$$144 = \frac{1}{3} \cdot 36 \cdot h$$

$$144 = 12h$$

$$12 = h$$

Choice B is the correct choice. Choices A, C, and D either do not use the correct formula or use it incorrectly.

38. **C** The next line in the solution of this equation must contain correct applications of the distributive property. Choice A is incorrect because $4 - (x + 1) = 4 - x - 1$. Choice B is incorrect because $4(x + 2) = 4x + 8$. Choice D is incorrect because $4(x + 2) = 4x + 8$, not $4x + 6$. Choice C is the correct answer.

39. **D** Choice D is correct because the rule for order of operations states that exponential expressions must be done first, $3^2 = 9$; then $9 \cdot 2 = 18$; and last, $7 + 18 = 25$. Choices A, B, and C are incorrect because the operations are done in an incorrect order.

40. **D** Choice D is correct because a 20% tip can be estimated by mentally rounding to the nearest ten dollars, taking 10% of that, and then double the 10%. $36.95 rounded to the nearest ten dollars is $40. 10% of $40 is $4. Doubling the $4, you get $8. Choice D is closest to that estimate. Choices A, B, and C are not close to the estimated tip.

41. **D** Choice D is correct because it would be reasonably easy to count the number of books written by Charles Dickens that were in the library. Choices A, B, and C are incorrect because it would be very difficult or nearly impossible to count the items indicated in each of the choices.

42. **A** Line segments have two endpoints and are named with those two endpoints. $\overline{AB}$ and $\overline{BA}$ are two different names for the same segment; the order in which you name the segments does not matter. Therefore, you really want to know the number of two letter groups (combinations) that you can make using the five letters in the drawing. The letters must be different; you cannot name a segment using the same letter twice. The formula for finding the number of combinations of n things taken r at a time is $_nC_r = \dfrac{n!}{(n-r)! \cdot r!}$. This can be found on the reference sheet, which you can use on the AIMS test. Therefore, we are evaluating the formula for $n = 5$ and $r = 2$:

$$_5C_2 = \frac{5!}{(5-2)! \cdot 2!} = \frac{5!}{3! \cdot 2!}$$

Choice A is the correct answer. The other choices either do not use the combination formula at all or they do not use it correctly.

43. **A** In the given formula $a_1 = 1$. The next term, a_2, can be found by applying the second part of the formula,

$$a_{n+1} = (a_n)^2 + 3; \text{ that is, } a_2 = (a_1)^2 + 3 = (1)^2 + 3 = 1 + 3 = 4.$$

Each of the next terms can be found in a similar manner:

$$a_3 = (a_2)^2 + 3 = (4)^2 + 3 = 16 + 3 = 19, \ a_4 = (a_3)^2 + 3 = (19)^2 + 3 = 361 + 3 = 364.$$

Therefore, choice A is the correct choice. Choices B, C, and D do not apply the formula correctly.

44. **D** The domain is the values that the independent variable (in this case, x) can be which eliminates choice B. Choice D is correct because the graph of the function shows that x is greater than or equal to negative one. Choices A and C are incorrect because the beginning x-value is incorrect.

45. **A** The solution to this problem is dependent on finding the slope of the two lines. The first equation is already in slope-intercept form ($y = mx + b$), so it is fairly easy to find its slope. The value of m, the coefficient of x, is the slope. Therefore, the slope of the first line is 3. The second equation is not in slope-intercept form, but you can get it into slope intercept form by dividing each of the terms of the equation by 3. This results in the equation, $y = 2x - 1$. This is exactly the same equation as the first equation. This means that the graph of the two equations is the same line. We say that the equations are coincident because the two graphs would coincide with each other. Therefore, choice A is the correct choice. If the slopes of the two equations were the same but the y-intercepts were different, the lines would have been parallel. If the slopes were different, then you know the lines intersect. But the lines may intersect in a special way—they may be perpendicular to each other. If the slopes are opposite reciprocals of each other $\left(e.g., \frac{2}{3} \text{ and } -\frac{3}{2} \right)$, then you know the lines are perpendicular to each other.

46.   **C**   If Charlie pays \$12 a night to bowl, he would pay $12n$ to bowl n nights. In addition he pays \$25 when he signs up for the league. Therefore, his total costs would be $12n + 25$. This is choice C. Choice B is incorrect because you are subtracting the initial fee of \$25. Choices A and D are incorrect because Charlie does not pay \$25 each night.

47.   **D**   To find the equation of a line given two points, you must first find the slope of the line. Using the slope formula from the reference sheet

$$m = \frac{y_2 - y_1}{x_2 - x_1}$$

$$m = \frac{4 - (-3)}{0 - 5}$$

$$m = \frac{7}{-5} = -\frac{7}{5}$$

The only choice that has a slope of $-\frac{7}{5}$ is choice D. Choice D is the correct choice.

48.   **B**   This equation is a radical equation. Radical equations are solved by first isolating the radical expression that contains the variable.

$$\sqrt{4x} + 5 = 9$$

$$\sqrt{4x} = 4$$

Then square both sides of the equation and solve the resulting equation.

$$\sqrt{4x} = 4$$

$$4x = 16$$

$$x = 4$$

Choice B is the correct solution.

49.   **A**   Choice A is the correct choice. Choice B is the net of a cube, choice C is the net of a rectangular prism, and choice D is the net of a triangular pyramid.

50.   **B**   Choices A, C, and D are incorrect because we know nothing about proportional sides and that is what the S stands for. Choice B is correct because you are given one pair of angles equal, and you can get the second pair from the vertical angles at point A.

51.   **D**   The given equation is a quadratic equation (function)—the exponent of 2 tells you this. Notice the coefficient of x^2 is positive; the graph opens upward. If $x = 0$, $y = -1$, the y-intercept is at $(0, -1)$. Note that this is also the vertex of the graph.

52.   **C**   Changing the 2 to a -3 will change the slope of the line from a positive slope to a negative slope. Choice B is incorrect because the slope of that line is positive. The problem does not indicate that you are to change anything else in the equa-

tion. The y-intercept is to remain at (0, 1). Choices A and D are incorrect because the y-intercepts changed. Choice C is the correct choice. The line has a y-intercept of 1 and a slope of –3.

53. **B** The sum of the angles of a polygon can be found by using the formula $S = (n - 2)180$, which can be found on the reference sheet. The variable, n, represents the number of sides in the polygon. The polygon in this problem has five sides. Therefore, $(5 - 2)180 = (3)180 = 540$. The angles given in the figure have a sum of 441. Therefore, $540 - 441 = 99$. Choice B is the correct answer.

54. **B** The tenth value in a sorted list of 19 values is the middle value. The measure of central tendency that is the middle value is the median. The mean is the average, the mode is the value that appears most often, and the range is the difference between the largest and the smallest values.

55. **A** Choice A is correct because you can count the number of three digit whole numbers. There are quite a lot of them but you can count them. Choices B, C, and D are incorrect because there are infinitely many numbers in each set; they each continue forever.

56. **B** Choice B is the correct choice. The expressions within the absolute value symbols should first be simplified, $6 + |-2| + |-5|$. Absolute value is always a positive number, so you are adding $6 + 2 + 5$. The correct answer is B. Choices A, C, and D either do not find the value of the expression within the absolute value symbols correctly or do not find the absolute values correctly.

57. **D** Choice D is correct because if you round what he charges each time he mows the lawn, you would get $12. To find the number of times Michael must mow the lawn, divide $240.00 by $12, which is 20 times. Choices A, B, and C do not generate enough money for Michael.

58. **B** Choice B is correct because the school choir is the group of students who will most likely be involved in the musical. Choices A, C, and D contain people who probably have very little interest or no interest in what musical is chosen.

59. **C** The expected probability of getting a 6 when you roll a die is one out of six. In this experiment, the number 6 was rolled four out of 18 times. That is more than the expected probability would indicate (1 out of 6 equals 3 out of 18). Choices A, B, and D are incorrect because they do not match the results of the sampling probability.

60. **C** Notice the pattern that the figures form: the 2nd, 4th, 6th, 8th, and 10th figures have the two top boxes shaded. The 1st figure has the top box shaded and the 3rd figure has the bottom box shaded. The 5th, 7th, and 9th figures have the top, then bottom, and then top boxes shaded. Therefore, continuing the pattern, the bottom box should be shaded in the 11th figure. Choice C is a figure with the bottom box shaded. Choices A, B, and D do not contain figures that follow the pattern.

61. **D** Choice D is the correct answer. The maximum height that the rocket will reach is at the maximum height of the curve. The ordered pair at that point is (3.5, 21). The y-coordinate of that point is the maximum height of the rocket. Choices A, B, and C are not reasonable choices for the maximum height.

62. **B** The expressions in choice A are equivalent because $4 \cdot 3^2 = 4 \cdot 9 = 36$ and $60 - 24 = 36$. The expressions in choice C are equivalent because when $a = 0$, $4(0^2 - 1) = 4(-1) = -4$ and $4(0 - 1) = 4(-1) = -4$. The expressions in choice D are equivalent because when $x = 1$, $5x^3 = 5(1)^3 = 5 \cdot 1 = 5$ and $3x + 2 = 3(1) + 2 = 3 + 2 = 5$. Choice B is the correct answer because the two expressions are not equivalent when $x = 2$ and $y = 3$, $3(2) + 4(3) = 6 + 12 = 18$ and $3(3) = 9$.

63. **A** You might notice that the cost for camping increases by $14 each one day. That means the slope of the equation is going to be 14. The equations in choices B and D do not have a slope of 14. The equation in choice C is incorrect because the values for n and C do not satisfy the equation. Choice A is the correct answer because the values for n and C do satisfy the equation.

64. **C** Systems of equations can be solved using substitution, elimination, or the graphical method. Generally graphing would be used only if you are given a graph in the question. Substitution or elimination would be appropriate for this question. if you use substitution, you would solve the second equation for y and then substitute that new name for y into the first equation.

$$2x-3y=-1$$
$$3x+y=-7$$
$$y=-3x-7$$
$$2x-3(-3x-7)=-1$$
$$2x+9x+21=-1$$
$$11x+21=-1$$
$$11x=-22$$
$$x=-2$$

Therefore, choice C is correct.

If you use the elimination method, you could multiply the second equation times 3 and add the two equations.

$$2x-3y=-1$$
$$3x+y=-7$$

$$2x-3y=-1$$
$$9x+3y=-21$$

$$11x=-22$$
$$x=-2$$

Again, the answer is –2 and choice C is correct.

65. **C** The definition for slope, taken from the reference sheet, is $m=\dfrac{y_2-y_1}{x_2-x_1}$, given, (x_1, y_1) and (x_2, y_2). Therefore, using the given points, $(6, -1)$ and $(-3, 5)$,

$m=\dfrac{5-(-1)}{-3-6}=\dfrac{6}{-9}=-\dfrac{2}{3}$. Choice C is the correct answer.

66. **A** The angles whose algebraic representations are given are vertical angles. Vertical angles are congruent. Therefore, the equation, $3x + 18 = 5x - 4$, is true. Solving the equation, $x = 11$. Choice A is the correct choice.

67. **C** The base angles of an isosceles triangle are congruent. Therefore, $\angle B$ and $\angle C$ have to be equal. The sum of the angles in any triangle is 180°. That means that $\angle B$ and $\angle C$ together are 78°. Since they are equal, they are each 39°. Choice C is the correct choice.

68. **D** The x-coordinate of the given ordered pair, $(3, -1)$, will be increased by 3 as a result of the move 3 units to the right. The y-coordinate of the given ordered pair,

(3, –1) will be decreased by 2 as a result of the move 2 units down. Therefore, the new ordered pair is (6, –3), Choice D is the correct choice.

69. **C** The distance formula, $d = \sqrt{(x_2 - x_1)^2 + (y_2 - y_1)^2}$ is contained on the reference sheet. Substituting the given points into the formula,

$$d = \sqrt{(-2 - [-5])^2 + (9 - 3)^2}$$
$$d = \sqrt{(-2 + 5)^2 + (9 - 3)^2}$$
$$d = \sqrt{(3)^2 + (6)^2}$$
$$d = \sqrt{9 + 36}$$
$$d = \sqrt{45}$$

Choice C is the correct choice.

70. **D** Choice A is incorrect because you must square the 4 before subtracting 5. Choice B is incorrect because 4 squared is 16, not 8. Choice C is incorrect because when moving from the fourth line to the fifth line, the multiplication of 2 times 11 must be completed before the addition. Choice D is correct because all of the rules for order of operations have been followed.

71. **A** Choice A is correct because the numbers in this set continue forever. Choices B, C, and D each have a finite number of values in the set; each set is countable.

72. **D** Choice D is correct because it is *not* an appropriate question to ask to determine the design of the label. Choices A, B, and C are appropriate questions.

73. **C** Choice C is correct. The number of degrees all the way around a circle is 360°. Region B is 135°. Probabilities related to geometric figures are found by using the total region or area as the denominator and the region desired as the numerator; therefore, the correct probability is $\frac{135}{360} = \frac{3}{8}$. Choices A, B, and D do not represent the correct fractional probability.

74. **B** Choice B is the correct answer because the number of possible numbers is eight and there are four prime numbers on the spinner (2, 7, 11, 23). Therefore, the probability is $\frac{4}{8} = \frac{1}{2}$. Choices A, C, and D may have the correct number of initial possible choices, but the numerators of the fractions are incorrect.

75. **C** The pattern would be as follows: Sunday—$1, Monday—$2, Tuesday—$4, Wednesday—$8, Thursday—$16, Friday—$32, Saturday—$64. Therefore, choice C is the correct answer. Choices A, B, and D do not contain choices that follow the pattern.

76. **C** To simplify and find an equivalent expression for the given expression, use the associative and commutative properties of multiplication to rearrange the product to be $(4 \cdot 5)(a^2 a^2)(b^2 b^4)$. Using arithmetic multiplication and the rules for multiplying exponential expressions, the product becomes $20a^4 b^6$, which is choice C. Choices A, B, and D incorrectly apply the rules for multiplying exponential expressions.

77. **D** Choices A, B, and C are incorrect because when 4 is substituted in for x, none of the equations is true. Substituting 4 in for x in the equation in choice D, $8(4) = 7(4) + 4$, $32 = 28 + 4$.

78. **B** The expression under the radical symbol in called the radicand. To multiply radical expressions, you multiply the radicands. Therefore,

$$\sqrt{12} \cdot \sqrt{3y^2} = \sqrt{36y^2} = 6y.$$

Therefore, choice B is the correct choice.

79.   **B**   Remember you are looking for a choice that is *not* true. Choice A is always true for any triangle. Choice C is true because the angles opposite the equal sides in an isosceles triangle are equal. Choice D is true because it is the Pythagorean theorem, which is true for every right triangle. Choice B is the choice that is not true. The hypotenuse of a right triangle is always longer than either of the two legs.

80.   **B**   The tangents drawn to a circle from the same point outside the circle are equal. Therefore, $3x - 1 = -4x + 13$. $7x = 14$ and $x = 2$. However, choice A is incorrect because the question asks you to find the length of $\overline{AB}$. $-4x + 13 = -4(2) + 13 = -8 + 13 = 5$. Therefore, choice B is the correct choice. Choices C and D come from making mistakes in solving the equation.

81.   **A**   The y-intercept of the boundary of the region is one. All of the equations have a y-intercept of one. The boundary of the shaded region is a dotted line whose slope is positive. Choices C and D are incorrect because they each have a slope of negative one. You must decide if the correct side of the boundary has been shaded. Pick a test point—say (0,0). Does it satisfy the inequality in choice B? $0 > 0 + 1$ No. Does it satisfy the inequality in choice A? $0 < 0 + 1$ Yes. Choice A is the correct choice.

82.   **D**   To find the value of y, you must set up a proportion using ratios of corresponding sides. Sides $\overline{DC}$ and $\overline{ZY}$ correspond, and $\overline{BC}$ and $\overline{XY}$ correspond. Therefore, the proportion that relates those corresponding sides is

$$\frac{4}{6} = \frac{8}{y}$$
$$4y = 48$$
$$y = 12$$

Choice D is the correct choice. Another pair of corresponding sides could have been chosen for the initial ratio.

83.   **A**   Choice A is correct because you can find what part of the 24 points that Carlos scored is of the total 88 points. The answer to $(24 \div 88)$ represents that decimal part. Multiplying by 100 changes the decimal to a percent. The other choices show a division or multiplication of the wrong numbers or the right numbers in the wrong order.

84.   **B**   Choice B is correct because the numbers in the 40s are 47, 48, and 48. Choice D is incorrect because there are two numbers in the 20s, 24 and 28. Choice C is incorrect because there are three numbers in the 40s. Choice A is incorrect because there are three numbers in the 50s.

85.   **A**   Choice A is correct because the income falls between the values $16,850 and $16,900, and the single column contains the tax amount of $2,601. Choices B, C, and D are values either in the wrong range of income or the wrong marriage status.

86.   **A**   Choice A is correct because the line is going down from D to E, which is an indication that the bank balance is decreasing. Choice A is the only choice that indicates something happening that would cause the balance to go down. Choice B is incorrect because a line that was going up would represent deposits. Choice C is incorrect because the line would be horizontal if the account neither increased nor decreased. D is incorrect because interest would increase the balance, and the line would be going up.

87. **C** You can find the *y*-intercept by letting $x = 0$ in the given equation. $3y - 2(0) = 12$. Therefore, $y = 4$, and the *y*-intercept is $(0, 4)$. Choice C is the correct answer. You can also rewrite the equation in slope-intercept form $(y = mx + b)$.

$$3y - 2x = 12$$
$$3y = 2x + 12$$
$$\frac{3y}{3} = \frac{2}{3}x + \frac{12}{3} \qquad \text{The value of } b, 4, \text{ is the } y\text{-coordinate of the } y\text{-intercept.}$$
$$y = \frac{2}{3}x + 4$$

88. **D** There are 360° in the sum of the angles at the center of every circle. In the diagram, the three angles shown have a sum of 220°. Therefore, the measure of $\angle APB$ is 140°. The measure of an arc of a circle is equal to the measure of the central angle that intercepts it. Therefore, $m\,\overset{\frown}{AB} = 140°$ and $m\,\overset{\frown}{CB} = 70°$. $\overset{\frown}{ABC}$ is the sum of $\overset{\frown}{AB}$ and $\overset{\frown}{CB}$. Therefore, $m\overset{\frown}{ABC} = 210°$. Answer D is the correct choice.

89. **B** The two-step reflection is shown on the graph below. A'' is at the point, $(3, -2)$. Choice B is the correct answer.

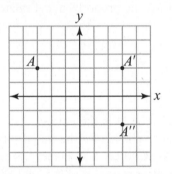

90. **C** This problem is solved by dividing the figure into rectangles and then finding the areas of those rectangles. There is more than one way to divide the given figure. Below is one possible choice.

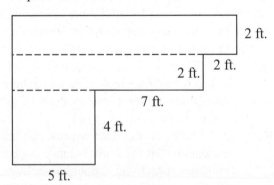

The length and width of the smallest rectangle (at the bottom) are 5 and 4. Its area is 20 sq. ft. The length and width of the middle rectangle are 12 and 2. Its area is 24 sq. ft. The length and width of the top rectangle are 14 and 2. Its area is 28 sq. ft. The sum of the three areas is 72 sq. ft. Choice C is the correct choice.

91. **B** The statement "Kelly does not take Algebra" satisfies the premise, "you do not take Algebra." Therefore, the conclusion, "Kelly will not take Geometry" is true. Choice B is the correct answer.

92. **A** Choice A is correct because Mary's change from $200 should be the money value of the coins needed to return the difference between $200.00 and $154.49, or $45.51. In coins and bills, that is one penny, two quarters, one $5 bill, and two $20 bills. Choices B, C, and D do not have a value of $45.51.

93. **A** Choice A is correct because the median score is very obvious in a box-and-whisker plot. It is the value indicated by the line within the box. The mode and the mean cannot be determined from a box-and-whisker plot. The range is not a measure of central tendency.

94. **C** Choice C is correct because each of the four sections (quartiles) of a box-and-whisker plot represent 25% of the scores. There are three sections, or 75%, below 60 inches. Choices A, B, and D do not represent the correct percentages.

95. **D** A good way to test for the correct answer in this question is to choose one of the x-y ordered pairs from the table and substitute it into the equations. If you choose the first ordered pair, $(2, 0)$, and substitute it into the equations, the only equation that it satisfies is $y = 4 - 2x$. Therefore, choice D is the correct answer. Using two of the ordered pairs in the table to find the slope of the line containing the points and then using one of the ordered pairs with the slope in the slope-intercept form of an equation, $y = mx + b$, could also solve the problem.

96. **C** Choice A is incorrect because $3^2 < 7^2$; that is, $9 < 49$. Choice B is incorrect because $\left(\sqrt{9}\right)^2 < \left(\sqrt{13}\right)^2$; that is, $9 < 13$. Choice D is incorrect because $(1.2)^2 < (3.5)^2$; that is, $1.44 < 12.25$. Choice C is the correct answer because $(-7)^2 > (-3)^2$; that is, $49 > 9$.

97. **D** Choice D is correct because if $A_n = 13$, that means $13 = |n - 2|$. Which of the choices for n would make this equation true? A is incorrect because $13 \neq |12 - 2|$. B is incorrect because $13 \neq |13 - 2|$. C is incorrect because $13 \neq |14 - 2|$. D is correct because $13 = |15 - 2|$.

98. **C** Choice C is correct using two different approaches to the problem. The first is that the y value is always one more than twice the x value. Or you might see the pattern in the column of y values—they increase by two each time x increases by one. Choices A, B, and D do not reflect the pattern created in the table.

99. **D** Choice D is the correct choice because the points in that scatterplot are the closest to being linear. Choices A, B, and C do not represent points that are somewhat arranged in a line.

100. **D** Choices A, B, and C are incorrect because each describes a second event that is dependent upon the first event. Choice D describes an event that does not depend on the first event. The results of getting a head or a tail on one flip do not affect the results of the next flip. This is the basis of independence.

Strand 1 | Number Sense and Number Operations

<hr>

When you complete your study of Strand 1, you will be able to:

- Classify real numbers as members of the natural numbers, the whole numbers, the integers, the rational numbers, or the irrational numbers.
- Identify properties of the real number system: commutative, associative, distributive, identity, inverse, and closure.
- Distinguish between finite and infinite sets of numbers.
- Understand and apply numerical operations
  - Solve word problems.
  - Simplify numerical expressions.
  - Compute using scientific notation.
- Use estimation strategies reasonably and fluently.

It is quite fitting that the first strand of the Arizona Mathematics Standard deals with number sense. The concepts that make up this standard lay the foundation for each of the other standards. Even though you have been exposed to these concepts from an early age and some of the concepts seem trivial, nonetheless it is important to have a firm grasp of number sense. Number sense helps to define and clarify the types of numbers used in any problem involving mathematics.

The Real Number System

Our starting point is to learn the relationships among numbers and the different number systems or sets. A set is a collection of objects. In our number systems, we categorize a number as belonging to a set or several sets. A subset is a part of a larger set. Its members are contained completely in the larger set. The members are often called elements. Using this concept of sets, let's define each of the number systems.

The Real Numbers
- Natural numbers
- Whole numbers
- Integers
- Rational numbers
- Irrational numbers

Natural (Counting) Numbers

The initial set of numbers is called natural numbers. Some textbooks refer to them as counting numbers. At an early age you begin by assigning numbers to objects in an effort to count them. Hence, our first set of numbers is the counting numbers.

$$\{1,2,3,4,5,6,7,\ldots\}$$

When a problem requires counting, the set of natural numbers is all that is needed. Did you notice the three dots in the set of numbers above? These dots mean that the set continues in the same pattern.

Whole Numbers

Adding the digit zero to the natural numbers forms a new set called whole numbers. The value zero adds the concept of "how many" of an object. When a set of numbers is part of (contained in) another set of numbers, we call it a subset. So, the set of natural numbers is a subset of the set of whole numbers.

$$\{0, 1, 2, 3, 4, 5, 6, 7, \ldots \}$$

Integers

In order to represent such things as temperatures below zero degrees or overdrawn bank accounts, negative numbers are used. These numbers are included in the set of numbers called integers. When we combine the counting numbers with their opposites along with zero, we form the set of integers. Often times, the set is listed in the following manner:

$$\{\ldots, -3, -2, -1, 0, 1, 2, 3, \ldots \}$$

Using the subset definition, the set of whole numbers is a subset of the set of integers.

Rational Numbers

The operation of division leads us to the next number set. This set of numbers is called rational numbers. These are fractions that are in essence "ratios of integers." We use fractional numbers to represent the part of a whole. You can take a fractional number and express it as a decimal number by dividing the denominator (bottom number) into the numerator (top number). The resulting division will come to an end (terminate) or will reveal a pattern of digits (repeating digits). So, terminating decimal numbers and repeating decimal numbers are rational numbers. Every member of the set of integers can be written as a rational number. Simply take the given integer and divide it by the integer 1, and you have a ratio of integers.

$$4 = \frac{4}{1} \qquad -16 = \frac{-16}{1} \qquad 0 = \frac{0}{1}$$

Using the subset definition, integers are a subset of rational numbers.

Irrational Numbers

There are numbers that cannot be expressed as a ratio of integers. In decimal form they are nonrepeating and nonterminating. These are called irrational numbers, meaning *not* rational. Some irrational numbers are radicals that are not perfect squares:

$$\sqrt{2}, \sqrt{20}, \sqrt{6}, \sqrt{5}, -\sqrt{95}, \sqrt{3}, \text{ and so forth.}$$

Note that $\sqrt{4}$ is not in this list of examples. Since 4 is a perfect square and its square root is 2, which is a rational number, $\sqrt{4}$ is a rational number. Since irrational numbers and rational numbers describe two different sets of numbers, we would state that irrational numbers are *not* a subset of rational numbers. A very important irrational number is π (*pi*). Although π is sometimes represented as 3.14 or $\frac{22}{7}$, these values are just approximations. As stated, the rational numbers and the irrational numbers do not have any common members.

Real Numbers

The real number system is created by combining (union) the irrational numbers with the rational numbers. Most of the mathematics classes that you will take deal primarily with the real number system.

Let's classify these numbers. (Remember that a number may be in more than one set.)

Value	Number system(s)
-8	Integer, rational, and real
1.25	Rational and real
25	Natural, whole, integer, rational, and real
$\sqrt{11}$	Irrational and real
$\dfrac{2}{3}$	Rational and real
π	Irrational and real
$4.\overline{6}$	Rational and real
$\sqrt{49}$	Natural, whole, integer, rational, and real
4.040040004 . . .	Irrational and real (This number may appear to be a repeating decimal, but the repeating blocks of numbers are different.)

Study this list. Did you notice something common about each number represented? Each belongs to the real number system.

Properties of the Real Number System

There are some properties or rules that are always true for our real number system. These properties are the basis for expressing and solving algebraic sentences, often called equations.

Properties of Equalities

The **reflexive property** of equality just says that $a = a$; anything is equal to itself. The equal sign is like a mirror, and the image it "reflects" is the same as the original.

The **symmetric property** of equality says that if $a = b$, then $b = a$. This is often used when displaying a solution. The equation $5 = x$ could also be written as $x = 5$.

The **transitive property** of equality says that if $a = b$ and $b = c$, then $a = c$. The transitive property is a property of real numbers, but it is also used in geometry to help us create deductive arguments. Consider the following example. If the measure of angle A is equal to the measure of angle B and the measure of angle B is equal to the measure of angle C, then we can say that the measure of angle A is equal to the measure of angle C.

If $m\angle A = m\angle B$ and $m\angle B = m\angle C$, then $m\angle A = m\angle C$.

The **addition property** of equality says that if $a = b$, then $a + c = b + c$. If you add the same real number to (or subtract the same real number from) both sides of an equation, the equation continues to be true. The basic rules for solving equations are centered around the equal sign. The equal sign in an equation is like the center of a scale: both sides, left and right, must be the same value in order for the scale to stay in balance and the equation to be true.

The **multiplication property** of equality says that if $a = b$, then $a \cdot c = b \cdot c$. If you multiply (or divide) by the same nonzero number on both sides of an equation, the equation continues to be equal. There is one exception to this property and that is dividing both sides by zero. Division by zero is not a defined operation.

The **associative property** allows you to group numbers in any way without changing the answer. It doesn't matter how you combine them; the answer will always be the same. The operations of addition and multiplication are both associative. Notice in the associative property that the order of the numbers does not change, but rather the grouping of the numbers.

Here is an addition example.

$$3 + (4 + 5) = (3 + 4) + 5$$
$$3 + \quad 9 \quad = \quad 7 \quad + 5$$
$$12 \quad = \quad 12$$

In general terms, $a + (b + c) = (a + b) + c$

Here is a multiplication example.

$$3 \cdot (4 \cdot 5) = (3 \cdot 4) \cdot 5$$
$$3 \cdot \quad 20 \quad = \quad 12 \quad \cdot 5$$
$$60 \quad = \quad 60$$

In general terms, $a \cdot (b \cdot c) = (a \cdot b) \cdot c$

Note that the associative property is not valid for the operations of subtraction and division. Consider the following example using subtraction.

$$13 - (5 - 2) = (13 - 5) - 2$$
$$13 - \quad 3 \quad = \quad 8 \quad - 2$$
$$10 \quad = \quad 6 \quad \text{This is a false statement.}$$

The **commutative property** allows you to change *the order* of the numbers in the operation without changing the result. Addition and multiplication are both commutative. Here are some examples of the commutative properties of addition and multiplication.

$$18 + 15 = 15 + 18 \qquad 8 \cdot 5 = 5 \cdot 8$$
$$33 \quad = \quad 33 \qquad\qquad 40 = 40$$

In general terms, $a + b = b + a$ and $a \cdot b = b \cdot a$

The order in which you add or multiply two numbers does not matter, but subtraction and division are not commutative. Look at the examples that follow.

$$18 - 15 = 15 - 18 \qquad 20 \div 5 = 5 \div 20$$
$$3 \quad \neq \quad -3 \qquad\qquad 4 \neq \frac{5}{20} \text{ or } \frac{1}{4}$$

The **distributive property** of multiplication over addition says that when a number is multiplied by the sum of two other numbers, it can be multiplied by each of the numbers in the sum separately and then added.

Here's an example of the distributive property of multiplication over addition.

$$3(4+5) = (3 \cdot 4) + (3 \cdot 5)$$
$$3 \cdot 9 \ = \ 12 \ + \ 15$$
$$27 \ = \ 27$$

Here's an example of the distributive property of multiplication over subtraction.

$$3(8-5) = (3 \cdot 8) - (3 \cdot 5)$$
$$3 \cdot 3 \ = \ 24 \ - \ 15$$
$$9 \ = \ 9$$

> In general terms, $a \cdot (b + c) = a \cdot b + a \cdot c$
> $a \cdot (b - c) = a \cdot b - a \cdot c$

Closure Property

The closure property requires that the answer from any operation on two numbers in a set of numbers is also included in the set. For example, if we take two real numbers and multiply them together, the answer is always another real number. Because this is always true, we say that the real numbers are "closed under the operation of multiplication." Real numbers are also closed under addition, subtraction, and division. However, consider the set of whole numbers and the operation of subtraction. This set is not closed under the operation of subtraction. The problem $7 - 10 = -3$ is an example of this because the answer, -3, is not a member of the set of whole numbers.

Identity Properties

An identity is a special kind of number. When you use an operation to combine an identity with another number, that number stays the same. Zero is the additive identity because adding zero to a number will not change it— the number stays the same.

> Identity Elements
> • $a + 0 = a$
> • $a \cdot 1 = a$

> In general terms, $0 + a = a$
> $a + 0 = a$

Since any number multiplied by 1 remains constant, the multiplicative identity is 1.

> In general terms, $1 \cdot a = a$
> $a \cdot 1 = a$

Inverse Property

A number's additive inverse is a number that you can add to the original number to get the additive identity. For example, the additive inverse of 15 is –15, because

Inverses
• $a + (-a) = 0$
• $a \cdot \left(\dfrac{1}{a}\right) = 1$

$15 + (-15) = 0$ (Remember, zero is the additive identity.)

> In general terms, $a + (-a) = 0$

Similarly, if the product of two numbers is the multiplicative identity, the numbers are multiplicative inverses. Since $5 \cdot \dfrac{1}{5} = 1$ (Remember, one is the multiplicative identity.), the multiplicative inverse of 5 is $\dfrac{1}{5}$.

> In general terms, $a \cdot \left(\dfrac{1}{a}\right) = 1$

Here is another example of the inverse property using mixed numbers:

$$3\frac{1}{2} \cdot \frac{2}{7} \text{ rewritten as } \frac{7}{2} \cdot \frac{2}{7} = 1$$

Zero is the only real number that does not have a multiplicative inverse, since no matter what you multiply it by; the answer is always 0, not the multiplicative inverse, 1.

Finite and Infinite Sets of Numbers

We have been using the terms "set" and "subset" to describe the various number systems. When you are able to count the number of elements in a set, the set is said to be a finite set. If the number of elements is not countable, the set is called an infinite set.

Finite—Countable
Infinite—Not Countable

The following are examples of finite sets:

Description of the set	Listing of the set	Count
The number of letters in your name	$\{a, n, n, i, e\}$	5 letters
The number of numbers between 500 and 700 inclusive	$\{500, 501, 502, 503, \ldots, 700\}$	201 numbers
The counting numbers less than five	$\{4, 3, 2, 1\}$	4 numbers

Sometimes the set contains many numbers, say the counting numbers less than 100. We use the "three-dot" notation, also called an ellipsis, to describe this finite set of numbers, $\{1, 2, 3, 4, \ldots, 99\}$. The three dots suggest that a pattern is formed from the preceding numbers. When the set contains an ending number the set stops at that value.

We use the three-dot notation for infinite sets also. Infinite sets are uncountable so they will not contain an ending number, $\{1, 2, 3, 4, \ldots\}$. This set is the infinite set of numbers that we have earlier defined as the counting numbers.

Description of the set	Listing of the set	Count
The set of even integers	$\{\ldots, -4, -2, 0, 2, 4, \ldots\}$	infinitely many
The number of fractions between 0 and 1	{Any proper fraction greater than 0 and less than 1}	infinitely many
The number of decimal places in the irrational number π	π is a nonterminating, nonrepeating decimal number	infinitely many

Appropriate Mathematical Terminology

The vocabulary used in a mathematics course is both similar to and different from that in other classes you take. Many of the words you use in mathematics will also be used in your science class or your English class. For example, the word "similar" used in your English class means having like characteristics. In your mathematics class, "similar" means having the same size and shape. Those terms are close in meaning. On the other hand, in music the word "chord" refers to a certain combination of notes, but in mathematics the word "chord" refers to a line segment whose endpoints are on a circle. Those two terms are very different.

Building your mathematical vocabulary takes time and work. It is often helpful to use a glossary specifically designed for mathematics; that is, the terms in the glossary are mathematical. At the end of this book, you will find such a glossary. Don't try to memorize the terms all at once—continue to work toward learning them as you develop your mathematical knowledge. Try to familiarize yourself with the words and allow the context of their usage to help you decide the meaning of the words.

Solving Word Problems

There is a good deal of effort and learning devoted to finding the solution to various problems. You learn step-by-step processes called algorithms to find answers. Some processes are straightforward and pretty easy to understand—some are not. Setting up and solving word problems is often difficult. The key to success is being able to translate word phrases into mathematical sentences. Once you have the problem written in mathematical symbols and numbers, the solution to the problem rests on your ability to solve a mathematical problem. Learning how to translate from words to mathematics takes lots of practice.

It is very important to begin by reading the problem through, carefully and completely. Do not try to solve the problem until you know for sure what you are being asked to find. Organize your thoughts—keeping yourself and your thoughts organized will help you think more clearly.

Here are some key words that will help you recognize the operation(s) you need to use to solve the problem.

Addition	Subtraction	Multiplication	Division
Sum	Minus	Of	Quotient
Increased by	Difference	Product	Divided by
More than	Decreased by	Times	Ratio of
Altogether	Less than	Multiply	Per
Total of	Fewer than	Twice (times 2)	Out of
Added to	Less	Doubled	Divisor

Word Phrases	Mathematical Expression
The ratio of a to b	$\dfrac{a}{b}$
5 less than y	$y - 5$
The sum of 4 and y	$4 + y$
3 more than x	$3 + x$
7 less q	$7 - q$
twice x	$2x$
m increased by 6	$m + 6$
half of x	$\dfrac{1}{2}x$
the product of a and b	ab
m divided by n	$\dfrac{m}{n}$

Mathematical sentences have their own "verbs" that represent certain words or phrases. The most often used mathematical verb is equals (=). Below is a list of words or phrases that could be replaced with an equal sign.

is	are	was
were	will be	has
have	gives	yields

Examples:

Mary (M) *is* 2 years older than Jim (J). $M = J + 2$
Adding 72 to some number x *yields* 100. $x + 72 = 100$
Five less W *will be* 89. $5 - W = 89$

Some other mathematical verbs are listed below.

Is greater than	>		Is less than	<
Is greater than or equal to	≥		Is less than or equal to	≤
Is at least	≥		Is at most	≤

This may be the time to talk about the possible confusion that can take place when using the following three phrases. They sound very similar and yet are represented by different mathematical phrases or sentences.

"7 less a number x" is represented as $7 - x$	The expression, "7 less a number x," is simply taking some value from 7.
"7 less than a number x" is represented as $x - 7$	The expression, "7 less than a number x," is simply taking 7 from some value x. The word "than" changes the meaning and order.
"7 is less than the number x" is represented as $7 < x$	The inclusion of the word "is" in this example indicates that one of the mathematical "verbs" should be used—notice the use of the symbol, <.

The material covered in word problems can involve a multitude of subjects:

Money	Geometry	Ages
Science	Business	Travel and/or Distance
Numbers	Percents	Sports
Weather	Measurement	Work

Having general knowledge in these areas is very helpful. Also, as mentioned earlier, learning how to do word problems takes lots of practice—the more practice, the greater the understanding.

Example: Juan installs car alarms. It takes him about 1 hour and 20 minutes to complete an installation. Estimate how many alarms he can install in during an 8-hour work day.

Installations	Time Needed
1	1 hour 20 minutes
2	2 hours 40 minutes
3	3 hours 60 minutes = 4 hours
4	5 hours 20 minutes
5	6 hours 40 minutes
6	7 hours 60 minutes = 8 hours

Juan can install 6 alarms during an 8-hour work day.

Example: Your restaurant bill was $56.30. What would be a 15% gratuity (tip)?

The gratuity can be determined by finding 15% of $56.30, which is approximately $8.45.

The gratuity can also be determined using estimation skills and some mental arithmetic by realizing that 15% is 10% + 5%, and that 10% of $56.30 ($5.63) can be found by moving the decimal point one place to the left. Then 5% of $56.30 is just half of what 10% was ($2.82). Therefore, the gratuity would be 10% ($5.63) plus 5% ($2.82) or $8.45.

Simplifying Numerical Expressions and Order of Operations

Being able to use the rules for operating with signed numbers is necessary for success in mathematics. Numbers may be positive or negative. There are concepts and rules related to working with as well as operating on these numbers.

Absolute Value

The absolute value of a number is defined as being the distance that number is from zero. That is, because –3 is three units from 0, its absolute value is 3. The number 3 also has an absolute value of 3 because its distance from 0 is 3.

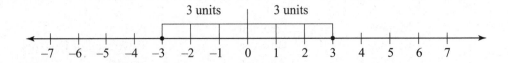

The symbol that is used to denote absolute value is $|\ |$. So the absolute value of –7 is written as $\left|-7\right|$ and is equal to 7.

Addition of Real Numbers

To add two numbers with like signs, add their absolute values and write the common sign.

Example: Find $-7 + (-4)$ $\left|-7\right| = 7$ $\left|-4\right| = 4$ $7 + 4 = 11$

Use the common negative sign.
Therefore, $-7 + (-4) = -11$

To add two numbers with unlike signs, find the difference of their absolute values and write the sign of the number with the greater absolute value.

Example: Find $3 + (-8)$ $\left|3\right| = 3$ $\left|-8\right| = 8$ $8 - 3 = 5$

Use the sign of the number with
the larger absolute value (– 8).
Therefore, $3 + (-8) = -5$

Subtraction of Real Numbers

Subtraction of real numbers is defined as "adding the opposite"; $a - b$ means $a + (-b)$. Change the operation to addition and change the sign of the second number to its opposite. Once the problem is written as an addition problem, the rules for addition of signed numbers apply. Let's look at some examples.

Examples:

$5 - 3 = 5 + (-3) = 2$	$-4 - 5 = -4 + (-5) = -9$
$12 - 15 = 12 + (-15) = -3$	$5 - (-3) = 5 + 3 = 8$
$6\frac{2}{5} - 8\frac{4}{5} = 6\frac{2}{5} + \left(-8\frac{4}{5}\right) = -2\frac{2}{5}$	$-2.6 - 1.2 = -2.6 + (-1.2) = -3.8$

Multiplication and Division of Real Numbers

Rules for multiplying signed numbers:

- The product of two numbers with like signs is positive.

- The product of two numbers with unlike signs is negative.

Examples:

$5 \cdot (-3) = -15$
$-2 \cdot (-5) = 10$
$-6 \cdot 2 = -12$

Rules for dividing signed numbers:

- The quotient of two numbers with like signs is positive.

- The quotient of two numbers with unlike signs is negative.

Examples:

$\dfrac{15}{-3} = -5$
$\dfrac{-12}{-3} = 4$
$-10 \div 5 = -2$

Order of Operations

The order in which mathematical operations are performed is defined by what mathematicians identify as the "Order of Operations" and is often remembered using the mnemonic, "**P**lease **E**xcuse **M**y **D**ear **A**unt **S**ally." The first letter of each of the words in this mnemonic represents an operation or group of operations.

P	**Parentheses**
E	**Exponents**
M, D	**Multiply or Divide**
A, S	**Add or Subtract**

P	Parentheses	Do the operations that are contained within parentheses, braces, brackets, or bars.
E	Exponents	Find the value of numbers with exponents (or powers).
M, D	Multiply or Divide	Then do all multiplications and divisions, from left to right.
A, S	Add or Subtract	Finally, do all additions and subtractions, from left to right.

It is very important to remember that the left to right rule applies to multiplication and division, as well as addition and subtraction.

Examples:

Parentheses
$7(2 + 3 + 5)$
$7 \cdot (10)$
70

Exponents
$2 + 3^2$
$2 + 9$
11

Multiplication and/or Division
$9 + 12 \div 4 \cdot 3$
$9 + \quad 3 \cdot 3$
$9 + \quad \quad 9$
18

Addition and/or Subtraction
$7 - 3 + 1 - 2$
$4 \quad + 1 - 2$
$5 \quad \quad - 2$
3

Scientific Notation

Scientific Notation
5.76×10^3

The format for writing a given decimal number in scientific notation is to write it as a product of two factors: (a decimal greater than or equal to 1 but less than 10) × (a power of 10). Here is an example of a number in scientific notation.

Scientific notation	Decimal factor	Power of ten
5.76×10^3	$1 \leq 5.76 < 10$	10^3 is a power of 10 that represents 1,000
5.3×10^{-4}	$1 \leq 5.3 < 10$	10^{-4} is a power of 10 that represents 0.0001

One of the skills that is required on the AIMS test is to convert a decimal number to scientific notation or convert a number in scientific notation to its decimal form. The tricky part in these conversions is dealing with the exponent.

Here is a quick review of the positive and negative powers of ten. Notice the powers of ten. Pay close attention to the exponents and the value of the number.

Positive powers of ten	Expanded decimal form
10^1	10
10^2	100
10^3	1,000
10^4	10,000
10^5	100,000

Negative powers of ten	Expanded decimal form
10^{-1}	0.1
10^{-2}	0.01
10^{-3}	0.001
10^{-4}	0.0001
10^{-5}	0.00001

Do you notice the pattern? As the powers of ten increased the number in expanded form got larger. The powers of ten that have positive exponents will always be greater than or equal to ten. The number of zeroes is the same as the exponent of ten. The powers of ten that have negative exponents are smaller than one. The number of decimal places is the same as the absolute value of the exponent.

There is a relationship between the exponent of ten and the moving of the decimal point (the direction and number of places). Let's look at how to write decimal numbers using scientific notation.

Examples: The problem: Rewrite 456,000 in scientific notation.

1. Find the decimal point. If the decimal is not displayed, it is at the end of the number. Move the decimal point to a position that makes the number's value greater than or equal to 1 but less than 10.

4.56 is the number between 1 and 10

2. Count the number of places you moved the decimal point. In this case, you moved the decimal point 5 places to the left.

3. The exponent on the power of ten is the same number as the number of places in step two. Because 456,000 is a large number, the exponent should be positive.

10^5

4. Solution.

$456,000 = 4.56 \times 10^5$

The problem: Rewrite 0.5632 in scientific notation.

1. Move the decimal point in the given number to a position that makes the number's value greater than or equal to 1 but less than 10.

$0.5.632$
$\rightarrow$

5.632 is the number between 1 and 10

2. Count the number of places you moved the decimal point. In this case, you moved the decimal point 1 place to the right.

$0.5.632$
1

3. The exponent on the power of ten is the same number as the number of places in step 2. Because 0.5632 is a number less than 1, the exponent should be negative.

10^{-1}

4. Solution.

$0.5632 = 5.632 \times 10^{-1}$

Changing a number in scientific notation to a number in decimal notation is the reverse process and, therefore, the decimal moves in the other direction. A positive exponent causes the decimal point to move to the right. Recall that positive exponents mean the decimal value will be greater than ten. A negative exponent causes the decimal point to move to the left. Negative exponents mean the decimal value will be less than one. You may need to add zeroes as placeholders.

If the power of ten is a positive number, move the decimal point to the right the same number of places as the exponent.	$2.378 \times 10^2 = 2.37.8 = 237.8$ $\rightarrow$	Notice by moving the decimal point to the right, the number increased in value.
If the power of ten is a negative number, move the decimal point to the left the same number of places as the exponent.	$4.37 \times 10^{-3} = .004.37 = 0.00437$ $\leftarrow$	Notice by moving the decimal point to the left, the number decreased in value.

Computing in Scientific Notation

Computing in scientific notation requires the use of the commutative and associative properties.

Example: Consider the following problem: $(2.5 \times 10^4)(3.7 \times 10^3)$

In this problem, the commutative and associative properties are used to group the decimal numbers 2.5 and 3.7 together and to group the powers of 10 together. The numbers 2.5 and 3.7 are multiplied and the base 10 numbers are multiplied.

Remember exponents count factors; therefore, 10^4 means you are multiplying four factors of 10, and 10^3 means you are multiplying three factors of 10. $10^4 \times 10^3$ is a total of seven factors of 10.

$$(2.5 \times 10^4)(3.7 \times 10^3) = (2.5 \times 3.7)(10^4 \times 10^3)$$ Commutative and associative properties
$$= 9.25 \times 10^7$$ Multiplication and properties of exponents

Example: Consider the following division problem: $\dfrac{8.4 \times 10^5}{2.0 \times 10^3}$

As in the previous example, the commutative and associative properties are used to group the decimal numbers 8.4 and 2.0 together and to group the powers of 10 together. The number 2.0 divides into 8.4 and the base 10 numbers are divided.

$$\frac{8.4 \times 10^5}{2.0 \times 10^3} = \left(\frac{8.4}{2.0}\right) \times \left(\frac{10^5}{10^3}\right)$$ Commutative and associative properties

Remember exponents count factors; therefore, 10^5 means you are multiplying five factors of 10, and 10^3 means you are multiplying three factors of 10.

$$10^5 \div 10^3 = \frac{10^5}{10^3} = \frac{\cancel{10} \cdot \cancel{10} \cdot \cancel{10} \cdot 10 \cdot 10}{\cancel{10} \cdot \cancel{10} \cdot \cancel{10}} = 10^{5-3} = 10^2$$

10^2 is two factors of 10.

$$\frac{8.4 \times 10^5}{2.0 \times 10^3} = \left(\frac{8.4}{2.0}\right) \times \left(\frac{10^5}{10^3}\right)$$ Commutative and associative properties

$$= 4.2 \times 10^2$$ Multiplication and properties of exponents

Example: Consider the following problem: $(2.6 \times 10^6)(4.2 \times 10^3)$

This time when multiplying the 2.6 and 4.2, you get a product that is not between 1 and 10. Complete the problem as usual and in the last step move the decimal point and change the exponent for the base 10 to make your answer in proper form.

$$(2.6 \times 10^6)(4.2 \times 10^3) = (2.6 \times 4.2)(10^6 \times 10^3)$$ Commutative and associative properties

$$= 10.92 \times 10^9$$ Multiplication and properties of exponents

$$= 10,920,000,000$$

$$= 1.092 \times 10^{10}$$ Final answer in scientific format

In the preceding example, 10.92 is not in scientific notation format. It must be changed to a number that is greater than or equal to 1 and less than 10. Let's look at the number, 10.92×10^9. Its value is 10,920,000,000. When you use the rules to change decimal numbers to scientific notation, it becomes 1.092×10^{10}.

Subscripts

a_1 = the first term
a_2 = the second term
a_3 = the third term
a_4 = the fourth term

$\vdots$

a_n = the nth term

Subscripts are used in mathematics to indicate a different value for a similar variable. For example, two ordered pairs might be represented as (x_1, y_1) and (x_2, y_2). The first ordered pair is read as "x sub 1, y sub 1." The 1 is a subscript for the variables x and y. Likewise, we use a 2 in the second ordered pair as a subscript, to distinguish the ordered pairs.

Subscripts are used in the following formulas.

Example:

$$\text{Slope} = \frac{y_2 - y_1}{x_2 - x_1}$$

If $(x_1, y_1) = A\,(1, 2)$ and $(x_2, y_2) = B\,(3, 5)$, what is the slope of $\overrightarrow{AB}$?

$$\frac{y_2 - y_1}{x_2 - x_1} = \frac{5 - 2}{3 - 1} = \frac{3}{2}$$

Example:

$$Distance = \sqrt{(x_2 - x_1)^2 + (y_2 - y_1)^2}$$

If $(x_1, y_1) = A(3, 5)$ and $(x_2, y_2) = B(1, 2)$, what is the length of $\overline{AB}$?

$$\begin{aligned} AB &= \sqrt{(x_2 - x_1)^2 + (y_2 - y_1)^2} \\ &= \sqrt{(1-3)^2 + (2-5)^2} \\ &= \sqrt{(-2)^2 + (-3)^2} \\ &= \sqrt{4+9} \\ &= \sqrt{13} \end{aligned}$$

Subscripts are also used in the study of sequences and series.

$$a_1 = 3$$
$$a_n = a_{n-1} + 2$$

In this example, 1, n, $n - 1$ are subscripts.

2nd term	3rd term	4th term
$a_2 = a_{2-1} + 2$	$a_3 = a_{3-1} + 2$	$a_4 = a_{4-1} + 2$
$a_2 = a_1 + 2$	$a_3 = a_2 + 2$	$a_4 = a_3 + 2$
$a_2 = 3 + 2$	$a_3 = 5 + 2$	$a_4 = 7 + 2$
$a_2 = 5$	$a_3 = 7$	$a_4 = 9$

In the above example, $a_1 = 3$ means the first term of the sequence is 3. To find the next term (a_2), use a_1, which is 3, and then add 2. To find the next term (a_3), use a_2, which is 5, and add 2. Continue in the same manner adding 2 to the previous term to find the next term.

This generates the sequence: 3, 5, 7, 9, 11, 13, . . .

Term	1	2	3	4	5	6
Sequence value	3	5	7	9	11	13

Notice that the sequence numbers are placed in the table with the corresponding term number. It is helpful to make a table like the one above when creating a sequence.

Estimation

A great skill to develop is the skill of estimation. In the problem-solving process, it is helpful to estimate your solution. It is like "looking at the big picture" and determining a reasonable guess. Consider finding the mean or average of five numbers. You have learned to add the five numbers together and then divide by 5 (the number of numbers). However, what if the numbers were very large or you had many more than just five numbers? The process of finding the average can be accomplished, but by using estimation, you may be able to arrive at a reasonable solution more quickly.

It is often helpful to rewrite fractions and decimals as approximate whole numbers. If you are buying eight baseballs for $8.78, a ballpark estimate would be 8 times $9.00 or approximately $72.00. Suppose that you need 15 pieces of wood, each piece being 5 foot 8 inches long. You could round the length to 6 feet and multiply by 15 and get 90 feet. This value is greater than the actual length needed, but would serve as a quick estimate.

If you are working with very large numbers, you can round to a power of ten. There may be 1,234,567 bananas in a truck. Choosing from the many values that you could round the number of bananas to, let's round it to 1,000,000 (one million) bananas. You have 10,029 monkeys to share the bananas, which could be rounded to 10,000 (ten thousand) monkeys. The estimate of how many bananas each monkey will get is found by dividing 1,000,000 by 10,000; therefore, each monkey would get approximately 100 bananas.

We use the process of estimation to help determine a reasonable range of values for a solution. This is not to say that "close enough" is "good enough," but rather is my answer "reasonable"?

Working with irrational numbers is a good example of a time when you would want to use approximations. Since an irrational number is a nonrepeating, nonterminating decimal number, we can only approximate its value with rational numbers. Consider the square root of 33. Since 33 is not a perfect square, its square root is an irrational number. If we recognize that the whole number 33 is between 25 and 36 (two numbers that are perfect squares), we can approximate the irrational square root of 33 to be bigger than 5 and smaller than 6.

$$\sqrt{25} < \sqrt{33} < \sqrt{36}$$
$$5 < \sqrt{33} < 6$$

An often-used irrational number is pi (π). If you are computing the area of a circle, you will recall the formula for the area of a circle is $A = \pi r^2$. Most students use 3.14 or $\frac{22}{7}$ as an approximation of π. If a circle has a radius of 5.27 inches, with rounding, the radius is 5 inches and squaring that value would equal 25. If you were to use just 3 as the approximation for π, you could estimate the area to be a little greater than 75 (25 times 3) square inches.

In conclusion, understanding how numbers work, or making "sense" of numbers, is the foundation for the study of any and all parts of mathematics. Therefore, it is important for you to have a good grasp of the concepts in this strand as you study the rest of the strands. You will see these concepts in action in Strands 2 through 5. Refer back to this strand as needed through the remainder of your preparation for the AIMS test.

Are you able to do each of the following?

❏ Classify real numbers as natural, whole, integers, rational, or irrational?
❏ Identify the properties of real numbers: commutative, associative, distributive, identity, inverse, closure?
❏ Distinguish between finite and infinite sets of numbers?
❏ Solve word problems?
❏ Compute using scientific notation?
❏ Simplify numerical expressions?
❏ Use estimation strategies reasonably and fluently?

Practice Problems—Strand 1

1. The set of real numbers shown below is a subset of which of the following?

$$\left\{-1, -1\frac{1}{2}, 0.6679, 3, 4\right\}$$

   A. The set of integers
   B. The set of counting numbers
   C. The set of rational numbers
   D. The set of irrational numbers

2. Which of the following is a set of rational numbers?

   A. $\left\{\frac{1}{2}, \sqrt{4}, 7.3\right\}$

   B. $\left\{\frac{1}{4}, \sqrt{3}, 3.6\right\}$

   C. $\left\{\frac{1}{5}, \sqrt{6}, 5.12\right\}$

   D. $\left\{\frac{1}{6}, \sqrt{9}, 2.010010001\ldots\right\}$

3. Which property is demonstrated below?

   $A = 2C$ and $2C = D$, so $A = D$

   A. Associative property
   B. Commutative property
   C. Reflexive property
   D. Transitive property

4. Which of the following illustrates the associative property of multiplication?

   A. $5(10+y)=5(10)+5(y)$

   B. $(4\cdot 5w)2=2(4\cdot 5w)$

   C. $6(15\cdot x)=(6\cdot 15)x$

   D. $3+(4+5z)=(3+4)+5z$

5. Which of the following sets of numbers is NOT finite?

   A. $\{1, 2, 3, \ldots, 100\}$
   B. $\{-8, 2, 0, 7, \sqrt{81}\}$
   C. {positive integers}
   D. {two digit natural number}

6. Which of the following is finite?

   A. Counting numbers less than 5
   B. Even integers less than 5
   C. Rational numbers less than 5
   D. Real numbers less than 5

7. A football is advertised at a special discount of ¼ off of the regular price of $52. What is the sale price of the football?

   A. $13.00
   B. $39.00
   C. $65.00
   D. $208.00

8. Nita worked 40 hours per week for 3 weeks. She earns $7.25 an hour. What will be her total gross pay (pay before deductions)?

   A. $290.00
   B. $507.50
   C. $870.00
   D. $2,900.00

9. Mars is about 142,000,000 miles from the sun. Pluto is about 26 times farther from the sun than Mars. What operation would be used to approximate the miles Pluto is from the sun?

   A. Addition
   B. Subtraction
   C. Multiplication
   D. Division

10. What operation would be used to determine the unit cost of an ounce of pineapple juice if you can buy 12 ounces for $0.96?

    A. Addition
    B. Subtraction
    C. Multiplication
    D. Division

11. What is the value of the expression below?

$$4+\left|-10+8\right|+\left|6-11\right|$$

    A. 3
    B. 11
    C. 27
    D. 39

12. What is the value of the following expression?

$$-2\left|4-7\right|+9^2\div 3$$

    A. −3
    B. 21
    C. 25
    D. 29

13. Consider the following number pattern.

$$5, 8, a_1, 14, a_2, \ldots$$

    What is the value of a_2?

    A. 16
    B. 17
    C. 18
    D. 19

14. Which list was generated from the statements:

$$a_1 = 5$$
$$a_{n+1} = 2a_n$$

    A. $\{5, 7, 9, 11, \ldots\}$
    B. $\{5, 10, 15, 20, \ldots\}$
    C. $\{5, 10, 12, 14, \ldots\}$
    D. $\{5, 10, 20, 40, \ldots\}$

15. What is the coefficient of the second term in the following polynomial?

$$P = 12x^3 + 4x^2 - 2xz + 9y$$

    A. −2
    B. z
    C. 4
    D. y

16. William declared the following numbers to be prime numbers.

$$\{3, 11, 39, 43\}$$

    He made an error. Which one in this list is **not** prime?

    A. 3
    B. 11
    C. 39
    D. 43

17. Determine the value of the following expression.

$$\frac{(3.6\times10^4)(2.8\times10^7)}{(1.6\times10^5)}$$

    A. 6.3×10^6
    B. 6.3×10^{16}
    C. 6.3×10^{23}
    D. 6.3×10^{33}

18. Bombay, India, has a population of about 11 million and a land area of only 96 square miles. Estimate the number of people per square mile in scientific notation.

    A. 1.15×10^3
    B. 1.15×10^5
    C. 1.15×10^7
    D. 1.15×10^9

19. What is the value of the following expression?

$$7+2\sqrt{4}-(9+5)\div 7$$

A. 0
B. 7
C. 9
D. 17

20. What is the value of the following expression?

$$\sqrt{9}-8\div 2+2$$

A. −3
B. −1.25
C. 0.5
D. 1

21. Jamie earned the following scores on his tests in his math class.

58%, 86%, 72%, 79%, 68%

Which estimation is closest to the actual mean of his test scores?

A. 62%
B. 73%
C. 86%
D. 94%

22. What is the best estimate for the following expression?

$$4.41\times 10^{-2} + 1.63\times 10^{-2}$$

A. 6.0×10^{4}
B. 6.0×10^{-4}
C. 6.0×10^{-3}
D. 6.0×10^{-2}

23. Jack worked for 27.35 hours at a pay rate of $5.15 per hour. What would be a reasonable estimate of his wages before taxes?

A. $28.00
B. $100.00
C. $135.00
D. $315.00

24. On a recent floating river raft trip, Josie was told the total distance traveled was $5\frac{2}{3}$ miles and the overall time in the water was 30 minutes. The raft did not have a motor so the speed at which they traveled was the same as the speed at which the water was moving. Which of the following is a reasonable approximation of the rate of the water?

A. 5 miles/hr
B. 7 miles/hr
C. 9 miles/hr
D. 11 miles/hr

25. Which value is closest to $\sqrt{84}$?

A. 8.1
B. 9.2
C. 9.9
D. 10.1

26. Point X is graphed on the number line as shown below.

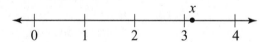

Which of the following numbers is closest to the coordinate of point X?

A. $\sqrt{6}$

B. $\sqrt{8}$

C. $\sqrt{11}$

D. $\sqrt{15}$

Answer Key

1.	C	6.	A	11.	B	16.	C	21.	B	26. C
2.	A	7.	B	12.	B	17.	A	22.	D	
3.	D	8.	C	13.	B	18.	B	23.	C	
4.	C	9.	C	14.	D	19.	C	24.	D	
5.	C	10.	D	15.	C	20.	D	25.	B	

Answers Explained

1. **C** The set of numbers contains fractions and decimals and these numbers are members of rational numbers. Notice that this list does include counting numbers (3 and 4), as well as the −1, which is from the set of integers. When describing the solution, the set that contains all of the given numbers is required. There are no irrational numbers in the set.

2. **A** Choices B, C, and D each contain an irrational number and must be excluded. The set of rational numbers is only the numbers from option A.

3. **D** $2C$ acts as the connector between A and D. This is an example of the transitive property.

4. **C** A and D are examples of the distributive property, and B is an example the commutative property. Notice in the associative property the grouping symbols are around different factors. The associative property allows different grouping of the factors while keeping the value of the expression the same.

5. **C** A finite set of numbers is a set of numbers whose members can be counted. Choices A, B, and D have members that can be counted. It is not possible to count the members in option C, the integers.

6. **A** The integers, rational numbers, and real numbers all include infinitely many numbers on the left side of 0 on the number line. However, the last counting number (or natural number) as you move toward the left on the number line is 1.

7. **B** Find the discount (¼ of 52) and subtract it from 52. You could also find ¾ of 52.

8. **C** Multiply 40 times 3 to find the total hours worked. Next, multiply 120 times $7.25 to find the pay.

9. **C** The second sentence contains the phrase "26 times farther," which indicates multiplication.

10. **D** Unit cost means cost per unit, in this case, cost per ounce. The word "per" indicates division.

11. **B** Simplify the absolute value expressions first and then combine terms from left to right by adding.

$$4+\left|-10+8\right|+\left|6-11\right|$$
$$4+\left|-2\right|+\left|-5\right|$$
$$4+2+5$$
$$6+5$$
$$11$$

12.   **B**   Use the order of operations (PEMDAS). Notice the absolute value bars are symbols of enclosure just as parentheses.

$$-2\left|4-7\right|+9^2\div 3$$
$$-2\left|-3\right|+9^2\div 3$$
$$-2(3)+9^2\div 3$$
$$-2(3)+81\div 3$$
$$-6+81\div 3$$
$$-6+27$$
$$21$$

13.   **B**   The second term is 3 more than the first. If the next term is 3 more than the second, it would be 11. The fourth term is 3 more than 11. Therefore, the fifth term, identified as a_2, would be 17.

14.   **D**   The second given statement is saying that each next term is found by taking 2 times the previous term; that is, the second term is found by taking 2 times the first term, and the third term is found by taking 2 times the second term.

15.   **C**   The second term is $4x^2$ and the coefficient in that term is 4.

16.   **C**   A prime number has only one set of factors and those factors must be unique. Because 39 has the factors 1 and 39 and 3 and 13, it is not prime.

17.   **A**   Multiply 3.6 times 2.8 and divide the result by 1.6. That is 6.3. Then add the exponents on 10 in the numerator and subtract the exponent on 10 in the denominator: $4 + 7 - 5 = 6$. Therefore, the exponent on 10 in the answer is 6.

18.   **B**   Eleven million, written in scientific notation, is 1.1×10^7. Ninety-six rounds to 100. One hundred, written in scientific notation is 1.0×10^2. Therefore, it would be reasonable for the answer to be close to 1.15×10^5.

19.   **C**   Using the order of operations (PEMDAS):

$$7+2\sqrt{4}-(9+5)\div 7$$
$$7+2\sqrt{4}-(14)\div 7$$
$$7+2(2)-(14)\div 7$$
$$7+4-(14)\div 7$$
$$7+4-2$$
$$11-2$$
$$9$$

20.   **D**   Using the order of operations (PEMDAS):

$$\sqrt{9}-8\div 2+2$$
$$3-8\div 2+2$$
$$3-4+2$$
$$-1+2$$
$$1$$

21.   **B**   There is more than one way to arrive at an estimate of this answer. Consider estimating each score individually to the nearest tens (60%, 90%, 70%, 80%, 70%) and then finding the mean by addition and dividing by 5 (370 ÷ 5 = 74). Or find the average by adding the given scores and dividing by 5 (58 + 86 + 72 + 79 + 68 = 363, 363 ÷ 5 = 72.6). Both of these solutions, 74 and 72.6, are closest to 73%.

22.   **D**   Because each of the addends contains 10^{-2}, they are like terms. That means you can add 4.41 + 1.63, which is 6.04 and is closest to choice D. Remember, when adding like terms, the base and exponent do not change. Simply combine the coefficients.

23.   **C**   Using estimates for each of the numbers, a reasonable estimate could be obtained by multiplying 27 times 5, which is 135.

24.   **D**   In looking for a reasonable answer, it's important that the answer make sense. You do not have to do a written calculation. You can solve the problem with a simple mental calculation. The distance traveled is close to 5½ miles and the time it took to travel those miles is ½ hour. Double the amount traveled in a half hour to find the distance traveled per hour.

25.   **B** $\sqrt{84}$ is between the perfect squares $\sqrt{81}$ and $\sqrt{100}$ which equal 9 and 10. Notice that 84 is closer to 81, so choose the estimate closer to 9 than 10. 9.2 is the best estimate.

26.   **C** X is between 3 and 4, which equal $\sqrt{9}$ and $\sqrt{16}$. Since X is closer to 3 than 4, the best estimate is $\sqrt{11}$.

Strand 2 | Data Analysis, Discrete Mathematics, and Probability

When you complete your study of Strand 2 you will be able to:

- Understand concepts related to data collection
- Represent data using a variety of graphs, charts, and plots
- Understand and use measures of central tendency
- Make reasonable predictions and draw inferences related to data presented in a variety of forms
- Draw and interpret a line of best fit

As you recall, the first strand dealt with number sense. Number sense helps to define and clarify the types of numbers used in any problem involving mathematics. It is very important to have a firm grasp of the vocabulary and the mathematical properties on which we base our calculations. We want to make sure that when we complete a calculation, it is correct. As we continue studying each of the strands, pay close attention to the vocabulary.

The focus of the second strand of the Arizona mathematics standard is statistics and probability. This strand, like the first strand, deals with foundations of mathematics. **Statistics** is defined as the collection, organization, description, and analysis of numerical data—in other words how to make sense of numerical information. As we collect data, we can use graphs, diagrams, and terms that summarize data to communicate with others. The concept of probability is also included because we want to know the likelihood of the occurrence of the data.

As we collect data (or information), we often will use graphs and pictures to describe the data. Shape, symmetry, and tendencies can become quite useful in conveying meaning about data. It is very easy to become overwhelmed with seemingly unrelated numbers. However, when you order, categorize, and form basic groups, you get a better handle on the data. An appropriate graph or chart will help communicate your findings to others.

Part of the process of drawing conclusions involves making a prediction. If we have reliable data, we can use this to reveal future data. Conclusions, predictions, and interpreting the data depends on having good data that are well organized and is presented using accurate concepts and calculations. If the data are not good, the conclusions are more apt to be incorrect. Sometimes we need to make conclusions based on data that are not the best but are all we have.

Although applications of mathematics can be found in all areas of the high school curriculum, the topics in Strand 2 are truly found "across the curriculum." The analysis of data can take place in a science class or a social studies class; charts and graphs appear in a child development class as well as a business class or a physical education class. The analysis of data allows you to make predictions related to population growth in foreign countries in your geography class. Analyzing data related to the number of Spanish-speaking people in the United States would be helpful in studying the number of bilingual people in the United States.

Discrete mathematics, the mathematics used in determining "how many," can be found in genetics problems in a biology class or in the auto mechanics class when determining the number of combinations of options available when purchasing a new car. A simple application of discrete mathematics can be found in any of your classes when your teacher creates groups for any collaborative activity you do with your classmates. How many different groups of three can your teacher make with the 30 students in your class?

The amount of statistical information available to help make decisions in business, politics, research, and everyday life is staggering. Consumer surveys guide the development and marketing of products. Experiments evaluate the safety and usefulness of new medical treatments. Statistics sway public opinion on issues and represent—or misrepresent—the quality and effectiveness of commercial products. Through experiences with the collection and analysis of data, students learn how to interpret such information.

As this chapter is developed and explained, keep thinking of all of the places and times when you have encountered an application of data analysis, probability, or discrete mathematics. We think you will realize how much these topics permeate your everyday life.

Data Analysis

The first major concept in Strand 2 is data analysis. The three areas of study are (1) collecting data, (2) displaying data, and (3) interpreting data.

Collecting Data

The first thing that needs to be considered as you collect data is the questions that will be used to collect the data or to determine what data need to be collected. These questions can be those asked in taking a survey, or they can be questions that a business or organization would need answered to appropriately and efficiently use collected data to make decisions related to further business operations.

The collection of data and the methods used to collect the data are very important. The value and usefulness of the collected data will be determined by several factors. The choice of questions asked on a survey and the individuals taking the survey contribute to the value of the data. How the collected data are organized helps convey the information to others.

Example: The questions asked for a survey should appropriately speak to the issues needing to be resolved.

- If your survey is trying to determine the favorite fast food choices of students in a high school cafeteria, be sure you limit the questions to specific fast food choices as opposed to asking questions related to overall food likes and dislikes.

- If a business is trying to predict future population growth in a certain area, the statistics they use should be related to that specific area as opposed to a more widespread area.

Example: The questions asked should not contain obvious bias.

- To ask survey participants for their preferences related to favorite movies and limit the choices to G or PG movies would be creating a built-in bias.

- The data a business gathers in making predictions about their future productivity should be very specific and should not seek to gather peripheral data. It may not be appropriate for a business seeking information related to the housing market in an area to include questions related to gas prices.

Example: A survey may be biased in terms of who is being surveyed.

- To conduct a survey about motorcycles outside a motorcycle business would instill bias in the survey.

• A survey conducted with the freshman class in a school to determine student preferences for the junior/senior prom might contain a built-in bias because of lack of interest or subject knowledge.

The questions asked in a survey must be measurable. The numerical results from the survey questions are used to create charts and graphs. A few examples of what the measurements could be are numbers of responses, an amount of money, or a preference taken from a given number of choices.

Displaying Data

Organizing and displaying data are done to get a "picture" of the data, to create a visual that makes the data more easily recognizable and understandable. The process of organizing the data involves tallying the responses, organizing the data in a table or graph, and calculating values of central tendency. Let's begin with the organization of the data.

The data must first be tallied. An example of a tally of the number of movies watched by 36 students during a specific week is shown at the right. It helps to put the numbers in order, either increasing or decreasing, so that you can recognize patterns or trends. The tally below is called a frequency distribution.

Number	Tally	Frequency
10	/	1
9	/	1
8	/	1
7	//	2
6	/	1
5	//////	6
4	////////	8
3	///////	7
2	////	4
1	///	3
0	//	2

In addition to a frequency distribution, data are sometimes tallied using a stem-and-leaf plot. Stem-and-leaf plots have become a popular way of tallying data because of the ease with which information can be identified. On the next page is a set of test scores for a geometry test. The scores are ordered and placed in a stem-and-leaf plot.

| Scores in random order:

93,92,92,82,85,84,96,94,95,93,75,
98,96,75,73,98,89,55,85,98,96

Scores in order largest to smallest:

98,98,98,96,96,96,95,94,93,93,92,
92,89,85,85,84,82,75,75,73,55 | Stem-and-leaf plot:

10\|
9\| 8 8 6 6 6 5 4 3 3 2 2 8
8\| 9 5 5 4 2
7\| 5 5 3
6\|
5\|5

key 9\|2 = 92 |

After the data are tallied, it is often valuable to put the data into a table or chart. A table or chart may be considered a matrix. A **matrix** is a rectangular array of numbers arranged in rows and columns. The plural of the word "matrix" is matrices. The matrix below on the left represents data tallied from a survey about movies. The graph below is a bar graph that shows the data in a visual format. Another way to show data visually is the box-and-whisker plot below the bar graph.

Number of movies watched	Number of students
0	2
1	3
2	4
3	7
4	8
5	6
6	1
7	2
8	1
9	1
10	1

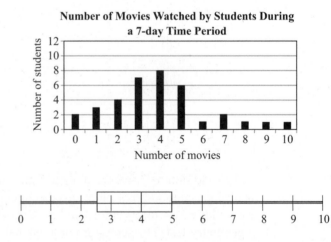

Notice the data related to the movie survey can be displayed in more than one way. This technique of displaying data in multiple ways provides the reader a more complete picture.

Several different types of graphs can be used to display data—some better than others for certain situations. The types of data graphs that we will discuss are bar graphs, line graphs, circle graphs, scatter plots, box-and-whisker plots, and stem-and-leaf plots.

Types of Graphs
• Bar Graph
• Line Graph
• Circle Graph
• Scatter Plot
• Box-and-Whisker Plot
• Stem-and-Leaf Plot

Bar Graph

A bar graph uses solid bars to represent numeric values related to the data. A simple glance at a bar graph allows you to see relationships between the data. Sometimes it is helpful to use a double bar graph to compare sets of data. The double bar graph that follows shows the home run leaders in the American and National Baseball Leagues from 1998 to 2005. Many interesting things are apparent from reading the bar graph. Notice that the National League leader in 1998, 1999, and 2001 had more than 60 home runs but the National League leader in 2002 only had 49 home runs.

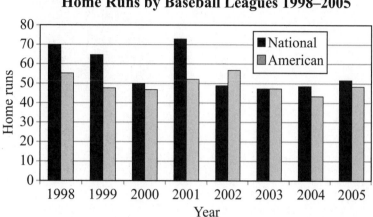

Home Runs by Baseball Leagues 1998–2005

Line Graph

A line graph is best at showing how things change over time. A line graph shows whether a value is increasing, decreasing, or staying the same; line graphs often have one axis showing a measure of time and the other showing some numeric value. The line graph below represents how the batting average for a baseball player on the Arizona Diamondbacks team changed over a period of time.

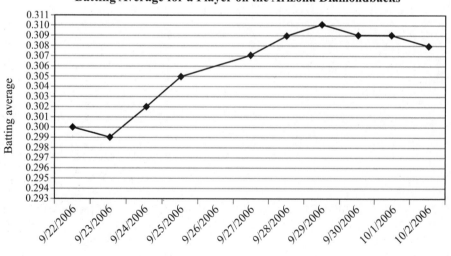

Batting Average for a Player on the Arizona Diamondbacks

Circle Graph

Circle graphs show how a part relates to the whole. Circle graphs are generally used when a large group is separated into smaller groups. The following circle graph shows the types of electric power produced in the United States.

Types of Electric Power Produced in the United States

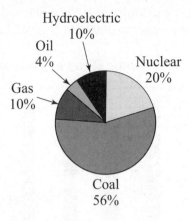

Scatter Plot

Scatter plots, or scattergrams, are used to show the relationship between two sets of data. The corresponding data values are paired together to form ordered pairs. Those ordered pairs are graphed on a coordinate plane containing horizontal and vertical axes. From the graph of these ordered pairs, you can tell if there is a positive or negative correlation between the two sets of data. The data represent a positive correlation if one value increases as the other increases; the data represent a negative correlation if one value decreases as the other increases. The scatter plot below illustrates a positive correlation.

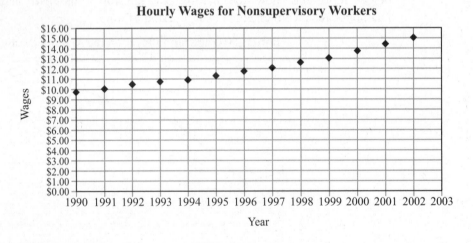

This chart and scattergram relate the weight of a vehicle with its average highway mileage.

Weight (lb)	Mileage (mpg)
2,000	42
2,150	40
2,350	38
2,500	37
2,700	35
3,100	28
3,500	23
3,900	19
4,100	11

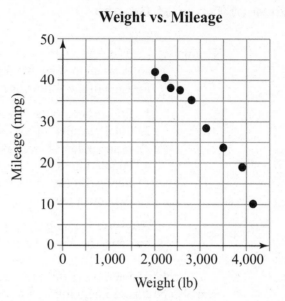

Weight vs. Mileage

The data suggest a negative correlation between the weight and the mileage. As the weight increases the mileage decreases.

Here are data values relating the student's shoe size with a quiz score.

Shoe Size vs. Quiz Score

Shoe size	Quiz score
5	62
5.5	89
7	72
8	90
9	57
9.5	73
10	61
11	98
12	69

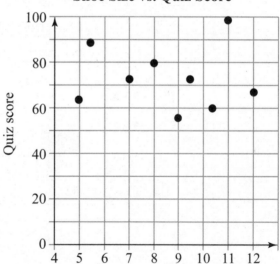

Notice that the data points do not form a pattern, but rather are scattered about. This would suggest that there is no correlation between shoe size and quiz scores; as a shoe size increases the quiz score both increases and decreases. Because there is no correlation positive or negative, shoe size is not a good predictor of quiz scores.

Box-and-Whisker Plot

Box-and-whisker plots are best used to show the concentration of data within a certain range. The median of the data as well as the upper and lower quartile of the data is used to create the box-and-whisker plots. Box-and-whisker plots divide the data into four regions, each representing 25% of the data.

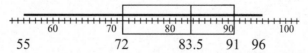

55   72   83.5   91   96

Review of Types of Graphs

Example: Let's look at a variety of topics and decide which type of graph would be most appropriate to show the data. Which type of graph would be most appropriate for each of the following?

1. Price of computers over a span of years
2. Total receipts for a weekend showing of five different movies
3. Comparison of salaries of men and women in five different careers
4. Population of Phoenix metro area by ethnicity
5. Representation of test scores by median and quartile placement
6. Average allowance for 15 year olds over the past 20 years

Answers: 1. line graph; 2. bar graph; 3. double bar graph; 4. circle graph; 5. box-and-whiskers plot; 6. line graph

Graphs can help us better understand the significance and meaning of data, but they can also be misleading. Graphs are misleading when they visually lead us to a false conclusion. For example, the following bar graph shows the number of tenants evicted from apartments in Phoenix, Arizona, in 1998–2001. The bar for 2001 visually appears to be almost three times (an increase of 300%) as tall as the bar for 1998 when, in fact, as the numbers indicate, the increase was only about 40%.

Eviction Filings Increase in Maricopa County

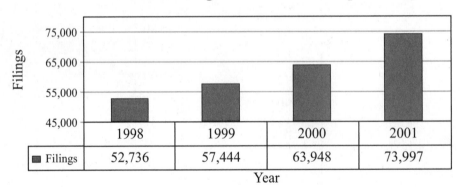

Source: *The Arizona Republic*, January 21, 2003.

There are other ways in which the reader can be misled by a graph. Oftentimes the scale on either the horizontal or vertical axis is incomplete. The vertical scale on the following Miles of Seacoast graph shows the numbers from 500 to 1,200 as missing. This makes the graph visually misleading, making you think that Florida only has a little more than twice as much seacoast as Texas. Florida actually has almost four times as much seacoast as Texas.

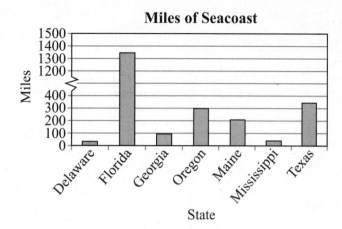

Sometimes the vertical or horizontal scale is intentionally cut to save space. It is very important for you to study the graphs on the AIMS test carefully so that you are sure you are interpreting them correctly.

Unequal intervals on the vertical scale can be misleading. The graph below starts with an interval of ten. You might assume each of the other intervals is also ten, which would be incorrect.

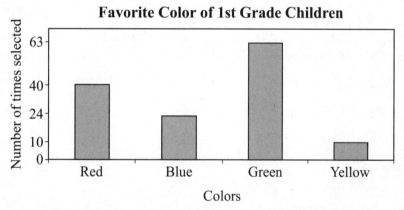

It might seem like a good idea to mark the vertical scale with the exact values for the data, but it can be misleading. To repeat, it is very important for you to study the graphs you are given on the AIMS test to be sure you are interpreting them correctly.

Interpreting Data

Measures of Central Tendency

A measure of central tendency is a value used to better understand the meaning of data. Three measures of central tendency are commonly used in mathematics. They are the mean (or average), the median, and the mode. The following table summarizes the appropriate usage of each of the measures of central tendency.

Measures of Central Tendency
• Mean
• Median
• Mode

Measure of central tendency	Definition	Example
Mean	The mean (or average) of a set of numbers is the sum of the numbers divided by the number of numbers.	The mean is most appropriate when data are continuous and spaced evenly. For example, the batting average of a baseball player represents performance over a time period.
Median	The median is the middle number of a set of numbers when the numbers are written in order.	The results of the earlier example of geometry test scores, with one student getting a score much lower than most of the students, can best be represented using the median as opposed to the mean.
Mode	The mode is the number that occurs most frequently in a group of numbers.	A shoe store that keeps records of the number of items of each style of shoes it sells would be most interested in the mode when collecting sales data. The mode would tell the company which shoe in their inventory was the most popular and, therefore, the one they would carry the most. The mode is best used when data are clustered in one or more specific areas.

The three measures of central tendency, mean, median, and mode, as well as range and quartiles, are probably the most important concepts necessary in summarizing data and data sets.

Let's begin by using the test scores from a previous example (93, 92, 92, 82, 85, 84, 96, 94, 95, 93, 75, 98, 96, 75, 73, 98, 89, 55, 85, 98, 96).

The **mean** (or average) is the sum of the numbers in the data set divided by the number of numbers in the data set. Using the list of scores on the geometry test shown above, the mean would be the sum of the scores (1,844) divided by 21 (number of scores) or 87.8.

The **median** would be the middle of the 21 scores when they are arranged in order (98, 98, 98, 96, 96, 96, 95, 94, 93, 93, 92, 92, 89, 85, 85, 84, 82, 75, 75, 73, 55), which is 92.

The **mode** is the score that appears most often. In this case there is a double mode because 96 and 98 each appear three times.

The **range** of the scores is the difference between the highest and the lowest score, which is 98 − 55, or 43.

Quartiles partition a group of scores into four quarters. The median is the second quartile (92), the median of the lower half is the first quartile (83), and the median of the upper half is the third quartile, (96). That partitioning is shown below.

55 73 75 75 82 | 84 85 85 89 92 (92) 93 93 94 95 96 | 96 96 98 98 98

Let's work with a different data set to practice these concepts. Suppose you work at a restaurant waiting tables. Your tips for twenty consecutive days are shown in the table below.

Days	1	2	3	4	5	6	7	8	9	10	11	12	13	14	15	16	17	18	19	20
Tips (in $)	18	21	33	15	16	22	24	16	13	36	56	40	32	16	24	21	58	40	16	24

What are the values of each of the measures of central tendency for this new set of data?

First, the data should be arranged in order from smallest to largest: 13, 15, 16, 16, 16, 16, 18, 21, 21, 22, 24, 24, 24, 32, 33, 36, 40, 40, 56, 58.

Data Summary

Mean (the average)	27 (rounded to the nearest whole number)
Median (the middle)	23 (the average between the two middle scores, 22 and 24)
Mode	16
Lower quartile	16
Upper quartile	34.5
Range	45

About half the time you can expect to take home at least $23 in tips. The average amount that you get for tips is $27.

Making Predictions and Drawing Inferences

As mentioned earlier, scatter plots are used to show trends in data and to make predictions about the future. Data sets displayed in tables and charts can also show trends over time and can be used to make predictions. For example consider the next problem.

The table shows the cost of a new sports utility vehicle. If the trend continues, how much will a new sports utility vehicle cost in 2006?

Year	Cost
1996	$ 5,000
1998	$10,000
2000	$15,000
2002	$20,000

The table shows that the cost of the vehicle is increasing $5,000 every two years. Since the table ends at 2002, there would two $5,000 increases by the year 2006. Therefore, the cost of the vehicle would be $10,000 more than $20,000, or a total of $30,000.

The scatter plot below represents new seedling growth on Mt. Bigelow over 12 years. Based on the information provided, what would be a reasonable prediction for seedling growth in 2005?

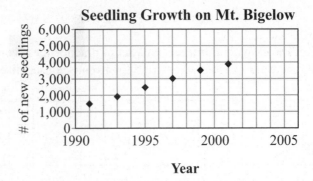

Notice the pattern of the data points is slowly going up. This trend suggests that about 5,000 seedlings would grow on Mt. Bigelow in 2005.

The patterns represented in the preceding examples were linear or near linear. Data patterns are not always linear but still can be used to make reasonable predictions. Shown below is a graph of the world population from 1950 to 2000. Notice the graph is definitely not linear, but we can infer from the graph that the population has increased much more rapidly in the past 50 years than in the 150 years before that.

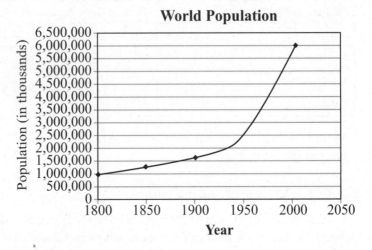

To draw an inference is to create meaning and applicable relationships about the given data. The meaning created is not necessarily stated explicitly but rather is implied.

A bar graph of the female/male percentage of the population of Arizona, Alaska, Utah, New Mexico, and Colorado is shown below. A data table is also shown on the next page. What are some inferences that can be made related to this data?

Female/Male Population Percents for Five Selected States

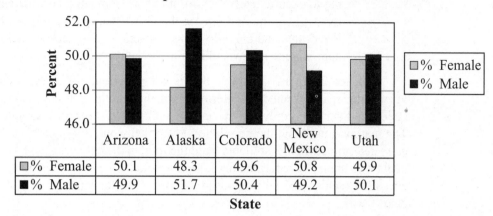

	Arizona	Alaska	Colorado	New Mexico	Utah
☐% Female	50.1	48.3	49.6	50.8	49.9
■% Male	49.9	51.7	50.4	49.2	50.1

State

Did you notice that the female population in Arizona is slightly larger than the male population? This may be attributed to the older population in Arizona and the fact that women tend to live longer than men. Alaska, on the other hand, has a much larger population of males. This may be due to the need for workers to support various industries such as oil, fishing, and timber. What other inferences would you make concerning the table and the data shown?

The process of drawing inferences involves going beyond what is obvious—making predictions, discussing what the statistics imply, creating meaning from the data. The analysis of our data must be reasonable. In the graph comparing the population of men and women, one might conclude that women dislike cold weather and therefore refuse to live in Alaska. Is this a reasonable conclusion?

Sometimes we make unreasonable conclusions based on what we think is true. For example, those of us who live in North American might think all swans are white. However, if you travel to the southern part of Australia you will find black swans.

Census vs. Sampling

Statistical studies often contain a large collection of data. The question becomes: "How many data items are needed to accurately yield useful conclusions?" Collecting data from a part of the total population is called **sampling**. Collecting data from every member of an identified population is called a **census**.

Sample → Part
Census → All

Examples: You are interested in finding how many students in your high school have a part time job. If you were to ask each student in your school, you would be taking a census to answer the question. In a small school, this would not be difficult to do. If your high school is very large, asking every student may be time consuming and not practical. You could choose to take a random sample of students and ask for their response. These responses are then used to make predictions about the entire population.

Sampling in some instances is a necessity. Suppose you are concerned with the number of chocolate chips in your favorite cookie. If you bought every box of chocolate cookies from your grocery store and inspected every cookie, the process would be costly and time consuming. In addition, when you completed the task, a new batch of cookies would arrive to the store and your conclusions may not transfer to future boxes of cookies.

If you wanted to know the number of fish in the lake, it would be impossible to capture every fish for a count. Game and fish officials will often capture, tag, count, and release the fish. Then after a period of time, the process is repeated and the ratio of tagged versus untagged fish helps to determine the approximate number of fish in the lake.

A **biased sample** is a sample that is not representative of the identified population. For example, if I wanted to determine whether motorcycles were safe to drive on city streets and surveyed only the owners of motorcycle shops, my sample would be biased. The method I used and the population I sampled would produce biased results. If I wanted to determine whether students liked the food in the cafeteria and I sampled only the students who ate in the cafeteria, my sample would be biased. Being able to recognize differences between biased and unbiased sampling will enhance your ability to get accurate data.

One way to help avoid bias is to take a random sample. If you were to ask just your friends and/or students that you see on a daily basis, you could draw conclusions, but the conclusions may be biased. By choosing to randomly select students for your sample, you would have an unbiased sample.

Historical Note

Would you be surprised to hear that the United States Armed Forces once had a problem with biased results? In 1970 the Armed Forces held a lottery for drafting young men into the military. Numbered marbles representing each day of the year were placed into a large container. The order that the marbles were withdrawn corresponded with the order in which you would be eligible for the draft. One by one the marbles were drawn, and the dates from the marbles were placed on a large board. At the conclusion of the process it became apparent that the December birthdays were predominantly among the first dates drawn. The dated marbles had been placed into the container by month, and the December dates were placed into the container last. The container was not mixed before the drawing, and consequently the December birthdays were selected early and often. Initially, the error was not detected. When the dates were placed on a scatter plot, the graph did not reveal any irregularities. With the help of statistics, the trend was discovered, and a new drawing was ordered. This time, with the help of statisticians and thorough mixing, two containers were used to determine the draft lottery order. One marble was drawn with the birth date and a second marble was drawn with a number for order placement. This procedure helped to ensure a "random" drawing.

Line of Best Fit

One of the skills necessary in making predictions related to data analysis is to be able to draw a line of best fit for a scatter plot. A **line of best fit** is a line that appears to most closely follow the pattern shown by a data set in a scatter plot. The line of best fit does not have to pass through any of the points in the data set. It is suggested that there should be as many points above the line as below the line, but the line must still fit the flow of the data. The following scatter plot has a line of best fit that is appropriate for the data.

Population of 16-Year-Olds in Arizona Who Are Unemployed

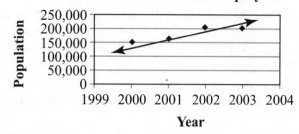

The line of best fit in the diagram below does not follow the flow of the data and would not be a good predictor. In Strand 3 we will talk about finding the equation of the line of best fit and using that equation to make predictions about the data.

Median Income for Male Full-Time, Year-Round Workers in Arizona

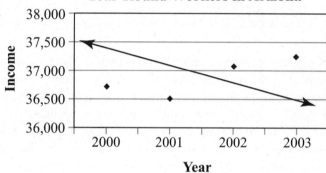

Correlation

The relationship between the two variables in a data set can represent positive, negative, or no correlation.

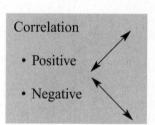

Correlation

• Positive

• Negative

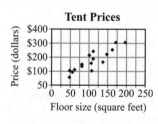

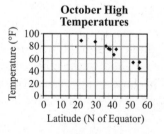

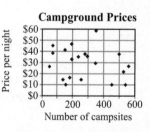

Positive correlation, as represented by the graph above on the left, takes place when one variable value increases as the other also increases. Lines that represent a positive correlation always have a positive slope. Negative correlation, as represented by the graph in the middle, takes place when one variable increases as the other decreases. Lines that represent a negative correlation always have a negative slope. A graph represents no correlation between the variables when there is no appropriate line of best fit, as shown by the graph on the right.

Normal Curve

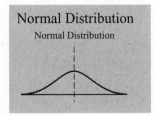

Normal Distribution

Normal Distribution

A graph can take on any overall shape when plotted as a bar graph or listed in a stem-and-leaf plot. The data obtained from many trials often follow a common pattern. This is called a normal distribution. The graph of the normal distribution is called the normal curve. The key characteristics of a normal distribution are

1. The mean, median, and mode are all equal.
2. The normal curve is bell-shaped and symmetrical about the mean value.

Here are some example graphs of the normal distribution:

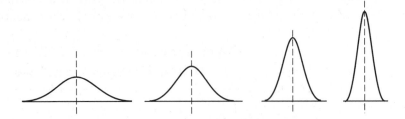

Notice that graphs can be tall and skinny or short and wide. The shape of the graph is determined by the range and spread of the data values. Extreme values, also known as outliers, stretch the graph and may cause the mean to be different than the median. The graph is "skewed" when the mean is different than the median.

An example of skewed data might be shown if you were to graph time in years on the horizontal axis and the number of cars for each year on the vertical axis. The collection of data may need to occur at a large sporting or community event. You might find a large number of recent models and as the years progressed, the number of older cars might begin to decrease.

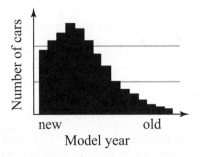

Another example of a graph where the data are skewed could come from a survey of ages. Suppose your youth group planned to visit a retirement home and you wanted to plot all of the ages of those in attendance (including your group). If your graph ranged from youngest to oldest, it would have quite a spread in data values and would be skewed left.

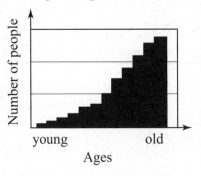

Probability

The second and third major concepts in Strand 2 are probability and discrete mathematics. Probability involves determining in how many ways an event can happen, and discrete mathematics is used in answering the question "how many"? The next section incorporates the concepts of discrete mathematics in the discussion of probability.

$$P(E) = \frac{\text{Desired outcome}}{\text{Possible outcomes}}$$

Probability has been called the Science of Chance. To some the word "chance" might bring up the notion of unpredictability. The following three sentences are typical statements using the word chance.

There is a good chance that something might happen.

There is a 40% chance of rain.

Buy a raffle ticket, and there is a chance that you might win.

Include the term "random," and you can get caught up in emotions and the notion of being "lucky" or "unlucky." So how does probability relate to chance? How can we use the scientific method to investigate the probability of an event?

As usual, we need to begin with defining and clarifying terms. Here are the key terms:

Probability—the ratio of a particular outcome to the total possible outcomes (the measure of the likelihood of an event occurring).

Random Sample—each item or element of the population has an equal chance of being chosen as part of a sample of the population.

Experimental Probability—relating to the outcomes of an actual performance of a probability activity or experiment.

Theoretical Probability—a predicted value that is based on the definition of probability; the probability of an event without doing an experiment or analyzing data.

Event—one of the many occurrences that can take place during a probability activity.

As we will see in this section, there are two types of probability. Gathering data through observation determines the experimental probability. The probability determined by using the definition of probability is called theoretical probability. We will compare the results of experimental data with the expected probability. This will enable us to draw conclusions about our current data and speculate about the nature of additional data.

Experimental Probability vs. Theoretical (Expected) Probability

What would you expect if you flipped a fair coin ten times and recorded the number of heads and tails? The results could be five heads and five tails. Would you be surprised if you received six heads and four tails or seven heads and three tails? What about eight and two tails? By flipping the coin, the actual results are called experimental data. A coin has two sides and therefore when it is flipped there will be two outcomes. Based on the theory of probability with a fair coin, the probability of flipping a coin and getting a head is ½. This is the theoretical expectation.

We are not overly alarmed when flipping a coin that we do not always record a 50-50 split between the heads and tails. It is also quite natural to flip a coin multiple times and get a string of same results. What do we expect after a great many number of trials? The number of heads will closely match the number of tails. We expect the results from our experiment to match the results from the theoretical probability. The more trials we perform, the closer the number of heads will match the number of tails.

With a few repetitions, it is easily shown that the results from an experimental model may not yield the same results as the theory of probability.

Testing the Theory

Want to see if this is true? Flip a coin and keep track of the results. Repeat the process in groups of ten flips. Here is an example performed by the authors; For ease of recording, we kept track of the number of heads.

Set	Heads	Set size	Ratio for each set	Running total of the number of heads	Running total of the number of flips	Overall Ratio
1	6	10	.6	6	10	.6
2	3	10	.3	9	20	.45
3	6	10	.6	15	30	.5
4	5	10	.5	20	40	.5
5	4	10	.4	24	50	.48
6	3	10	.3	27	60	.45
7	7	10	.7	34	70	.486
8	2	10	.2	36	80	.45
9	6	10	.6	42	90	.467
10	6	10	.6	48	100	.48

These results shows the variation of the probability for ten flips. Two sets of flips were a 50-50 split for heads and tails, but the rest were not an even split. Note that as we were performing this experiment, there were several runs of heads only, as well as tails only. This is to be expected and is part of the nature of random events. Notice that the total ratio of heads to tails fluctuated but is slowly approaching the theoretical value of .50 or ½.

The unpredictability or random nature of experimental events is the basis for games of chance. After just a few trials, any outcome can occur. In the long run, the experimental probability outcome matches the theoretical probability. Theoretical probability is not based on experimental data, but instead is determined by the definition of probability.

Probability is the measure of the likelihood of an event occurring. In mathematics, we express this measure as the ratio of the number of actual occurrences to the number of possible occurrences. Simply put, probability is the desired outcomes compared to total possible outcomes. It is sometimes stated as the ratio of success to total. As you recall, a ratio is nothing more than a fraction where you are comparing one number to another. The fraction can be converted to a decimal or to a percentage by dividing the numerator by the denominator. Due to the nature of some decimals with repeating digits or long strings of digits, probability answers are often left in fractional form.

Probability of Rolling a Single Die

Let's take a closer look at the probability fraction. The numerator corresponds to the outcomes that you are focusing on. The value for the denominator represents the total possible outcomes. Consider the roll of a single six-sided die. Each side of the die has dots corresponding to the numbers one through six:

Spots						
Number	1	2	3	4	5	6
Probability	$\frac{1}{6}$	$\frac{1}{6}$	$\frac{1}{6}$	$\frac{1}{6}$	$\frac{1}{6}$	$\frac{1}{6}$

The die is called a fair die if each side is equally likely to be on the top. (The term "loaded die" comes from the fact that a given number will come up more than the other numbers, due to the uneven weight or construction of the die.)

The probability of rolling the die and having four spots showing can be determined by noting that there is just one side with four spots, so the numerator of the probability fraction is one. The denominator of the probability problem represents the total number of possible outcomes. In our die probability example, since we are rolling a die with a possible six sides with spots, the denominator of the probability fraction is six. Let's put the two parts together. To find the probability of rolling a four from a fair die, the numerator is one and the denominator is six.

$$P(Rolling \ a \ four) = \frac{1}{6}$$

Notice the capital P in the above statement. This represents the word "probability," and the parenthetical expression is the event. Therefore, here is another way to write this: Let E = the event of rolling a four on a six-sided, fair die:

$$P(E) = \frac{1}{6}$$

In general terms,

$$P(E) = \frac{\text{Number of ways your desired outcomes can occur}}{\text{Total number of possible ways}}$$

Examples: In rolling a single die, let E = event of the die showing an even number. There are 3 even numbers (2, 4, and 6) and there are six total possible outcomes (1, 2, 3, 4, 5, 6). So, $P(E) = \frac{3}{6} = \frac{1}{2}$.

In rolling a single die, let E = event of the die showing a number greater than 4. There are two numbers greater than 4 (5 and 6), and there are six total possible outcomes (1, 2, 3, 4, 5, 6). So, $P(E) = \frac{2}{6} = \frac{1}{3}$.

There are two special cases:

In rolling a single die, let E = event of the die showing a number greater than 6. There are no numbers greater 6 and there are six total possible outcomes (1, 2, 3, 4, 5, 6). So, $P(E) = \frac{0}{6} = 0$.

In rolling a single die, let E = event of the die showing any number (1, 2, 3, 4, 5, 6). There are 6 numbers, and there are six total possible outcomes (1, 2, 3, 4, 5, 6). So,

$$P(E) = \frac{6}{6} = 1.$$

These last two examples demonstrate the two extreme values for probability problems. The largest value for a probability fraction will equal one and the smallest value will be zero. A probability answer will never exceed one. The desired outcomes will never exceed the total possible outcomes.

Probability and a Pair of Dice

Let's look at the probabilities formed by counting the spots from two dice. A good place to start is to determine the total possible sums when you roll two dice. We saw the outcomes for just one dice was the numbers from 1 to 6. The sum of the number of the spots for two dice can result in the numbers from 2 through 12. A challenging part to this problem will be to find the total possible ways that we can obtain these sums. When the possible outcomes are relatively small, we can list them. In the terms of formal mathematics, when you list all the possible outcomes of an activity, you are defining the sample space or outcome set. Here is the sample space for a pair of dice. (We distinguish the dice by having one red and the other green. The red die will be listed across the top of the chart and the green die will be listed down the left side.)

RED

Number of spots	1	2	3	4	5	6
1	1,1	1,2	1,3	1,4	1,5	1,6
2	2,1	2,2	2,3	2,4	2,5	2,6
3	3,1	3,2	3,3	3,4	3,5	3,6
4	4,1	4,2	4,3	4,4	4,5	4,6
5	5,1	5,2	5,3	5,4	5,5	5,6
6	6,1	6,2	6,3	6,4	6,5	6,6

(G R E E N labeled down the left side)

From the table it is easy to see that there are 36 different results. This is helpful. Not only will we be able to calculate the various sums, but we have the denominator for the probability questions.

Example: In rolling a pair of dice, let E = event of the sum of the dice is 2. There is just one combination from the table (1,1), and there are 36 total possible outcomes. So, $P(E) = \frac{1}{36}$.

Example: In rolling a pair of dice, let E = event of the sum of the dice is 7. There are 6 combinations from the table (1,6) (2,5) (3,4) (4,3) (5,2), and (6,1) and there are 36 total possible outcomes. So,

$$P(E) = \frac{6}{36} = \frac{1}{6}.$$

Example: In rolling a pair of dice, let E = event of the sum of the dice is 10. There are 3 combinations from the table (4,6) (5,5), and (6,4), and there are 36 total possible outcomes. So,

$$P(E) = \frac{3}{36} = \frac{1}{12}.$$

Example: In rolling a pair of dice, let E = event of the sum of the dice is 12. There is just one combination from the table (6,6), and there are 36 total possible outcomes. So, $P(E) = \frac{1}{36}.$

Probability and a Spinner

Oftentimes in various board games instead of using a pair of dice to play the game, a spinner is used.

Notice that the eight regions are equal in size. So, the pointer is equally likely to land on each of the numbers. (Oftentimes, games will include the rule that if the spinner is on the line, the player simply respins.)

For the event of landing on a 1, the probability would be

$$P(E) = \frac{1}{8} \qquad P(E) = \frac{\text{Number of 1's}}{\text{Total number of values on spinner}}$$

For the event of landing on a 1 using the new spinner shown below, the probability would be:

$$P(E) = \frac{3}{8} \qquad P(E) = \frac{\text{Number of 1's}}{\text{Total number of values on spinner}}$$

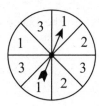

Probability and Flipping Coins

Let's look at the outcomes for flipping two fair coins. To help distinguish the outcomes we will flip a quarter and a nickel. As usual, we need to know the total possible outcomes for the denominator of our probability fraction.

Here is the sample space:

Heads	Heads	Heads	Tails
Tails	Heads	Tails	Tails

As you can see, we can flip two heads, a head and a tail, a tail and a head, and two tails. Since we have one outcome of both heads out of four possible outcomes, we would say that the probability of flipping two coins and getting both heads is $\frac{1}{4}$. If the event were two tails, the probability would be the same. If the event was to flip one head and one tail, you can see from the sample space that can occur two different ways for a probability of $\frac{2}{4} = \frac{1}{2}$.

Probability and the Number of Children

A husband and wife would like to have three children. If we assume the likelihood of having a girl is the same as having a boy, we can answer some basic probability questions and see some helpful ways to organize the outcomes.

First, let's list the sample space. Remember that the sample space is a listing of all possible outcomes. We will use a G for the girl and B for the boy. It is also important to note that the order of listing the children represents the birth order for the family. So, GBG would represent a girl was first born, then a boy, and then lastly a girl. This would be different for GGB. Although, both use the same letters, the second sample represents that the first two children born to the family were girls and the youngest was the boy. For three children, we can list the following:

All girls	Two girls and one boy	Two boys and one girl	All boys
GGG	GGB	GBB	BBB
	GBG	BGB	
	BGG	BBG	

Note the attempt to develop and follow a pattern with writing the letters. The list started with all girls and then replaced one girl with a boy into each place in the birth order. Next two boys were introduced to replace two girls. Then finally three boys were listed.

It is easy to miss some of the combinations when listing them. Another organization graph is a tree diagram. Here each pair of branches represents the outcomes of either a girl or a boy. The ordering of the tree can also display the birth order. Here is the tree depicting the three children family.

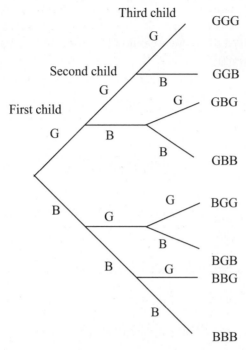

Let's use the tree to answer these probability problems.

Let E = the event of having exactly three girls. This is just the top branch on the tree.

$$P(E)=\frac{1}{8} \qquad P(E)=\frac{\text{One branch (topmost)}}{\text{Total number of branches}}$$

Let E = the event of having two girls and one boy. Look carefully at the tree. You will find three branches that fit this requirement. (You can also look at the sample space listed earlier.)

$$P(E)=\frac{3}{8} \qquad P(E)=\frac{\text{Three branches}}{\text{Total number of branches}}$$

Let E = the event of having an odd number of girls. With some quick thinking, we see that this occurs when you have three girls and when you have one girl (and two boys). Notice from the tree that all these occurrences are separate branches. So

$$P(E)=\frac{1}{8}+\frac{3}{8}=\frac{4}{8}=\frac{1}{2}$$

$$P(E)=\frac{\text{One branch (topmost)} + \text{Three branches of one girl and two boys}}{\text{Total number of branches}}$$

Let E = the event of having the same number of girls as boys. This time as you look at the tree, you will find there are no branches to represent this. Since three children is an odd number, it cannot be split evenly in half to have an equal number of girls as boys. Having a probability of zero comes from events that cannot happen.

$$P(E)=\frac{0}{8}=0 \qquad P(E)=\frac{\text{No branches}}{\text{Total number of branches}}$$

Probability and Colored Marbles

Suppose we have a bag of marbles. Each of the marbles is the same shape and texture and is only distinguishable by color. When you withdraw a single marble from the bag, each marble is equally likely to be selected. Initially, we count the colored marbles and set up the following chart. Since the total number of different marbles is 30, the denominator for our probability fraction will be 30. Since we are conducting a single draw of a marble and will replace the marble after noting its color, the total marbles for each drawing will always be 30.

Bag of Marbles

Red	White	Blue	Total
5	10	15	30

What is the probability that you will draw a red marble? The outcome that we are looking for is a red marble. Note that there are 5 red marbles in the bag. The total number of marbles is 30. The probability of drawing a red marble is $P(Red) = \dfrac{5}{30} = \dfrac{1}{6}$. If we repeat the process (replacing the previously drawn marble), the probability of drawing a white marble is $P(White) = \dfrac{10}{30} = \dfrac{1}{3}$, and the probability of drawing a blue marble is $P(Blue) = \dfrac{15}{30} = \dfrac{1}{2}$.

Probability with a Standard Deck of Cards

Summary table for cards:

- There are 52 cards in a standard deck.

- There are two colors—red and black.

- There are four suits—Hearts, Diamonds, Clubs, and Spades.

- Each suit has 13 cards—numbered from one (Ace) to ten, a Jack, a Queen, and a King.

- There are 12 face cards—4 suits and 3 face cards in each suit.

Let's begin by drawing one card from a well-shuffled standard deck of cards. We use the term "well-shuffled" to ensure that each card has an equally likely chance of being drawn. Remember, one of the key components for probability is that each outcome is equally likely.

Let E = the event of receiving a black card. There are 52 cards in a deck, and half of the cards are black (spades and clubs). Half of 52 is 26. So, there are 26 black cards.

$$P(E) = \frac{26}{52} = \frac{1}{2} \qquad P(E) = \frac{26 \text{ black}}{\text{Total number of cards}}$$

You could also answer this question by just looking at the colors only and not concern yourself with the total number of cards. With the deck of cards having the same number of red cards as black, you can say that for the desired outcome this is one color out of two possible colors.

$$P(E) = \frac{1}{2} \qquad P(E) = \frac{\text{One color (black)}}{\text{Total number of colors}}$$

You could also answer this question by just looking at the suits only. With the deck of cards having four suits, two of which are black, you can say that for the desired outcome this is two suits out of four possible suits.

$$P(E) = \frac{2}{4} = \frac{1}{2} \qquad P(E) = \frac{\text{Two suits (Spades and Clubs)}}{\text{Total number of suits}}$$

Let E = event of drawing a Queen from the deck of cards. Since there are four Queens in a standard deck of 52 cards, we find the probability to be

$$P(E) = \frac{4}{52} = \frac{1}{13} \qquad P(E) = \frac{\text{Number of Queens}}{\text{Total number of cards}}$$

Let E = event of drawing the Queen of Hearts from the deck of cards. Since there is only one Queen of Hearts in a standard deck of 52 cards, we find the probability to be

$$P(E) = \frac{1}{52} \qquad P(E) = \frac{\text{Number of Queens of Hearts}}{\text{Total number of cards}}$$

Let E = event of drawing a Diamond from the deck of cards. Since there are 13 Diamonds in a standard deck of 52 cards, we find the probability to be

$$P(E) = \frac{13}{52} = \frac{1}{4} \qquad P(E) = \frac{\text{Number of Diamonds}}{\text{Total number of cards}}$$

Independence vs. Dependence

Many of the examples shown so far have been about the probability of a single event. You have seen what happens when you repeat the experiment of flipping a coin and multiple children in a family. Let's go back through each probability example and discuss the event happening in

Independent events → Replacement
Dependent events → No replacement

more than one trial. This is called a compound event. This will show probability solutions from repeated trials and introduce the terms "independent events" and "dependent events" in light of probability.

Independent events—two events in which the outcome of the second event is not related to the outcome of the first event.

Dependent events—two events in which the outcome of the second event is affected by the outcome of the first event.

Probability model	Independent or dependent	Reasoning
Flipping a coin	Independent	Each time you flip the coin there are always 2 options.
Rolling a single die	Independent	Each time you roll a die there are always 6 options.
Rolling a pair of dice	Independent	Each time there are 11 sums available. The number of spots in the dice is the same for each roll.
Children in a family	Independent	This is generally assumed to be true.

Probability model	Independent or dependent	Reasoning
Drawing marbles from a bag	Could be either	It is independent if before the next marble is selected, the first marble is returned to the bag. The total number and proportions of color of marbles must be the same. It is dependent when the first marble drawn is not replaced. This will change the probability on the second draw.
Selecting a card from a deck of cards	Could be either	Here again the answer of independence and dependence is based on whether the first card is replaced into the deck. If the card is replaced, the total number of cards from which to select is the same as the first event. If the card is not replaced, then the total number of cards will be one less and the probabilities will change.

Drawing Marbles—Without Replacement

In our marble bag, we had 30 marbles. Five of the marbles are red, 10 marbles are white, and 15 marbles are blue. For the event of drawing the first marble of any color the probabilities are

$$P(Red) = \frac{5}{30} = \frac{1}{6} \qquad P(White) = \frac{10}{30} = \frac{1}{3} \qquad P(Blue) = \frac{15}{30} = \frac{1}{2}$$

Suppose the first marble drawn was red. If we draw another marble without replacing the first red marble, two numbers will have changed. First, the bag no longer contains 30 marbles, and the number of red marbles is decreased by one. Here are the new probabilities showing that a red marble previously chosen is no longer in the bag.

$$P(Red) = \frac{4}{29} \qquad P(White) = \frac{10}{29} \qquad P(Blue) = \frac{15}{29}$$

Suppose that we drew another red marble and did not replace either the first red or the second red marble. Again, all the probabilities would change. Here are the new probabilities showing that two red marbles previously chosen are no longer in the bag.

$$P(Red) = \frac{3}{28} \qquad P(White) = \frac{10}{28} \qquad P(Blue) = \frac{15}{28}$$

Simply stated, when you do not replace the marbles after you select one, the probability values change for the next draw.

Standard Deck of Cards—Without Replacement

Let E = event of drawing a Queen from the deck of cards. Since there are four Queens in a standard deck of 52 cards, we find the probability of the first draw to be

$$P(E) = \frac{4}{52} = \frac{1}{13} \qquad P(E) = \frac{\text{Number of Queens}}{\text{Total number of cards}}$$

If we did draw a Queen and did not replace the card back into the deck, the probability of drawing another Queen on the next draw is

$$P(E) = \frac{3}{51} = \frac{1}{17} \qquad P(E) = \frac{\text{Number of Queens} - 1}{\text{Total number of cards} - 1}$$

Now, if we did not draw a Queen on the first draw and did not replace the card into the deck, the probability of drawing a Queen on the next draw is

$$P(E) = \frac{4}{51} \qquad P(E) = \frac{\text{Number of Queens}}{\text{Total number of cards} - 1}$$

In the previous examples when there is replacement, the probabilities of subsequent events are not affected. The replacement of a marble or card creates an independent event. When items are not replaced, the outcome of the first event affects the probability of the next event. If the marble or card was not replaced, the event is dependent.

Probabilities Using Geometric Shapes

The calculations for probability can also extend to geometric shapes. When a figure is equally divided, you can determine the probability by counting the number of shaded regions and the number of total regions. Using the definition of probability, we make a ratio of the shaded regions to the total number of regions. Check out these shapes and corresponding probabilities.

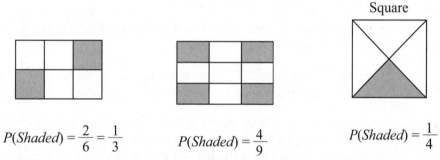

$$P(Shaded) = \frac{2}{6} = \frac{1}{3} \qquad P(Shaded) = \frac{4}{9} \qquad P(Shaded) = \frac{1}{4}$$

In the following examples, finding the probability of selecting the shaded region from the geometric figure requires you to determine the area of the shaded region and total area.

In this target, the radius of the smallest circle is 2 feet. The radius of the middle circle is 4 feet, and the radius of the largest circle is 6 feet.

To calculate probabilities related to this geometric figure, first compute the areas of the circles. Remember the formula for area of a circle is $A = \pi r^2$.

Circle	Radius	Area = πr^2
Small	2 feet	4π
Middle	4 feet	16π
Large	6 feet	36π

Let's first find the probability of hitting the bull's eye. The area of the small circle is 4π, and the area of the target is 36π.

$$P(Bull's\ eye) = \frac{4\pi}{36\pi} = \frac{1}{9}$$

Next, let's look at the probability of hitting the white region of the target. The area of the middle circle is 16π. The area in the white portion of the target is the middle circle minus the bull's eye ($16\pi - 4\pi = 12\pi$).

$$P(White\ ring) = \frac{12\pi}{36\pi} = \frac{1}{3}$$

Next, let's look at the probability of hitting the outside ring of the target. The area of the outside circle is 16π. The area of the middle circle is 36π. Again we subtract the areas ($36\pi - 16\pi = 20\pi$).

$$P(Outside\ ring) = \frac{20\pi}{36\pi} = \frac{5}{9}$$

Sample Space

In our quest to understand probability, we have defined probability as a ratio (fraction) comparing the desired outcome to the total possible outcomes. Remember the term "outcome set" defines the total possible outcomes. In some problems, it is possible to list the elements of a sample space rather quickly.

- Consider the possible outcomes of rolling a typical six-sided die. The sample space is the numbers from one to six, inclusive.

- In a marble bag probability problem, just count the number for each color of marbles and find the total number of marbles, and you have the sample space.

Not all probability problems have a sample space that can be quickly determined. In an earlier example, the sample space of rolling a pair of dice was displayed in a matrix. With the outcomes of one die listed across the top and the outcome of the other die listed down the side, the resulting matrix displayed all possible outcomes. Probability questions can be more easily answered when the data are required.

The tree diagram is also quite useful in displaying the sample space. This was shown in the earlier problem dealing with the number of ways three children might be born to a couple. You can list all the possible outcomes as you follow each branch combination. We could make a tree diagram for the rolling of a pair of dice instead of a chart, but this might become confusing and hard to follow. How many branches would you have?

On the first die you would list six branches. On each of these branches you would list six additional branches, for a total of 36 branches. Quite cumbersome! It looks like the chart is the way to go. What if you were to roll three dice instead of a pair of dice? A three-dimensional matrix now is a challenge to draw or envision. In mathematics there is a counting principle that allows us to calculate the number of ways. This counting principle will not list the individual outcomes but instead aids in counting the items. Since the basis for probability is a ratio of the desired outcomes to the total outcomes, knowing these values will allow you to set up the probability fraction.

Counting Principle

> If you do one task in x number of ways and a second task in y number of ways, then both tasks can be done in x times y number of ways.

The counting principle is also called the multiplication principle for sets.

In the dice problem, the first task (rolling a die) has six outcomes. The second task is the same, six outcomes. The total number of possible outcomes is $6 \cdot 6 = 36$. Looking back at the two-dimensional matrix, there were 36 entries. To answer the question of rolling three dice, the solution of the total outcomes can now be easily answered ($6 \cdot 6 \cdot 6 = 216$).

Suppose your favorite ice cream parlor has five new flavors of ice cream, four toppings, and three sizes. How many different treats could you make if you select one size, one topping, and one new flavor? Using the counting principle, the solution would be 60 different ways ($3 \cdot 4 \cdot 5 = 60$).

In a city election, there are four candidates for mayor, six candidates for vice mayor, two candidates for secretary, and three for treasurer. In how many ways can these four offices be filled?

$$4 \cdot 6 \cdot 2 \cdot 3 = 144 \text{ ways}$$

A certain combination lock can be set to open to any three-letter sequence. How many such sequences are possible? Since there are 26 letters in the alphabet, there are 26 choices for each of the three letters. By the counting principle, there are $26 \cdot 26 \cdot 26 \cdot = 17{,}576$ different sequences.

Counting problems are related to two general concepts: permutations and combinations. Before looking at these two concepts, we will review a useful notation called factorial.

Factorial

The symbol for factorial is the exclamation point (!). The expression 5! would be read five factorial. Here is a table of sample factorials:

> Factorial
> $n! = 1 \cdot 2 \cdot 3 \cdot 4 \bullet \cdots \bullet n$

Factorial	Expanded form	Product
· 1!	1	1
2!	$2 \cdot 1$	2
3!	$3 \cdot 2 \cdot 1$	6
4!	$4 \cdot 3 \cdot 2 \cdot 1$	24
5!	$5 \cdot 4 \cdot 3 \cdot 2 \cdot 1$	120
6!	$6 \cdot 5 \cdot 4 \cdot 3 \cdot 2 \cdot 1$	720
7!	$7 \cdot 6 \cdot 5 \cdot 4 \cdot 3 \cdot 2 \cdot 1$	5,040
0!	Special case	1

In general terms, factorial is expressed as

$$n! = n \cdot (n-1) \cdot (n-2) \cdots \cdot 3 \cdot 2 \cdot 1$$

This notation is a compact way to express a product. The factors start with the factorial number and decrease by one down to one. Since multiplication is commutable, you could start with one and list the factors consecutively up to and including the factorial number. Factorial notation is used in both permutations and combinations.

Permutation

Permutations → Order matters
Combinations → Order does not matter

A permutation is an ordered arrangement of a set of events or items. Consider the set of three letters $\{a,b,c\}$. If we wanted to arrange the three letters where each list of letters is unique, we are forming permutations.

$$\{\{a, b, c\}, \{a, c, b\}, \{b, a, c\}, \{b, c, a\}, \{c, a, b\}, \{c, b, a\}\}$$

Notice that none of the letters are repeated within any permutation; that is, $\{a, a, a\}$ is not a permutation of the three letters.

The total number of permutations of three letters is fairly small. In other cases, there are so many permutations that listing each permutation to count them is time consuming. Consequently, we use a formula to determine the number of permutations. In a permutation problem you are given two values. One value represents the total number of objects (n), and the second number represents the number of objects (r), that you are selecting.

Permutations of n objects taken r at a time: $\quad {}_nP_r = \dfrac{n!}{(n-r)!}$

Going back to the previous example of the permutations with the three letters, $\{a, b, c\}$, the number of permutations could be determined using the formula.

$$ {}_3P_3 = \frac{3!}{(3-3)!} = \frac{3!}{(0)!} = \frac{3 \cdot 2 \cdot 1}{1} = 6 \text{ ways} $$

When finding the number of ways events can happen, it is often stated that the events cannot be repeated and that the ordering of their occurrence is important. This is the key characteristic of a permutation.

Example: You have ten horses in a race with prizes for the first, second, and third place finishers. (Notice the result is an ordering.) To find the number of different ways that the results might occur, you would use the permutation formula. In the equation, n would be 10, which is the total number of horses. The r would be 3, which represents the number of prizes.

$$ {}_{10}P_3 = \frac{10!}{(10-3)!} = \frac{10!}{(7)!} = \frac{10 \cdot 9 \cdot 8 \cdot 7 \cdot 6 \cdot 5 \cdot 4 \cdot 3 \cdot 2 \cdot 1}{7 \cdot 6 \cdot 5 \cdot 4 \cdot 3 \cdot 2 \cdot 1} $$

This can be simplified:

$$ \frac{10 \cdot 9 \cdot 8 \cdot \cancel{7} \cdot \cancel{6} \cdot \cancel{5} \cdot \cancel{4} \cdot \cancel{3} \cdot \cancel{2} \cdot \cancel{1}}{\cancel{7} \cdot \cancel{6} \cdot \cancel{5} \cdot \cancel{4} \cdot \cancel{3} \cdot \cancel{2} \cdot \cancel{1}} = 720 $$

Instead of listing all of the factors, you may choose to reduce common factorials from the fraction.

$$ {}_{10}P_3 = \frac{10!}{(10-3)!} = \frac{10!}{(7)!} = \frac{10 \cdot 9 \cdot 8 \cdot \cancel{7!}}{\cancel{7!}} = 10 \cdot 9 \cdot 8 = 720 $$

Example: Suppose you are judging an art show with eight entries. You are planning to award prizes to the top four places. In how many different ways can you rank the entries?

$$_8P_4 = \frac{8!}{(8-4)!} = \frac{8!}{(4)!} = \frac{8\cdot7\cdot6\cdot5\cdot \cancel{4!}}{\cancel{4!}} = 8\cdot7\cdot6\cdot5 = 1{,}680$$

Example: Suppose that there are six students who wish to be student body officers: president, vice president, and secretary. In how many ways can the six students fill the three offices?

$$_6P_3 = \frac{6!}{(6-3)!} = \frac{6!}{(3)!} = \frac{6\cdot5\cdot4\cdot \cancel{3!}}{\cancel{3!}} = 6\cdot5\cdot4 = 120$$

Combinations

Just as in permutations, a combination is the number of ways an event can happen. The difference between a permutation and a combination is the requirement for ordering and arranging of the items; in combinations, order doesn't matter. If you are forming a group or collection where order doesn't matter, you will use the combination formula to count the number of ways.

Let's go back to the set of three letters {a, b, c}. If order doesn't matter, the three letters form one group. In a combination problem, you are given two values. One value represents the total number of objects (n), and the other value represents the number of objects (r) you are selecting.

Combinations of n objects taken r at a time: $_nC_r = \dfrac{n!}{(n-r)!\,r!}$

In our problem of three letters taken three at a time we let the variable n be three and the variable r is also three. The setup using the formula is

$$_3C_3 = \frac{3!}{(3-3)!\cdot3!} = \frac{3!}{(0)!\cdot3!} = 1 \text{ way}$$

Example: How many committees of three people can be formed from a group of eight people? Note here we are forming a committee. This is a group of individuals, and there is no concern with the selection order. So we use the combination formula.

$$_8C_3 = \frac{8!}{(8-3)!\cdot3!} = \frac{8!}{(5)!\cdot3!} = \frac{8\cdot7\cdot6\cdot \cancel{5!}}{\cancel{5!}\cdot3!} = \frac{8\cdot7\cdot\cancel{6}}{\cancel{3}\cdot\cancel{2}\cdot1} = 8\cdot7 = 56$$

Example: A manager is planning to select four employees for promotion. There are 12 employees eligible. In how many ways can four be chosen?

$$_{12}C_4 = \frac{12!}{(12-4)!\cdot4!} = \frac{12!}{(8)!\cdot4!} = \frac{12\cdot11\cdot10\cdot9\cdot \cancel{8!}}{\cancel{8!}\cdot4!} = \frac{\cancel{12}\cdot11\cdot\cancel{10}^{5}\cdot9}{\cancel{4}\cdot\cancel{3}\cdot\cancel{2}\cdot1} = 11\cdot5\cdot9 = 495$$

Example: A manager is planning to place four employees into different jobs—office technician, administrative assistant, receptionist, and data entry specialist. There are 12 employees eligible. Note that in this example the jobs are distinct and placement of an individual in one job would be different than placement in another job. In how many ways can four be arranged?

$$_{12}P_4 = \frac{12!}{(12-4)!} = \frac{12!}{(8)!} = \frac{12 \cdot 11 \cdot 10 \cdot 9 \cdot \cancel{8}!}{\cancel{8}!} = 12 \cdot 11 \cdot 10 \cdot 9 = 11,880$$

Summary—Permutations or Combinations

If you were counting the number of groups, you would use the combination formula. If the order of the objects matters, you would use the permutation formula to count the number of distinct arrangements. The number of combinations for a given set is always smaller than the number of permutations for that set.

Permutation	Combination
Different orderings or arrangements of r objects are different permutations.	Each choice or subset of r objects gives one combination. Order within the group of r objects does not matter.
$_{n}P_r = \dfrac{n!}{(n-r)!}$	$_{n}C_r = \dfrac{n!}{(n-r)!\,r!}$
Clue Words: arrangement, order, schedule	Clue Words: group, committee, set, sample
Order Matters!	Order does not matter!

Are you able to do each of the following?

❏ Understand concepts related to data collection

❏ Represent data using a variety of graphs, charts, and plots:
- Bar graph
- Line graph
- Circle graph
- Scatter plot
- Box-and-whisker plot
- Stem-and-leaf plot

❏ Understand and use measures of central tendency:
- Mean
- Median
- Mode

❏ Make reasonable predictions and draw inferences from data presented in a variety of ways

❏ Draw and interpret a line of best fit

Practice Problems—Strand 2

1. From the graph below, which would be the correct interpretation of how fast the runner is running?

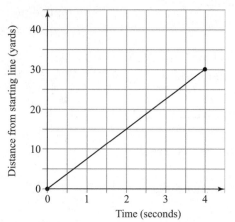

A. $\dfrac{10 \text{ yd}}{1 \text{ sec}}$ **B.** $\dfrac{30 \text{ yd}}{4 \text{ sec}} = \dfrac{15 \text{ yd}}{2 \text{ sec}}$ **C.** $\dfrac{20 \text{ yd}}{3 \text{ sec}}$ **D.** $\dfrac{25 \text{ yd}}{2 \text{ sec}}$

---

2. Given the following scatter plots of the weight of a vehicle and the miles per gallon, which graph displays the line of best fit for the graph?

A.
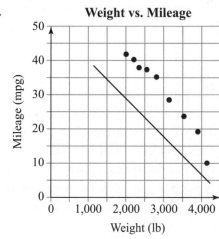

B.
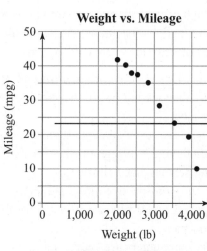

C.
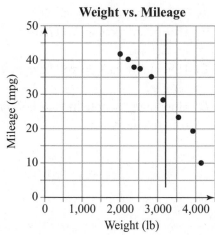

D.
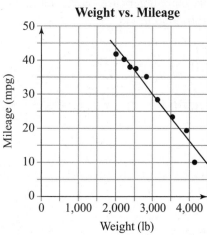

GO ON ➡

3. Census and sampling are two methods for gathering information. Which of the following statements are examples of sampling?

   I. Asking every third student in your history class if they are registered to vote.
   II. Buying a bag of potato chips from three different stores to find the freshest brand.
   III. Making an inventory of each music CD you own.
   IV. Measuring the length of the five pencils that you own.
   V. Reading the first and last pages of a book.

   A. I, III
   B. I, IV,V
   C. III, IV
   D. I, II,V

4. For which of the following would you use the combinations formula to determine the number of ways the event can happen?

   I. You place eight names in a hat and draw three names and award the same prize to each.
   II. There are seven students running a race and there are different prizes for first- and second-place finishers.
   III. Twenty prospective jurors are interviewed, and twelve are selected for the jury panel.
   IV. Six hamburgers are tasted and rated from high to low.

   A. I, II
   B. I, III
   C. I, IV
   D. I, II, III, IV

5. A laundry basket contains six black socks and ten white socks. A sock is pulled out of the basket at random, its color is noted. It is returned to the basket. Another sock is then pulled out. What is the probability that the second sock is white?

   A. $\dfrac{1}{8}$

   B. $\dfrac{3}{8}$

   C. $\dfrac{3}{5}$

   D. $\dfrac{5}{8}$

6. From a standard deck of cards, one card is drawn. What is the probability the card is an Ace or a black Queen?

   A. $\dfrac{8}{52} = \dfrac{2}{13}$

   B. $\dfrac{6}{52} = \dfrac{3}{26}$

   C. $\dfrac{6}{26} = \dfrac{3}{13}$

   D. $\dfrac{3}{14}$

7. The city is considering building a new skate park. You have been asked to survey residents to determine if there will be community support for this park. Which of the following is an appropriate question to ask to collect data?

   A. What is your yearly salary?
   B. How many pets do you have in your household?
   C. How many cars do you have in your family?
   D. How many people in your household are rollerblade skaters?

8. Tom, Dick, Larry, and Mary all ran the 2 mile race. In how many orders can the four people finish the race?

   A. 8
   B. 16
   C. 20
   D. 24

---

9. Find the mode of the set of data shown in the stem-and-leaf plot below.

Stem	Leaf
2	4 5
3	0 1 1 3 4 8
4	1 2 4 8 9
5	
6	7

   Key:
   2|4 = 24

   A. 31
   B. 34
   C. 36
   D. 67

---

10. Susan has started a summer reading program for herself, to improve her reading speed. After reading 1,000 pages, she begins to graph her progress. Predict Susan's reading speed after reading 5,000 pages of various books.

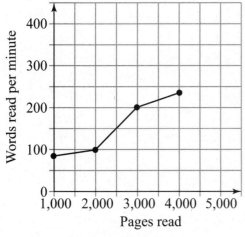

   A. 100 words per minute
   B. 200 words per minute
   C. 300 words per minute
   D. 400 words per minute

---

11. Which stem-and-leaf plot correctly displays the data given in the table?

Ages of People at a Diner					
25	17	30	7	16	27
34	56	10	21	20	62
17	12	22	45	42	53

   A.
Stem	Leaf
0	7
1	2 6 7 7
2	1 2 5 7
3	4
4	2 5
5	3 6
6	2

   3 | 4 = 34 years

   B.
Stem	Leaf
0	7
1	0 2 6 7 7
2	0 1 2 5 7
3	0 4
4	2 5
5	3 6
6	2

   3 | 4 = 34 years

   C.
Stem	Leaf
0	7
1	0 2 6 7
2	0 1 2 5 7
3	0 4
4	2 5
5	3 6
6	2

   3 | 4 = 34 years

   D.
Stem	Leaf
1	0 2 6 7 7
2	0 1 2 5 7
3	0 4
4	2 5
5	3 6
6	2

   3 | 4 = 34 years

---

12. At a deli, you have a choice of turkey, roast beef, or ham on your sandwich and a choice of potato chips or corn chips on the side. You can also choose a small, medium, or large drink? How many different ways can you order a sandwich, chips, and drink?

   A. 8
   B. 11
   C. 12
   D. 18

GO ON ➡

13. What is the probability that a randomly thrown dart will land in the shaded region of the dartboard?

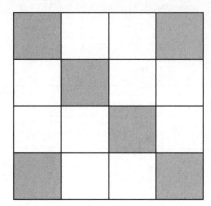

A. $\dfrac{1}{2}$

B. $\dfrac{6}{10} = \dfrac{3}{5}$

C. $\dfrac{6}{16} = \dfrac{3}{8}$

D. $\dfrac{10}{16} = \dfrac{5}{8}$

14. In how many different ways can you arrange the letters in the word QUALITY?

A. 4!

B. 5!

C. 6!

D. 7!

GO ON ➡

15. The data below are related to the percent of the population using the Internet.

U.S. region	Percentage using Internet
New England	71.3
Mid-Atlantic	53.3
National Capital	64.7
Southeast	45.8
South	51.1
Industrial Midwest	53.2
Upper Midwest	59.0
Lower Midwest	56.4
Border States	61.0
Mountain States	63.7
Pacific Northwest	72.2
California	64.6

Which of the following seems to be the most effective manner to display this data?

A.

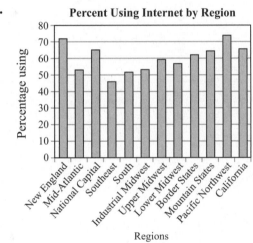

B.

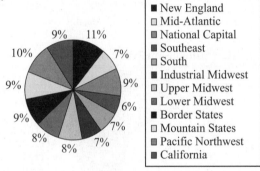

C.

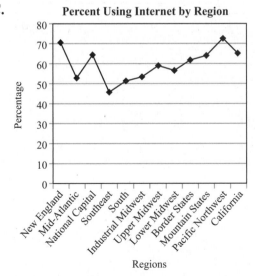

D.

```
45 | .8
51 | .1
53 | .2  .3
56 | .4
59 | .0
61 | .0
63 | .7
64 | .6  .7
71 | .3
72 | .2
```

Key
45.8 = 45 | .8

GO ON ➡

16. Which of the following is an equivalent display of the data related to total CD sales of the Centennial Music Store shown in the table below?

Month	Sales
January	126,000
February	782,000
March	238,712
April	81,970
May	328,877

A.

B. Millennium Music Store Sales

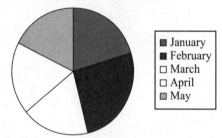

C. Millennium Music Store Sales

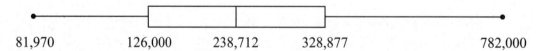

81,970   126,000   238,712   328,877   782,000

D.

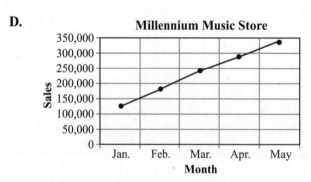

17. The box-and-whisker plot shown below represents the heights, in inches, of the members of the Ridge High School girls' basketball team.

    What height represents the maximum height of the smallest 75% of the members of the team?

    Height of Girls (in inches)

    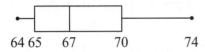

    64 65   67   70   74

    A. 64 inches
    B. 65 inches
    C. 67 inches
    D. 70 inches

18. Each of the events below is performed randomly. Which includes a dependent event?

    A. Twenty differently numbered marbles are put into a bag. One marble is drawn, the number is marked, the marble is set aside, and a second marble is drawn.
    B. A fair coin is flipped, the side it landed on is marked, and the coin is flipped again.
    C. A spinner with six congruent sectors is spun, the number is marked, and the spinner is spun again.
    D. A card is drawn from a deck of playing cards, replaced in the deck, and a second card is drawn from the deck.

19. Anthony took a survey of the "Happy Center" residents income. Their responses were

    $35,000 $61,000 $19,000
    $480,000 $22,000 $29,000 $16,500
    $13,000 $35,000 $23,000

    Find the median income.

    A. $26,000
    B. $73,400
    C. $81,600
    D. $101,250

20. Your scores on 8 math tests are listed below.

    74%, 68%, 78%, 82%,
    80%, 92%, 80%, 75%

    Your teacher gives you the choice of using the mean, median, or mode for your numeric grade. Assuming you want the highest grade possible, which measure of central tendency would you choose?

    A. Mean
    B. Median
    C. Mode
    D. Range

21. How many different four-digit numbers can be formed from using the digits 2, 4, 7, and 9, if each digit is used only once in each arrangement?

    A. 4
    B. 12
    C. 20
    D. 24

22. A driver training class collected the information in the table below. Estimate the stopping distance if a car is traveling at 80 mph.

Speed (mph)	Stopping distance (in feet)
10	15
20	40
30	75
40	120
50	175
60	240
70	315

    A. 325 feet
    B. 380 feet
    C. 400 feet
    D. 555 feet

GO ON ➡

23. A shoe store is considering carrying a new line of shoes. The manager wants to be sure to order the most popular size. Which measure of center would be used to measure the most popular shoe size?

A. Mean
B. Median
C. Mode
D. Range

24. Using the chart below as your basis for predicting, what do you think would be the total calories in a serving of ice cream that contained 15 fat grams?

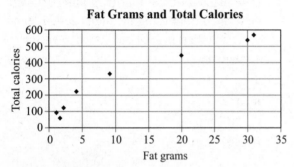

Fat Grams and Total Calories

A. 550 calories
B. 450 calories
C. 350 calories
D. 250 calories

25. A marketing company is conducting a survey to determine America's most popular professional football team. Where would they be most likely to obtain a unbiased random sample?

A. At a college football game
B. At an Arizona Cardinals game
C. By calling people from around the country on the telephone
D. At Phoenix Sky Harbor Airport

26. Use the scatter plot to predict the mileage of an engine with 250 horsepower.

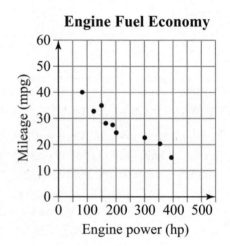

Engine Fuel Economy

A. 35 mpg
B. 30 mpg
C. 25 mpg
D. 15 mpg

GO ON ➡

Answer Key

1. B	6. B	11. B	16. A	21. D	26. C
2. D	7. D	12. D	17. D	22. C	
3. D	8. D	13. C	18. A	23. C	
4. B	9. A	14. D	19. A	24. C	
5. D	10. C	15. A	20. C	25. C	

Answers Explained

1. **B** The question is asking how fast. This is a question about finding a rate of change or slope. The graph relates time and distance. Determine the vertical change and the horizontal change to form the fraction of yards per second. The correct answer is B. The other answers do not correctly determine the rate of speed.

2. **D** The line of best fit is best displayed in graph D. The line attempts to pass through a majority of the points. The other graphs have lines, but they do not follow the trend of the points. Graph A does follow the trend, but notice that it does not include any of the given data points.

3. **D** Options I, II, and V each mention situations involving just a portion of a larger group. This is the concept of a sample. A census is an attempt to gather information for every member of a group.

4. **B** The key concept for a combination arrangement or collection is that order does not matter. Options I and III each have an arrangement, but there is no distinction within the group. Options II and IV each have an order, either high to low or first place, second place, and so on.

5. **D** The correct probability fraction is D. Since the first drawn sock was replaced, the total number of socks and the respective number of color socks do not change from the initial situation. The probability of drawing a white sock is 10 (number of white socks) divided by 16 (total number of socks). Notice the fraction was reduced. This is a common practice for the AIMS test.

6. **B** The probability numerator is the desired outcome. In this problem there are four Aces and two black Queens. The denominator is 52, the number of cards in a standard deck.

7. **D** The information gathered should be about rollerblading and the frequency of usage of a rollerblade park.

8. **D** The problem is asking for the number of different ordering of the race. Since order matters, this is a permutation problem using four for the variables n and r. Four permutations of four people is

$$\frac{4!}{(4-4)!} = \frac{4!}{0!} = \frac{4 \cdot 3 \cdot 2 \cdot 1}{1} = 24$$

Remember, you can use the formula sheet when taking the AIMS test.

9. **A** When reading a stem-and-leaf plot, look first at the key. The key tells you the place value for the stems and leaves. In this problem, the stem represents the tens place and the leaves the units. The mode is the measure of central tendency that occurs most often in the data set. The number 31 occurs twice.

10. **C** To predict the reading speed look at the trend of the data. As Susan's number of pages read increases, her words per minute also increase. Of the four possible answers, answer C is the most reasonable predicted reading speed.

11.   **B** To determine the correct stem-and-leaf plot, first organize the data. Since the data has two digits, let the stem represent the tens place and the leaves represent the units. Choice B correctly accounts for each number in the data set.

12.   **D** This problem is asking for the total number of ways to place an order. It uses the counting principle. Multiply the number of sandwiches by the number of side chips, and by the number of drink sizes ($3 \cdot 2 \cdot 3 = 18$).

13.   **C** For this probability problem count the number of shaded squares (6) and count the total number of squares (16). The ratio of $\frac{6}{16} = \frac{3}{8}$ accurately represents the probability. Remember, probability is the desired outcome divided by the total possible outcomes.

14.   **D** This arrangement involves different, distinguishable letters. The number of different arrangements is a permutation of seven letters taken seven letters at a time.

$$\frac{7!}{(7-7)!} = \frac{7!}{0!} = \frac{7!}{1} = 7!$$

Remember, $0! = 1$.

15.   **A** The bar graph is a more appropriate display of the data in the table. The bar graph allows you to see the relationship between the different regions more easily. Line graphs are used more often when data are continuous over time. Also, the line graph is difficult to read. The difference in the size of the regions in the circle graph is difficult to decipher. The stem and leaf graph organizes the data but does not show the relationships of the regions.

16.   **A** Choice A is correct. The data displayed in each of the other graphs is incorrect.

17.   **D** Each part of a box-and-whisker plot relates to a percentage of the population sample. The whiskers each represent 25 percent. The box represents 50 percent and the line in the box divides the 50 percent in half (25 percent for each part). This question asks for the lower 75 percent. Starting from the left end, the whisker represents 25 percent. The first quartile (left edge of box) to the median is another 25 percent. The next 25 percent is from the median to the third quartile (right end of the box). Using the scale below the graph, the heights up to 70 inches represent 75 percent of the population sample.

18.   **A** In probability, a dependent event occurs when a previous event affects the next event. Because the marble is not replaced after the first draw, the total number of marbles will change as will the amount of each color. This changes the probability values, and the answer to the second draw of a marble depends on what was drawn first. In the other examples, there is independence. In choice D, the first card is replaced into the deck. In choices B and C, the total outcomes (numbers on the die or spinner) as well as the possible outcomes do not change. The dependent probability is choice A.

19.   **A** The median income is the middle income. First, sort the data either smallest to largest or vice versa. Count to the center data value. Remember, when you have an even number of data values, you find the average of the two middle values.

20.   **C** The correct answer is C. The mean is 78.6, the mode is 80, and the median is 79. Although not a measure of central tendency, the range is 24.

21.   **D** The counting principle is used in this problem. Any of the four given digits can be in the first place, any of the three remaining digits can be placed in the second position, and of the two remaining digits can be selected for the third posi-

tion. The one last digit will be placed in the fourth place. The total number of ways that this can be represented is by the product of 4 • 3 • 2 • 1 = 24 total ways.

22. **C** The rate of change is not linear; it accelerates every 10 miles per hour. Between 10 and 20 it is 25 feet, between 20 and 30 it increases to 35 feet, between 30 and 40 it increases to 45 feet, between 40 and 50 it increases to 55 feet and so on. Following this pattern, between 70 miles per hour and 80 miles per hour the stopping distance increases to 85 feet. Choice C is the correct answer.

23. **C** The measure of central tendency that represents the most is the mode. The manager will determine the most popular by knowing the shoe size that she sells the most.

24. **C** Picture a line of best fit passing through the points. The trend appears to be between 300 and 400 calories. The best estimate would be 350 calories.

25. **C** The basic idea of an unbiased sample is to gather data from a wide range of people. In this problem, the unbiased sample could be taken by calling people from around the country. A biased sample would ask individuals from a single region. Choices A and B only question current, local football fans. Choice D extends the questions to a more diverse group of people but falls short if you are interested in "America's favorite." Option C gives broad coverage and asks both fans and nonfans alike.

26. **C** Imagine a trend line through the data. It would have a negative slope and pass through as many points as possible. To estimate the missing value, you would estimate that the line would correspond somewhere between 20 and 30 mph. The closest answer from the choice would be 25 mpg.

Patterns, Algebra, and Functions

When you complete your study of Strand 3 you will be able to:

- Identify patterns
- Recognize a function and determine its domain and range
- Describe, identify, and sketch graphs of real-world situations
- Express and interpret the relationship between data
- Determine if the graphs of two linear equations are
  - Parallel
  - Perpendicular
  - Coincident
  - Intersecting, but not perpendicular
- Solve maximum/minimum problems
- Manipulate mathematics expressions and equations
  - Evaluate algebraic expressions
  - Simplify algebraic expressions
    - Multiply and divide monomials
    - Add, subtract, and perform scalar multiplication with matrices
    - Calculate powers and roots of real numbers
  - Translate
    - Words to algebra
    - Algebra to words
  - Write a linear equation
    - Using a table of values
    - Using a data set that models a real-world situation
    - Given two points on the line
    - Given the slope and point on the line
    - Given the graph of the line
  - Solve
    - Linear equations
    - Linear inequalities
    - Algebraic proportions
    - Square root radical equations
    - Systems of equations
    - Quadratic equations
- Identify the sine, cosine, and tangent ratios
- Determine slope, x-intercept, and y-intercept of a linear equation
- Solve formulas for specified variables

As you recall, the first strand dealt with number sense. Number sense helped to define and clarify the types of numbers used in any problem involving mathematics. It is very important to have a firm grasp of the vocabulary and the mathematical properties on which we base our calculations. We want to make sure that when we complete a calculation, it is correct. The emphasis of the second strand developed the concepts of making sense of numerical information. Guidelines for collecting, organizing, and representing data were listed. The trends and patterns in data helped to bring meaning and predictability. The uncertainty of conclusions was explained by probability and chance.

A significant proportion of the AIMS test focuses on Strand 3—Patterns, Algebra, and Functions.

Patterns can be found in wallpaper, in jewelry, in music, in a block fence—patterns are everywhere. In mathematics, we look at patterns to find out what comes next. For example, what would be the tenth number in the pattern, what color would come next in the pattern?

Algebra has been called the challenge of finding the unknown. Like patterns, algebra can be found in many areas. It is often called the language of science, and businesses use algebra to predict future business needs.

A **function** is the relationship between two sets of numbers in which a value in the first set is paired with one and only one element in the second set. The first set of numbers is called the domain and the second set is called the range.

Functions written as algebraic expressions are often used to determine values based upon given conditions. Those values often form some type of numerical pattern.

Patterns

Mathematics involves the study and usage of patterns. When you first learned to count, you were learning a pattern—each number is one more than the previous number.

$$1, 2, 3, 4 \ldots$$

This is a pattern of consecutive counting numbers.

Mathematics is especially useful when it helps you predict; number patterns are all about prediction. What will the 50th number of this pattern be? How many cupcakes would we need if we gave a party for our neighborhood instead of just our class? The stock trader looks for trends—patterns—in the stock market in order to make wise decisions concerning an investment.

Sometimes you will be asked to generate a pattern from its description—starting with the number 2, generate a pattern in which each term in the pattern is 2 more than the preceding number.

$$2, 4, 6, 8, 10, 12, 14, \ldots$$

This is a pattern of consecutive even numbers. Notice we used the word term in the describing the pattern. We think of the "terms" in order, first term, second term, third term, and so on. The sixth term of the pattern just previously described, 2, 4, 6, 8, 10, 12, 14, . . . is 12.

The first performance objective in patterns is to communicate an iterative or recursive pattern using numbers or symbols. **Recursive** patterns are patterns that are generated with a given first number and a formula for generating each subsequent number from the preceding number. The pattern description mentioned earlier in this section—start with a number and generate a pattern in which each term is 2 more than the preceding term— describes a recursive pattern.

Another simple recursive pattern could be "starting with the number 2, generate each subsequent term by doubling the term before it." This pattern would be

$$2, 4, 8, 16, 32, 64, \ldots$$

The first performance objective in patterns also contains the word "iterative." **Iterative** means repetitive—that is, the pattern repeats some operation. The operation may not be numerical—it may be defined in some other manner. For example, the arrangement of blocks shown below represents an iterative pattern, a pattern in which each row contains two more blocks than the preceding row. The typical AIMS question related to this pattern might be, "How many blocks would be in row eight?"

One solution would be to continue the pattern by drawing down to the eighth row.

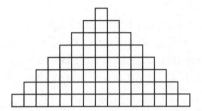

The answer would be (count them) 15.

However, this method can be time consuming. Another method would be to generate an algebraic rule to help determine the number of blocks in the eighth row.

Row number	1	2	3	4	5	6	7	8
Number of blocks	1	3	5	7	9	11	13	15

Notice that each row has two more blocks than the preceding row. However, you want to be able to find the number of blocks in any row without know the preceding rows. Notice that if you multiply the row number times 2, that value is always one more than the number of blocks. Therefore, the number of blocks in a row would be 1 less than 2 times the row number. Algebraically this could be represented by the expression $2r - 1$, where r is the row number. In the preceeding question, the eighth row would have 15 blocks because 1 less than 2 times 8 is 15.

Still another way mathematicians communicate the rule for generating a pattern is by using the term number and an algebraic equation that contains subscripts. Study the problem below.

What expression will define the nth term of the linear pattern contained in the table?

Term number	1	2	3	4	5	n
Term value	3	6	9	12	15	?

$$T_1 = 3 = 3(1) \quad T_2 = 6 = 3(2) \quad T_3 = 9 = 3(3) \quad \text{and so on}$$

The term value can be defined as $T_n = 3n$. Notice that the subscript number, n, represents the term number and is the same n as the n in the expression $3n$.

The pattern shown in the preceding table can also be communicated with a recursive rule. The definition begins by defining the first term of the pattern. The value of the next term is determined using that first term value. Subsequent terms follow the pattern of using the previous term:

$$T_1 = 3 \text{ and } T_n = T_{n-1} + 3$$

That means that when n is 2 (term #2), $T_2 = T_{2-1} + 3 = 3 + 3 = 6$, and when n is 3, we get $T_3 = T_{3-1} + 3 = 6 + 3 = 9$. Another way of saying this is the second term is 3 more than the first; the third term is 3 more than the second; each term is 3 more than the term before it. Notice the definition of this pattern uses subscripts. **Subscripts** are small numbers written to the right of and slightly below a term; they are used to designate placement in sequences, series, lists, and formulas.

Study the problem below. It is a pattern described using a recursive formula. This problem asks you to write the pattern of numbers from the formula.

Using the following recursion formula,

$$a_1 = 4$$
$$a_n = 2a_{n-1} + 6$$

what are the first four terms of this pattern?

When n is 1, you are describing the first term, which is given as $a_1 = 4$. This means the first term of the pattern is 4. When n is 2, a_{n-1} in the second equation of the formula is referring to $a_{2-1} = a_1$. If $n = 3$, a_{n-1} in the second equation of the formula is referring to $a_{3-1} = a_2$.

Here is the work for the development of the pattern:

$$a_1 = 4$$
$$a_2 = 2a_{2-1} + 6 = 2a_1 + 6 = 2(4) + 6 = 14$$
$$a_3 = 2a_{3-1} + 6 = 2a_2 + 6 = 2(14) + 6 = 34$$
$$a_4 = 2a_{4-1} + 6 = 2a_3 + 6 = 2(34) + 6 = 74$$

The first four terms in this pattern are 4, 14, 34, and 74.

A Challenging Pattern

Given the following sequence of values, write the next three terms in the pattern.

$$5, 6, 9, 14, 21, 30, \ldots$$

In this pattern, there is no constant difference between the terms, nor is there a common factor in the terms. The differences between consecutive terms display a recognizable pattern of odd consecutive numbers.

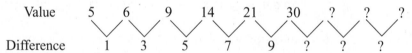

Value   5   6   9   14   21   30   ?   ?   ?

Difference   1   3   5   7   9   ?   ?   ?

Therefore, the next three differences should be the consecutive odd numbers, 11, 13, and 15. As a result, the next three terms in the pattern are 41, 54, and 69; 11 more than 30 is 41; 13 more than 41 is 54; 15 more than 54 is 69. Recursively, the pattern would be written as

$$a_1 = 5; \ a_n = a_{n-1} + (2n - 3)$$

Repetitive movements, reflections, rotations, and the like can be used in developing patterns. Can you guess what the fourth block should look like if the pattern continues?

The second block is a 90° clockwise rotation of the first block. The third block is a 90° clockwise rotation of the second block. Therefore, the fourth block should be a 90° clockwise rotation of the third block. The dots in the fourth block will be in the shape of a block letter U.

Functions

Recall that a function is a special relationship between numbers. It is a pairing of two or more values by a rule, table, graph, or diagram. The rule changes the input x into the output y. Each input value is paired with one and only one output value.

We use a specific format to write the rules used to describe the functions. The equation, $y = 5x$, written in function notation is $f(x) = 5x$, and is read, "f of x is equal to 5 times x." This notation shows that each output value is related to each input value by the function, f. The f in $f(x)$ is the name of the function; function names can be letters, such as $g(x)$ and $h(x)$, or sometimes they can be words, such as $\tan(x)$ and $\log(x)$.

It's easy to confuse the notation $f(x) =$ and $y =$. Typically, when graphing on a grid we use $y =$ and when identifying a function by name, we use $f(x) =$.

The types of functions that are most often studied in preparation for the AIMS test are linear, quadratic, absolute value, and square root (radical) functions.

Function	General Format	Examples
Linear	$y = mx + b$ $f(x) = mx + b$	$y = x$ $f(x) = -2x + 6$
Quadratic	$y = ax^2 + bx + c$ $f(x) = ax^2 + bx + c,\ a \neq 0$	$y = x^2 + 6x + 8$ $f(x) = -0.6x^2 + 6x$
Absolute Value	$y = |x|$ $f(x) = |x|$	$y = |x|$ $f(x) = 2|x| - 3$
Square root (radical)	$y = \sqrt{x}$ $f(x) = \sqrt{x}$	$y = 2\sqrt{x}$ $f(x) = \sqrt{x - 3}$

Let's look at a function in table format.

x	f(x)
1	5
2	10
3	15
4	20
5	25
6	30

Note that the value of *f(x)* is always 5 times *x*.

The equation *f(x)* = 5*x* is a simple in-and-out machine

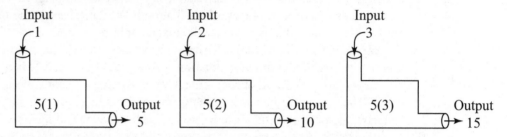

The *x* represents the input values, which as a set is called the **domain**. The output values are called the **range**. Informally, the domain is the set of all the numbers you can plug into the function, and the range is the set of all the numbers you can get out of the function.

Domain and Range

Domain—Values of *x*
Range—Values of *y*

You might be asked to find the domain and range of a function on the AIMS test. To find the domain, ask yourself, "What is *x* allowed to be?" There are some very specific things to check and watch for in answering that question. As we discuss different functions, we will sometimes use the *f(x)* = notation and sometimes the *y* = notation.

Function	Domain and Range
$y = \dfrac{4}{5-x}$	Ask yourself, "What is *x* allowed to be?" Notice the function is asking you to divide by 5 − *x*. We can divide by any number except 0. Therefore, we must determine if and when the denominator, 5 − *x*, would be 0. $5 - x = 0$ $-x = -5$ $x = 5$ So the domain of this function is all real numbers except 5. We write this as {all real numbers where *x* ≠ 5}. To find the range of this function, note that no matter how large or small *x* gets, the value of the function will never be 0. So the range is all real numbers except 0.

Function	Domain and Range
$f(x) = x^2$	What is x allowed to be? x is a real number and since we can square any real number, the domain of $f(x) = x^2$ is all real numbers. To find the range, think about what the results will be when we square any real number. Squaring any number always results in 0 or a positive number; therefore, the range is all positive real numbers greater than or equal to 0.
$y = 2x + 2$	What is x allowed to be? We can double any number and add 2 to the product. The result could be any number, positive, negative, or zero. The domain and range of this function is all real numbers.
$f(x) = \sqrt{x-2}$	What is x allowed to be? To find the square root of a number, the expression under the radical symbol, called the radicand, must be 0 or a positive number. $x - 2 \geq 0$ $x \geq 2$ Therefore, the domain is all real numbers greater than or equal to 2. Since $\sqrt{x-2}$ is always a nonnegative real number, the range of the function is all real numbers greater than or equal to 0.

The equation $y = 2x + 1$ can be represented as a simple in-and-out machine; it is the pattern of how one number changes into another. Every input value is doubled, and the product is increased by one. The equation is a function, and here is the table of values.

x	$2x+1$	y
1	$2(1)+1$	3
2	$2(2)+1$	5
3	$2(3)+1$	7
4	$2(4)+1$	9

When every input value generates a unique output value and every output value is paired with a unique input value, the function is called a **one-to-one function**. There is a one-to-one correspondence between input numbers and output numbers.

The equation, $y = x^2$, is also a function. Look at the table below. The input number 3 generates 9 and the input number, -3, also generates 9.

x	x^2	y
3	$(3)^2$	9
2	$(2)^2$	4
1	$(1)^2$	1
0	$(0)^2$	0
-1	$(-1)^2$	1
-2	$(-2)^2$	4
-3	$(-3)^2$	9

In this case, there are multiple x values that yield the same y value. The input-output relationship is not one-to-one, but is many-to-one; it still satisfies the requirements for a function. Every input value has only one output value paired with it.

The equation, $x = y^2$, is not a function. A value of x can correspond to two different values of y. When $x = 9$, y can be either 3 or -3; $9 = (3)^2$ and $9 = (-3)^2$. This contradicts the definition of a function that states that every x value corresponds to a unique value of y. The input-output relationship is a one-to-many relationship, which is not a functional relationship. A mapping diagram is a way of showing this relationship. In this diagram, the values in the oval on the left represent input and the values in the oval on the right represent output.

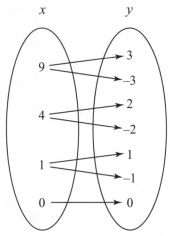

An input of 9 will generate both 3 and -3; an input of 4 will generate 2 and -2, and so on. In this case, the input-output relationship is a one-to-many relationship. This equation is not a function.

Here is another example of a relation, given by its mapping and its graph. Notice the input value 3 maps into two different output values. Also, notice that the vertical line at $x = 3$ touches two points of the relation. This is a one-to-many mapping, which means the relation is not a function.

$$\{(0, -2),(1,3),(2, 1),(3, 3)(3, 4)\}$$

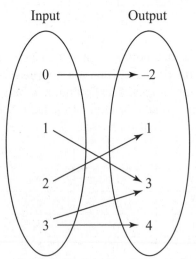

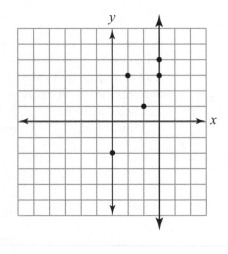

Comparing Mapping Diagrams

Let's compare the mapping diagram for each of the three relations we discussed earlier. Notice that in the first column of the first two relations all of the input values are different—they are both functions. In the first column of the third relation some of the input values are the same—it is not a function. This is another way of checking to see if a relation is a function.

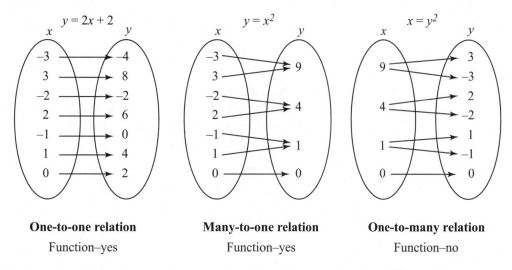

One-to-one relation	Many-to-one relation	One-to-many relation
Function–yes	Function–yes	Function–no

Vertical Line Test

Another way to look at each of these equations is to look at their graphs.

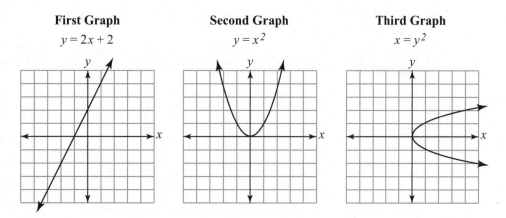

The first and second graphs represent functions. The third graph does not represent a function. One of the ways to test to see if a graph is the graph of a function is by using the vertical line test. If at any place on the coordinate plane a vertical line would touch the graph more than once, the equation is not a function. Look at the third graph drawn on the next page with a vertical test line inserted. Notice that the vertical line touches the graph more than once. This test visually shows $x = y^2$ is not a function.

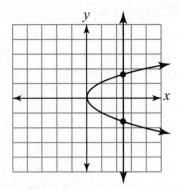

Can you find the table that matches the relationship below? If the relationship between *x* and *y* is expressed using the rule "To get *y*, multiply *x* by 2, then add 4." Which of the following tables contains values of *x* and *y* that satisfy the rule?

x	*y*
5	6
3	2
0	-4
–1	–6

x	*y*
–1	–6
–2	–8
–3	–10
–4	–12

x	*y*
1	5
3	7
6	10
12	16

x	*y*
–1	2
0	4
2	8
5	14

The fourth table on the right is the only table that contains values of *x* and *y* that satisfy the rule, $y = 2x + 4$.

When a diver goes underwater, the weight of the water exerts pressure on the diver. The following table shows how the water pressure on the diver increases as the diver's depth increases. What would be an equation with *D* for depth as the input and *P* for pressure as the output?

Diver's Depth *D*	*P* = 0.4*D*	Water pressure, *P*
10	0.4(10)	4
20	0.4(20)	8
30	0.4(30)	12
40	0.4(40)	16
50	0.4(50)	20

The equation, $P = 0.4D$, would represent this relationship.

Lines and Their Properties

Slope

The function, $y = 3x + 2$, is a linear function because when the set of values that satisfy the equation are graphed, they form a straight line. Notice the word "line" is the base of the word "linear."

$$y = 3x + 2$$

Table		**Graph**

x	$y = 3x + 2$	y
–2	$3(–2) + 2$	–4
–1	$3(–1) + 2$	–1
0	$3(0) + 2$	2
1	$3(1) + 2$	5
2	$3(2) + 2$	8
3	$3(3) + 2$	11

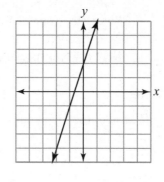

As you move from one point to another on the line, the x- and y-values change proportionally. That proportional relationship is called the slope of the line (or the rate of change) and is defined as the change in the y-values divided by the change in the x-values. This is also described as the difference in the y-values divided by the difference in the x-values; symbolically slope is can be written as

$$m = \frac{y_2 - y_1}{x_2 - x_1}$$

$$\boxed{\text{Slope} = \frac{\text{rise}}{\text{run}}}$$

In the previous example, using $y = 3x + 2$, with each increase of y by 3, x increases by 1.

You can find the slope of a line by using points on the line and the definition of slope. Pick two points from the table, such as $(–1, –1)$ and $(2, 8)$. If we use the definition of slope, the computation would be as follows:

$$m = \frac{y_2 - y_1}{x_2 - x_1} = \frac{8 - (–1)}{2 - (–1)} = \frac{9}{3} = 3$$

Let's use two points on a different line and go through the same process. Take the line whose equation is $y = 3x – 1$. The following table shows the process of finding two points.

Input x	Rule – $3x – 1$	Output y	Ordered Pair
–2	$–3 (–2) – 1$	5	$(–2, 5)$
1	$–3 (1) – 1$	–4	$(1, –4)$

Now use the slope formula with these two points.

$$m = \frac{y_2 - y_1}{x_2 - x_1} = \frac{–4 - (5)}{1 - (–2)} = \frac{–9}{3} = \frac{–3}{1}$$

Slope calculation	$y = mx + b$	Observation
$m = \frac{y_2 - y_1}{x_2 - x_1} = \frac{6 - 0}{2 - (–1)} = \frac{6}{3} = \boxed{2}$	$y = \boxed{2}x + 2$	m and the slope are the same
$m = \frac{y_2 - y_1}{x_2 - x_1} = \frac{–4 - (5)}{1 - (–2)} = \frac{–9}{3} = \frac{–3}{1} = \boxed{–3}$	$y = \boxed{–3}x – 1$	m and the slope are the same

On the line $y = –3x – 1$, the y-value decreases by 3, and the x-value increases by 1. Another way to state this same relationship is that the value of y increases by 3 as the value of x

decreases by 1. Did you notice in the equation above the coefficient of *x* (the value of *m* in $y = mx + b$) is the same as the slope of the line?

Since slope comes from a formula that uses a fraction, it is often helpful to write all slopes as fractions. This will help you identify the vertical and horizontal changes.

Here are two special cases of slope.

Horizontal Line

Consider the graph below with the points $A(-4, 3)$ and $B(2, 3)$.

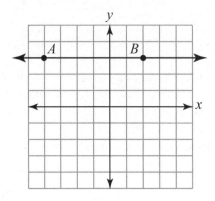

Notice $\overrightarrow{AB}$ is a horizontal line. Its slope is zero.

$$m = \frac{3-3}{-4-2} = \frac{0}{-6} = 0$$

Vertical Line

Consider the graph below with the points $C(3, 4)$ and $D(3, -1)$.

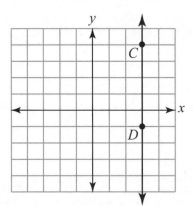

Notice $\overrightarrow{CD}$ is a vertical line. The denominator is zero, and division by zero is not a defined operation. Therefore, there is no slope for a vertical line.

$$m = \frac{4-(-1)}{3-3} = \frac{5}{0} = \text{undefined}$$

Summary of All Slopes

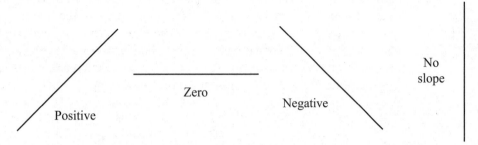

An easy way to remember the slope of lines is to imagine riding a bicycle along the graph above from left to right. You can ride uphill (positive slope), you can ride on flat land (zero slope), you can ride downhill (negative slope), but you cannot ride straight up (no slope).

Using the words "rate of change" for slope sometimes makes it easier to understand the concept of slope in real-world application problems, (e.g., change in wages over a period of years or change in the balance of a bank account over a period of time). Shown below is an example of rate of change (slope) applied to the change in hourly wages over a period of time. Over the time period from 1990 to 2002, the slope of the trend line (the equation of the trend line is shown) is 0.43, is the average increase in wages each year ($0.43/year).

Year	Year value	Hourly wages
1990	0	$9.64
1991	1	$10.04
1992	2	$10.42
1993	3	$10.71
1994	4	$10.91
1995	5	$11.22
1996	6	$11.61
1997	7	$12.07
1998	8	$12.54
1999	9	$13.17
2000	10	$13.63
2001	11	$14.31
2002	12	$15.01

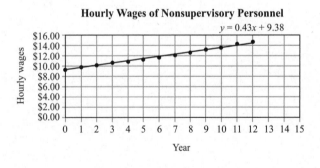

Interesting Note

There has been much discussion as to the reason for using m to represent slope. Many people think that m comes from the French word, *monter,* to climb. Others have said it comes from the word modulus, meaning the essential parameter determining a result. Therefore, m might have been used in the early days to mean modulus of slope, or parameter determining change. You might think of m as standing for move and b for begin. This relates to the way you graph linear equations by hand. You can use the b-value to plot the "beginning" point $(0, b)$. Then the m value instructs you where to move from point $(0, b)$ to plot the next point, thus giving you the graph of line for the equation.

x- and y-Intercepts

x-intercept $(x, 0)$
y-intercept $(0, y)$

Another very important bit of information that comes directly from the equation of a line in slope-intercept form ($y = mx + b$) is the y-intercept of the line.

Let's talk about intercepts in general, both the x- and the y-intercepts. The graph of the equation, $y = 2x + 2$ shown below, crosses both the x- and the y-axis. The point at which the graph crosses the x-axis $(-1, 0)$ is called the x-intercept and the point at which the graph crosses the y-axis $(0, 2)$ is called the y-intercept.

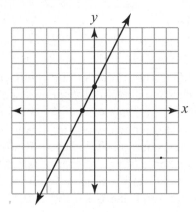

The intercept(s) of any equation can be found from the equation. Notice in the ordered pair for the x-intercept of the example, y is equal to 0, and in the ordered pair for the y-intercept, x is equal to 0. Look at the following examples. The equations do not have to be in any particular form for this process to work.

Equation	x-intercept	y-intercept
$3x + 4y = 12$	Let $y = 0$ $3x + 4(0) = 12$ $3x = 12$ $x = 4$ x-intercept is $(4, 0)$	Let $x = 0$ $3(0) + 4y = 12$ $4y = 12$ $y = 3$ y-intercept is $(0, 3)$
$y = -4x + 8$	Let $y = 0$ $0 = -4x + 8$ $4x = 8$ $x = 2$ x-intercept is $(2, 0)$	Let $x = 0$ $y = -4(0) + 8$ $y = 8$ y-intercept is $(0, 8)$
$2x = 3y + 6$	Let $y = 0$ $2x = 3(0) + 6$ $2x = 6$ $x = 3$ x-intercept is $(3, 0)$	Let $x = 0$ $2(0) = 3y + 6$ $-3y = 6$ $y = -2$ y-intercept is $(0, -2)$

A way to remember the process of finding intercepts is to always think of letting the other variable equal 0; that is, when you are finding the x-intercept, you let $y = 0$ and when you are finding the y-intercept you let $x = 0$.

You can also determine the y-intercept of a line by looking at the equation in slope-intercept form ($y = mx + b$). When an equation is in this form, the value of b is the y-value of the y-intercept.

Let's look again at the three equations.

Equation	Slope-intercept form	y-intercept $(0, b)$
$3x + 4y = 12$	$\begin{aligned} 3x + 4y &= 12 \\ 4y &= -3x + 12 \\ y &= \frac{-3}{4}x + 3 \end{aligned}$	The value of b is 3. Therefore, the y-intercept is $(0, 3)$.
$y = -4x + 8$	The equation is already in slope-intercept form.	The value of b is 8. Therefore, the y-intercept is $(0, 8)$.
$2x = 3y + 6$	$\begin{aligned} 2x &= 3y + 6 \\ 2x - 6 &= 3y \\ \frac{2}{3}x - 2 = y \quad &or \quad y = \frac{2}{3}x - 2 \end{aligned}$	The value of b is -2. Therefore, the y-intercept is $(0, -2)$.

Before leaving the topic of slope, there are two relationships when working with two or more lines that need to be defined. **Parallel lines** are lines in the same plane that never intersect and are always the same distance apart. The slopes of these lines are the same.

> Parallel—equal slopes
> Perpendicular—negative reciprocal slopes

Example:

Same slopes:
$y = 2x + 3$
$y = 2x - 4$

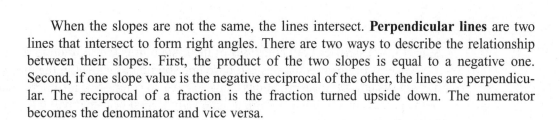

When the slopes are not the same, the lines intersect. **Perpendicular lines** are two lines that intersect to form right angles. There are two ways to describe the relationship between their slopes. First, the product of the two slopes is equal to a negative one. Second, if one slope value is the negative reciprocal of the other, the lines are perpendicular. The reciprocal of a fraction is the fraction turned upside down. The numerator becomes the denominator and vice versa.

Example:

Perpendicular Lines

Slopes

$y = 2x + 3$

$y = \dfrac{-1}{2}x - 4$

Product equal to negative one.

$m = 2$ and $m = \dfrac{-1}{2}$

$2 \cdot \dfrac{-1}{2} = -1$

Turn upside down and opposite sign.

$\dfrac{2}{1}$ and $\dfrac{-1}{2}$

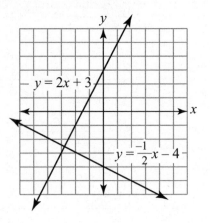

Systems of Equations

A **system of linear equations** is a set of two or more equations using the same variables.

Example 1	Example 2	Example 3
$3x - 4y = 23$	$x - 2y = 3$	$5x + 2y = 3$
$x + y = 3$	$3y = 9$	$y = 2x + 5$

Each of these systems is a linear system in two variables; the graph of each of the equations is a line (therefore linear). When the solution to a system is an ordered pair (x, y), you have a unique point. The ordered pair for this point is the only one that makes both equations true at the same time. Graphically speaking, this is where the two lines intersect.

Is the ordered pair, $(5, -2)$ a solution to the system in Example 1?

$$3x - 4y = 23 \qquad x + y = 3$$
$$3(5) - 4(-2) = 23 \qquad 5 + (-2) = 3$$
$$23 = 23 \qquad 3 = 3$$

Since $(5, -2)$ satisfies both equations of the system, it is the solution to the system.

Systems of linear equations can have one solution as shown in the preceeding example, but systems of linear equations can also have no solution or infinitely many solutions. In the process of solving a system, telltale signs will appear that will tell you whether the system has one solution, no solution, or infinitely many solutions. As we discuss the methods for solving systems (we study three of them), we will discuss those signs in detail, but the following table states what you are looking for in general in each case.

Number of Solutions

Method	One solution (x, y)	No solution— $\emptyset$	Infinitely many solutions —all (x, y) ordered pairs satisfy both equations
Substitution Elimination	The algebraic solution will generate one value for each of the variables. For example, $x = 3$ and $y = 7$.	The algebraic solution will generate a false statement that contains no variables. For example, $0 = -3$.	The algebraic solution will generate a true statement that contains no variables. For example, $0 = 0$.
Graphing	The lines will intersect at one point.	The lines will be parallel.	Graph of two equations will be the same line; the lines are said to coincide (be coincident).

It is possible to determine how many solutions you will have and how the graphs of the lines will be related to each other just by comparing their slopes and y-intercepts.

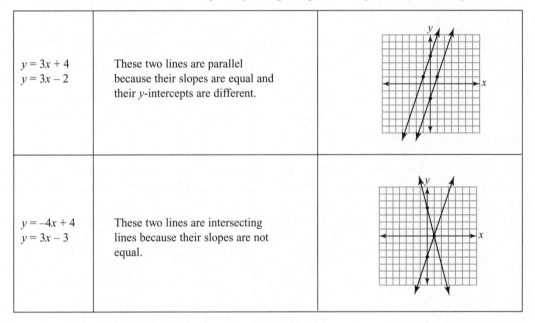

$y = 3x + 4$ $y = 3x - 2$	These two lines are parallel because their slopes are equal and their y-intercepts are different.	
$y = -4x + 4$ $y = 3x - 3$	These two lines are intersecting lines because their slopes are not equal.	

Sometimes you will encounter two lines that look like they probably have different slopes but when you simplify the equations you find they in fact are two equations for the same line. Consider the following equations.

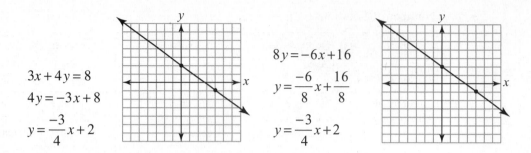

$$3x+4y=8$$
$$4y=-3x+8$$
$$y=\frac{-3}{4}x+2$$

$$8y=-6x+16$$
$$y=\frac{-6}{8}x+\frac{16}{8}$$
$$y=\frac{-3}{4}x+2$$

The graphs of these two equations are really the same line. Their slopes are the same, and their *y*-intercepts are the same. The two lines are said to coincide; that is, they are coincidental lines.

Real-World Graphs

Trends
• Positive /
• Negative \
• Constant –

The following graph represents a bicycle trip. During the trip, the bicycle got a flat tire. What intervals represent the time that the flat tire was being repaired?

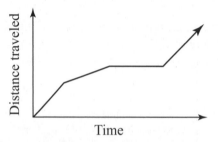

Because the bicycle will not be traveling during the time that the flat tire is being repaired, the horizontal portion of the graph (which shows no change in distance) correctly answers the question.

Real-world graphs are used to create visual displays of relationships between variables. They allow us to describe and analyze relationships in our world. They can be simple like the bicycle trip graph above, or they can be more complicated as shown in the next graph.

Suppose the path of a diver in the Olympics can be shown by this graph.

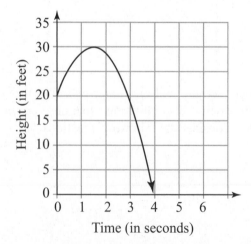

If the height of the dive is compared to the length of the time of the dive, approximately how many seconds does the judge have to view the dive in order to score it for competition? From the time at the beginning of the dive, which represents the height of the diving board, to the end of the dive, which represents the water level, the judge would have about 4 seconds to view the dive.

Maximum/Minimum Problems

The following graph shows the number of milligrams of a medication in the bloodstream from the time it was administered to 300 minutes after administration.

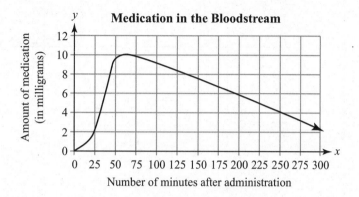

Notice that the graph shows a maximum (highest) value for the number of milligrams of medication in the bloodstream at approximately 55 minutes.

It is often very helpful to be able to determine from their graphs maximum and minimum values of relations. Here is the graph of the path that a diver could follow after jumping and diving into the water that we discussed earlier. The maximum value of the graph represents the height (30 feet) the diver reaches after jumping from the board.

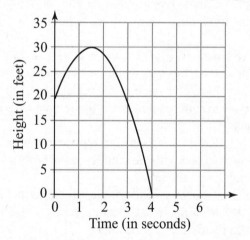

Later in Strand 4 we will discuss the graph of a quadratic function. On the next page you will see the graph of two different quadratic equations. One of them has a maximum value and the other one has a minimum value.

Equation	Graph	Maximum/minimum value
$y = x^2 + 4x + 2$		This graph has a minimum value at $(-2, -2)$.
$y = -x^2 + 4x + 2$		This graph has a maximum value at $(2, 6)$.

Algebra

An **algebraic representation** is a mathematical statement that contains variables and possibly numerical values. The variables are represented by letters (or symbols).

A numerical value is said to be a **constant**. Take for example the number 5. It can represent five dollars, five minutes, or five hamburgers. Placing the unit label with the number helps to identify what you are talking about. The value of five means just that—five things.

If you were to read that there were x dollars on the table, you would know what is being described, but you would not know how many dollars. The variable x can take on different values. The value of the variable can vary. Typically, in a mathematical situation, you are given additional information to help determine the value that the variable x is representing.

Any letter(s) can be used to represent the unknown value(s). Often the choice of the letter is associated with the context of the problem. You could use d for the unknown number of dollars on the table.

Finding answers to algebraic problems begins with understanding the key action words. The meaning of these action words will tell you what operation or procedure to use and when the problem is finished.

Mathematical Action Words

Let's start with a brief explanation of the action words and follow with several examples.

Evaluate—perform the various mathematical operations (addition, subtraction, multiplication, etc.) with the result of the problem being a numerical value. Other words fitting this category include "calculate" and "operate."

Translate—rewrite an English statement into a mathematical expression or equation. Occasionally a problem will require translating from math symbols to English statements.

Write—process the given information into a standard mathematical form. This most often will require you to find an equation.

Simplify—change a complex expression into a simple, concise form. The final solution may or may not be a numeric value.

Identify—recognize and state the requested solution based on a definition.

Solve—determine the value(s) of the variable(s) that makes the equation or inequality true.

The mathematical action words help to determine which procedure to follow. The word "algorithm" is often used in place of "procedure"; **algorithm** is defined as a step-by-step process or set of rules to follow to accomplish a goal. These procedures use the basic rules and definitions of algebra. Inasmuch as the vocabulary is found in several algorithms, the following is a list of some basic algebra definitions.

Algebra Vocabulary

Term—a quantity joined to other quantities by the operation of addition or subtraction.

Factor—a quantity joined to other quantities by the operation of multiplication or division.

Constant term—a numerical value.

Coefficient—a numerical factor of a term (e.g., $4x$, $4xy$, $4x^2y^3$—each term contains the coefficient of 4).

Like terms—terms that have the same base and exponents (e.g., $3x^2$ *and* $4x^2$ are like terms).

The next few pages contain information on how the action words—evaluate, translate, write, simplify, identify, and solve—are explored.

Evaluate

The goal in evaluating is to get an answer that is a numeric value.

Algebraic expression—an expression containing variables, numerals, grouping symbols, and operation signs, such as $+$, $-$.

Algorithm
Replace variables with the given values and simplify using the order of operations from Strand One (PEMDAS).

> Evaluate: $2a^2 + (b + 4c) - d$
>
> when $a = 3$, $b = -4$, $c = 5$, and $d = 7$
> $$2(3)^2 + ((-4) + 4(5)) - (7)$$
> $$2(3)^2 + ((-4) + 20) - (7)$$
> $$2(9) + (16) - (7)$$
> $$18 + (16) - (7)$$
> $$34 - (7)$$
> $$27$$

Absolute value expression—an expression that contains the absolute value symbol.

Algorithm

Begin as you would for an algebraic expression. Use the definition of absolute value as needed.

Evaluate: $|a - 5|$ when $a = -3$
$|-3 - 5|$
$|-8|$
8

Square root—an expression that contains a radical symbol.

Algorithm

Begin as you would for an algebraic expression. Use the definition of square root.

Evaluate: $\sqrt{a^2 + b^2}$ when $a = 4$ and $b = -3$

$\sqrt{(4)^2 + (-3)^2}$

$\sqrt{16 + 9}$

$\sqrt{25} = 5$

Simplify

The goal in simplifying is to make a simple expression.

Algebraic expression—an expression containing variables, numerals, grouping symbols, and operation signs, such as $+$, $-$.

Algorithm

Combine like terms. Use the rules for signed numbers. Remember to operate only on the coefficients.

Simplify: $5a^2 + 7 - 8b^3 - 3b^3 + 2a^2 - 8$
$5a^2 + 2a^2 - 8b^3 - 3b^3 + 7 - 8$
$7a^2 - 11b^3 - 1$

Powers

One of the most important parts of algebra involves working with exponents and performing basic operations using exponents.

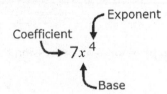

In writing powers, the number in front is called the coefficient. It is a factor of the expression. There is an "unshown" operator of multiplication between this coefficient and the base. The same is true between the coefficient and the radical symbol (e.g., $7x^4$ means 7 times x^4).

The exponent tells you the number of factors of the base. Exponential form gives you the opportunity to express a number in factored form and at the same time shorten the expression by grouping common factors.

$$\text{Exponential form } 7x^4 \qquad \text{Expanded form } 7 \cdot x \cdot x \cdot x \cdot x$$

Study the following chart to review the concepts related to working with exponents.

Operator	Expression	Process
Addition	Must have similar terms in order to add. The bases must be the same, and the exponents on each of the bases must be same.	Add the coefficients of the similar terms. Do not change the base or exponent.
Example	$2x^3 + 5x^2 + 4x^3 + 6x^2$	$6x^3 + 11x^2$

Operator	Expression	Process
Subtraction	Must have similar terms in order to subtract. The bases must be the same, and the exponents on each of the bases must be same.	Subtract the coefficients of the similar terms. Do not change the base or exponent.
Example	$2x^3 - 5x^2 - 4x^3 + 6x^2$	$-2x^3 + 1x^2$

Operator	Expression	Process
Multiplication	Contains factors.	Multiply the coefficients together. Keep the base(s) the same and add the corresponding exponents.
Example	$2x^3 \cdot 5x^2$ in expanded form $2xxx \cdot 5xx$	$10x^5$

Operator	Expression	Process
Division	Contains factors.	Divide the coefficients. Keep the base(s) the same, and subtract the corresponding exponents.
Example	$\dfrac{24x^5}{6x^2}$ in expanded form $\dfrac{24xxxxx}{6xx} = \dfrac{\overset{4}{\cancel{24}}\,xxx\,\cancel{x}\,\cancel{x}}{\cancel{6}\,\cancel{x}\,\cancel{x}}$	$4x^3$

Operator	Expression	Process
Raising to a power	Factor being raised to a higher power.	Raise the coefficient to the power. Keep the base(s) the same. Multiply the exponent(s) by the power.
Example	$(3x^4)^3$ in expanded form is $(3x^4)(3x^4)(3x^4)$	$3^3x^{12} = 27x^{12}$

Negative exponents	Factor being raised to a negative power	If a factor has a negative exponent, it can be moved from the numerator to the denominator and the exponent will change to positive. A factor in the denominator with a negative exponent can likewise be made positive by moving the factor to the numerator.
Example	$3^{-4} = \dfrac{1}{3^4}$, $\dfrac{1}{x^{-2}} = x^2$	$\dfrac{2x^{-4}y^2}{3x^2y^{-5}} = \dfrac{2y^2y^5}{3x^2x^4} = \dfrac{2y^7}{3x^6}$

Mixed Practice with Exponents

Problem	Display rule	Solution
$(2x^5y^2z^3)(3xyz)(5x^3y^4z^4)$	$= (2 \cdot 3 \cdot 5x^{5+1+3}y^{2+1+4}z^{3+1+4})$	$= (30x^9y^7z^8)$
$\dfrac{32x^4y^2z}{6xy^2z^3}$	$= \dfrac{\not{2} \cdot 16x^{4-1}y^{2-2}}{\not{2} \cdot 3z^{3-1}}$	$= \dfrac{16x^3}{3z^2}$ **Note:** Because the exponent of z is larger in the denominator, the subtraction is done in the denominator.
$(5x^3y^2z^4)^3$	$= (5^3x^{3\cdot3}y^{2\cdot3}z^{4\cdot3}) = (5^3x^9y^6z^{12})$	$= (125x^9y^6z^{12})$
$(2xy^3)^4\,(2x^2y)$	$= (2^4x^4y^{12})(2x^2y) = (2^{4+1}x^{4+2}y^{12+1})$	$= (32x^6y^{13})$
$\dfrac{2^3x^{-5}y^{-2}z^3}{2^{-2}x^6y^3z^{-4}}$	$= \dfrac{2^32^2z^3z^4}{x^6x^5y^3y^2} = \dfrac{2^5z^7}{x^{11}y^5}$	$= \dfrac{32z^7}{x^{11}y^5}$

Roots

Let's begin by reviewing the terminology and vocabulary.

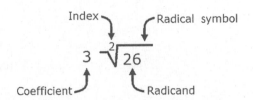

Radicals can have a coefficient also. It is the number in front of the radical sign. There is an "unshown" operator of multiplication between this coefficient and the radical.

To express a large number in factored form, a process of prime factorization is used. The prime factorization process breaks down a number into all prime factors. This can be shown using a factor tree.

A **prime number** is a positive integer that has exactly two different positive factors, itself and one; one is not a prime number.

Examples: 2 → prime

3 → prime

4 → not prime, factors 2 and 2

5 → prime

6 → not prime, factors 2 and 3

7 → prime

8 → not prime, factors 2 and 4

9 → not prime, factors 3 and 3

10 → not prime, factors 2 and 5

Can you see from the pattern above that every even number after 2 is not a prime?

Recognizing prime numbers is a good AIMS skill. Here is a list with the nonprime numbers crossed out. The nonprime numbers are called **composite numbers**. They are composed of factors other than one and the given number.

$$2 \quad 3 \quad \cancel{4} \quad 5 \quad \cancel{6} \quad 7 \quad \cancel{8} \quad \cancel{9} \quad \cancel{10}$$
$$11 \quad \cancel{12} \quad 13 \quad \cancel{14} \quad \cancel{15} \quad \cancel{16} \quad 17 \quad \cancel{18} \quad 19 \quad \cancel{20}$$
$$\cancel{21} \quad \cancel{22} \quad 23 \quad \cancel{24} \quad \cancel{25} \quad \cancel{26} \quad \cancel{27} \quad \cancel{28} \quad 29 \quad \cancel{30}$$
$$31 \quad \cancel{32} \quad \cancel{33} \quad \cancel{34} \quad \cancel{35} \quad \cancel{36} \quad 37 \quad \cancel{38} \quad \cancel{39} \quad \cancel{40}$$
$$41 \quad \cancel{42} \quad 43 \quad \cancel{44} \quad \cancel{45} \quad \cancel{46} \quad 47 \quad \cancel{48} \quad \cancel{49} \quad \cancel{50}$$

A factor tree is a diagram that helps you find the prime factors of a given number. Here are three different factor trees for the number 500.

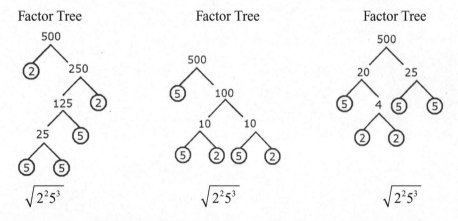

As you can see there are several different ways to make the factor tree for 500. Look closely at the prime factors at the end of each factor tree branch. The final answer in factored form is the same.

On a separate sheet of paper, try to make a factor tree for each of these values.

64   98   280

The results for the factor trees should be

$$64 = 2^6, \quad 98 = 2 \cdot 7^2, \text{ and } \quad 280 = 2^3 \cdot 5 \cdot 7$$

Let's use this new skill and turn our attention to radicals. When working with radicals you are sometimes asked to simplify or evaluate them.

Simplifying radicals

Square Roots
For a radical expression to be considered simplified, the factored radicand must not contain any perfect square factors. This can be accomplished by recognizing that a number is a perfect square, or you can use the factor tree method and remove prime factors in groups of two.

Examples:

Problem	Process	Solution
$\sqrt{16}$	Perfect square	4
$\sqrt{25}$	Perfect square	5
$\sqrt{32}$	$\sqrt{2^5} = \sqrt{2^2 \cdot 2^2 \cdot 2}$	$4\sqrt{2}$
$\sqrt{27}$	$\sqrt{3^3} = \sqrt{3^2 \cdot 3}$	$3\sqrt{3}$
$\sqrt{54}$	$\sqrt{2 \cdot 3^3} = \sqrt{2 \cdot 3^2 \cdot 3}$	$3\sqrt{2 \cdot 3} = 3\sqrt{6}$

Did you catch the step where $\sqrt{3^2}$ became 3? (Remember 3^2 is 9, a perfect square, and the square root of 9 is 3.

In square roots this happens to each pair of like factors. The square root removes the square power, and the result becomes a coefficient factor to the remaining part of the problem.

Examples:

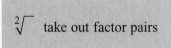

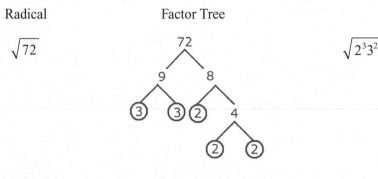

Radical

$$\sqrt{72}$$

Factor Tree

72

$$\sqrt{2^3 3^2}$$

Factored Form

$$\sqrt{2^3 3^2} = 2 \cdot 3\sqrt{2} = 6\sqrt{2}$$

Radical	Factor Tree	
$\sqrt{125}$	125	$\sqrt{5^3}$

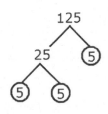

Factored Form

$$\sqrt{5^3} = 5\sqrt{5}$$

Radical	Factor Tree	
$\sqrt{198}$	198	$\sqrt{2 \cdot 7^2}$

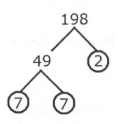

Factored Form

$$\sqrt{2 \cdot 7^2} = 7\sqrt{2}$$

Simplifying cube roots uses the same process, except instead of making pairs; you are making groups of threes.

$\sqrt[3]{}$ take out factor triples

Example:

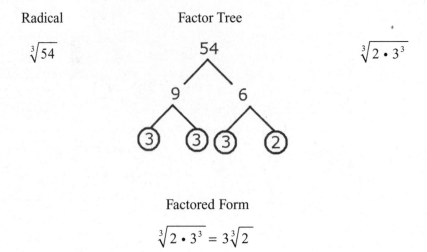

Radical   Factor Tree

$$\sqrt[3]{54}$$
$$\sqrt[3]{2 \cdot 3^3}$$

54

Factored Form

$$\sqrt[3]{2 \cdot 3^3} = 3\sqrt[3]{2}$$

Perfect square and perfect cube roots—expressions containing values and variables that are perfect squares or perfect cubes.

Algorithm
Write the radicand in factored form. In a square root problem, each perfect square factor is removed from under the radical symbol and its square root is placed as a single factor in the answer. When simplifying cube roots you need to find groups of three identical factors. (Note, when dealing with radicals and variables, the variables are typically representing positive numbers.)

Simplify: $\sqrt{25a^2} = 5a$

$\sqrt{49x^2y^4} = 7xy^2$

$\sqrt[3]{8x^6} = \sqrt[3]{(2)^3 \, x^2x^2x^2} = 2x^2$

Translate

The goal in evaluating is to change the words into an algebraic expression.

Algebraic expression—an expression containing variables, numerals, grouping symbols, and operators.

Algorithm
Apply operators from the key words as mentioned in Strand 1.

Translate:
- √ Some number increased by 8: $x + 8$
- √ Twice a number decreased by 5: $2x - 5$
- √ The square root of the sum of a number and 2: $\sqrt{x+2}$

Algebraic sentence—an equation or inequality containing variables, numerals, grouping symbols, operators, and an equal sign.

Algorithm

Apply operators from the key words to form an equation. Look for the key word representing the equal sign. We referred to several different words in Strand 1, but typically it is the word "is." The sentence may be an inequality, in which the inequality symbol separates the expressions. An equation or inequality equation is said to have a left and a right side with the equal (or inequality) sign as the separator.

Translate:
- √ Some number increased by 8 is 12 : $x + 8 = 12$
- √ Twice a number decreased by 5 is three times a number: $2x - 5 = 3x$
- √ The square root of the sum of a number and 2 is greater than the number squared: $\sqrt{x+2} > x^2$
- √ The first four test scores were: 88, 94, 72, and 80. What score would you have to obtain to average an 85? $\dfrac{88+94+72+80+x}{5} = 85$

Write

The goal of writing is to write a linear equation.

Write a linear equation from a given slope and a point

Algorithm

    a. Use the slope and one point to find the y-intercept.
    b. Write the equation using the slope and y-intercept.

Write the line:
Given: $m = 3$ and $(4, 10)$

    a. $y = mx + b$ use the point $(4, 10)$ and $m = 3$

    $\Rightarrow 10 = 3(4) + b \qquad \Rightarrow \qquad 10 = 12 + b \qquad -2 = b$

    b. $y = 3x - 2$

Write a linear equation that represents a table of values.

Algorithm

    a. Pick two points and compute the slope of the line.
    b. Use that slope and one given point to find the y-intercept.
    c. Write the equation using the slope and y-intercept.

Write the line:

$$\begin{array}{c|c} x & y \\ \hline 1 & 5 \\ 3 & 9 \\ 2 & 7 \end{array}$$

Given:

a. Slope $= \dfrac{y_2 - y_1}{x_2 - x_1} = \dfrac{9-5}{3-1} = \dfrac{4}{2} = 2$

b. $y = mx + b$ (use the point $(1, 5)$ and $m = 2$)

$\Rightarrow 5 = 2(1) + b \qquad \Rightarrow \qquad 5 = 2 + b \qquad 3 = b$

c. $y = 2x + 3$

Write a linear equation from a contextual problem

Algorithm

a. Identify two points and compute the slope of the line.
b. Use that slope and one given point to find the y-intercept.
c. Write the equation using the slope and y-intercept.

Given: A city's diving pool is being drained. If the pool is 14 feet deep and the water goes down 3 feet every 2 hours, write an equation that shows how fast the water is being drained from the pool.

Let $x =$ number of hours the pool has been draining, and $y =$ the depth of the water in the pool. The level of the water is at 14 feet at the beginning and goes down three feet after two hours. Therefore, two data points are $(0, 14)$ and $(2, 11)$.

a. Slope $= \dfrac{y_2 - y_1}{x_2 - x_1} = \dfrac{14-11}{0-2} = \dfrac{3}{-2} = \dfrac{-3}{2}$

b. y-intercept is $(0, 14)$, or calculated with the point $(0,14)$

$14 = \dfrac{-3}{2}(0) + b \implies 14 = b$

c. $y = mx + b \qquad y = \dfrac{-3}{2}x + 14$

Write a linear equation from a graph

Algorithm

a. Determine the vertical change and the horizontal change between two points on the line.
b. Use the graph to find the point at which the graph crosses the y-axis. This value is the y-intercept (b).
c. Write the equation using the slope and y-intercept.

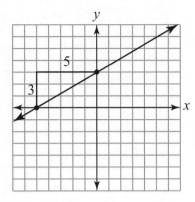

$$y = \frac{3}{5}x + 3$$

Solve

The goal of solving is to find the value(s) that makes the equation true.

Solve a linear equation (or inequality) with one variable

Algorithm

To find the solution of the equation, isolate the variable by using algebraic operations to get the variable by itself on one side of the equal (inequality) sign and everything else on the other side.

   a. Remove any parentheses using the distributive property.

   b. On each side of the equation, combine like terms.

   c. Isolate the variable by first moving unwanted terms using addition or subtraction. This is often called undoing. Use the opposite operation to move terms. (If the expression is adding 5, then subtracting 5 from both sides will move the 5 to the other side.)

   d. If the coefficient of the variable is not 1, you can make it a 1 by using either multiplication or division. You can multiply both sides of the equation by the reciprocal of that coefficient or divide both sides by that coefficient. This is true because dividing by a number is the same as multiplying by its reciprocal.

   e. Check the solution.

Solve this equation.

$8(x + 2) + 2x = 44 + 2\,(x - 2)$	Distribute to remove parentheses.
$8x + 16 + 2x = 44 + 2x - 4$	Gather like terms.
$10x + 16 = 40 + 2x$	Move terms by undoing.
$8x = 24$	Remove coefficient by dividing both
$x = 3$	sides by eight.

Check:

$$8(3 + 2) + 2(3) = 44 + 2\,(3 - 2)$$
$$8(5) + 6 = 44 + 2(1)$$
$$40 + 6 = 44 + 2$$
$$46 = 46 \checkmark$$

Solve this equation:

$$2|x| - 12 = 8$$
$$2|x| = 20$$
$$|x| = 10$$
$$x = 10 \ or \ x = -10$$

Move the term 12 by adding.
Remove coefficient by multiplying by ½.
Apply concept of absolute value.

Check:

$$2|10| - 12 = 8 \ or \ 2|-10| - 12 = 8$$
$$20 - 12 = 8$$
$$8 = 8 \checkmark$$

Solve this inequality:

$$-6(x - 2) > 2(1 - 2x)$$
$$-6x + 12 > 2 - 4x$$
$$-2x > -10$$
$$x < 5$$

Distribute to remove parentheses.
Move terms by undoing.
Remove coefficient by dividing by a negative two (−2). Remember to switch the inequality sign when dividing (or multiplying) by a negative number.

Check:

$$-6(4 - 2) > 2(1 - 2(4))$$
$$-6(2) > 2(1 - 8)$$
$$-12 > 2(-7)$$
$$-12 > -14 \checkmark$$

Select any number that satisfies the solution. This check uses 4.

Solve an algebraic proportion

Algorithm

A proportion is an equation where one ratio equals another ratio.

    a. Simplify each side to a single fraction, if needed.
    b. Apply the means-extremes product rule.

$$\frac{a}{b} = \frac{c}{d} \ \Rightarrow \ ad = bc$$

    c. Solve as usual for equations.

Solve the proportion.

$$\frac{x}{x+5}=\frac{2}{3}$$ Means-extremes product rule.

$$3x=2(x+5)$$ Distribute to remove parentheses.

$$3x=2x+10$$ Move terms and solve as usual.

$$x=10$$

Check:

$$\frac{10}{10+5}=\frac{2}{3}$$

$$\frac{10}{15}=\frac{2}{3} \quad ✓$$

Solve the proportion:

$$\frac{x}{8}=\frac{x+2}{24}$$ Means-extremes product rule.

$$24x=8(x+2)$$ Distribute to remove parentheses.

$$24x=8x+16$$ Move terms.

$$16x=16$$ Remove coefficient by dividing both sides by 16.

$$x=1$$

Check:

$$\frac{1}{8}=\frac{1+2}{24}$$

$$\frac{1}{8}=\frac{3}{24} \quad ✓$$

Solve an algebraic system of equations

Algorithms

There is a variety of ways to find the solution to a system of equations. The Math Standard emphasizes three ways: substitution, elimination, and graphing.

Substitution Method: The goal is to replace two equations with one equation that contains a common variable.

    a. Choose one equation and isolate a single variable.

    b. Use the result from the preceding step to replace the variable occurrence in the second equation.

    c. Solve the remaining equation as usual for a single variable.

    d. Substitute the solution into the first equation to find the value of the remaining variable.

    e. Check the solution in each equation of the system.

Solve a linear system using substitution.

$$x - y = 8$$
$$4x + y = 12$$

Given system of linear equations.

$$x - y = 8$$
$$\underline{+y = +y}$$
$$x = y + 8$$

Solve the top equation for x by adding y to both sides.

$$4(y + 8) + y = 12$$

Substitute this expression into the second equation in place of x.

$$4y + 32 + y = 12$$
$$5y + 32 = 12$$
$$5y = -20$$
$$y = -4$$

Solve for y by distributing and gathering similar terms.

$$x - (-4) = 8$$
$$x + 4 = 8$$
$$x = 4$$

Replace the value for y into either given equation and solve for x. The solution to the system is often written as an ordered pair. The solution to this system is $(4, -4)$.

$$4x + y = 12$$
$$4(4) + (-4) = 12$$
$$16 - 4 = 12$$
$$12 = 12 \checkmark$$

Check by placing your solution into both equations.

$$x - y = 8$$
$$4 - (-4) = 8$$
$$8 = 8 \checkmark$$

Elimination Method: The goal is to combine (add or subtract) the equations in such a way that one of the variables drops out and you are left with a single equation in a single variable. You can eliminate either variable when combining the equations.

a. Make the coefficients for a single variable opposite by multiplying by a scalar constant. (A scalar constant is a nonzero value.)
b. Add the two equations, using the rules for signed numbers.
c. Solve the resulting equation as usual for a single variable.
d. Substitute this value into either original equation to find the value of the remaining variable.
e. Check the solution in each equation of the system.

Solve the linear system using elimination.

$x - 3y = 3$
$4x + y = 12$

Given system of linear equations.

$x - 3y = 3$
$3(4x + y) = 3(12)$

Multiplying the second equation by 3 will make the coefficients for the y variable opposites.

$x - 3y = 3$
$12x + 3y = 36$

Combine the equations.

$13x + 0y = 39$
$13x = 39$
$x = 3$

Solve for x.

$3 - 3y = 3$
$-3y = 0$
$y = 0$

Replace your value of x into either original equation. This example used the first equation. The solution to this system is (3, 0).

$x - 3y = 3$
$3 - 3(0) = 3$
$3 = 3$ ✓

Check by placing your solution into both equations.

$4x + y = 12$
$4(3) + (0) = 12$
$12 = 12$ ✓

Here are two problems to give you additional practice in setting up the elimination method.

System of equations	Scalar to make x coefficients opposites	or	Scalar to make y coefficients opposites
$x - 2y = 4$ $3x + y = 12$	Multiply the top equation by –3.		Multiply the bottom equation by 2.
$2x - 3y = 8$ $4x + 2y = 7$	Multiply the top equation by a –2.		Multiply the top equation by 2 and the bottom equation by 3.

Graphing: The third method used for solving a system of equations will be discussed in coordinate geometry in Strand 4.

Solve formulas for specified variables

Solving a formula for a specified variable is an algebraic process in which you follow the rules of equality (addition, subtraction, multiplication, division) to isolate the specified variable. For example, distance is defined to be rate times time.

$$d = r \bullet t$$

Often a question asks us to find a rate given distance and time traveled. Therefore, it is handy to have the equation solved for the rate, r.

$r \cdot t = d$	Distance formula
$\dfrac{r \cdot \cancel{t}}{\cancel{t}} = \dfrac{d}{t}$	Solve for rate by dividing by time.
$r = \dfrac{d}{t}$	Rate equals distance divided by time. *Notice that rate is the ratio of distance to time just as we often describe rate as miles/hour.

| Given the formula for changing temperatures from Centigrade to Fahrenheit, solve for Fahrenheit:

 $C = \dfrac{5}{9}(F - 32)$

 $9C = 5(F - 32)$

 $9C = 5F - 160$

 $9C + 160 = 5F$

 $\dfrac{9}{5}C + 32 = F$ | Given the formula for the perimeter of a rectangle, solve for length:

 $P = 2L + 2W$

 $P - 2W = 2L$

 $\dfrac{P - 2W}{2} = L$ |
| Given the general equation of a line, solve for y:

 $Ax + By = C$

 $By = -Ax + C$

 $y = \dfrac{-A}{B}x + \dfrac{C}{B}$ | Given the formula for finding the area of a triangle, solve for height:

 $A = \dfrac{1}{2}bh$

 $2A = bh$

 $\dfrac{2A}{b} = h$ |

Solve radical equations

Algorithm

The key to solving a radical equation is to treat the radical as an unknown variable. When solving for a variable, the goal is to isolate the variable by the undoing processes. The same is true in a radical equation—isolate the radical. When the radical is by itself, you square both sides. This squaring process removes the radical symbol, and you·can continue to isolate the variable as needed.

Solve the radical equation:

$2\sqrt{x}+5=19$ Given a radical equation.

$2\sqrt{x}=14$ Isolate the equation for the radical by subtracting 5 from both sides.

$\sqrt{x}=7$ Divide each side by 2

$\left(\sqrt{x}\right)^2=\left(7\right)^2$ Solve for x by squaring both sides.

$x=49$ Solution

$2\sqrt{\left(49\right)}+5=19$ Check by placing your solution into the original equation.

$2\left(7\right)+5=19$

$14+5=19$ ✓

It is very important to check the solution to a radical equation. Squaring both sides of an equation may introduce a false solution. Study the example below.

Solve the radical equation:

$2\sqrt{x}+10=8$ Given a radical equation.

$2\sqrt{x}=-2$ Isolate the equation for the radical by subtracting 10 from both sides.

$\sqrt{x}=-1$ Divide each side by 2.

$\left(\sqrt{x}\right)^2=\left(-1\right)^2$ Solve for x by squaring both sides.

$x=1$ Solution

$2\sqrt{\left(1\right)}+10=8$ Check by placing your solution into the original equation. In checking the solution a false statement was produced. There is no solution to this radical equation.

$2\left(1\right)+10=8$

$12\neq8$

Solve quadratic equations

Algorithm

The quadratic formula, shown below, can be used to solve quadratic equations.

$$x=\frac{-b+\sqrt{b^2-4ac}}{2a}$$

The first step is to rewrite the quadratic equation in standard form. A quadratic equation is in standard form when the equation is equal to zero, and the terms are listed in descending order from highest to lowest power for the variable.

$$ax^2 + bx + c = 0$$

This first term, ax^2, is called the quadratic term. Every quadratic equation has this term. The next term is called the linear term, bx, and the last term is called the constant term, c. A quadratic equation might not have the linear and/or constant terms and yet still be a quadratic.

After placing the quadratic in standard form, match the coefficient values from your equation with the coefficients, a, b, and c, from the standard form. If the coefficient is negative, be sure to include the minus sign with the coefficient.

$$2x^2 - 7x + 3 = 0$$
$$a = 2, \ b = -7, \text{ and } c = 3$$

The next step requires you to replace the letters a, b, and c in the formula with the values from the problem. It is strongly recommended that you first replace each letter a, b, and c with empty parentheses and then place your values into the formula. It is easy to make a sign error when doing arithmetic with these numbers.

$$x = \frac{-(\ )\pm\sqrt{(\ )^2-4(\ )(\ )}}{2(\ )}$$

$$x = \frac{-(-7)\pm\sqrt{(-7)^2-4(2)(3)}}{2(2)}$$

Now, simplify the expression. You may want to start with the radicand. The terms $b^2 - 4ac$ are called the discriminant. Since the discriminant is under the radical symbol, it is of particular interest.

• If the discriminant value is negative, stop the process and declare that the quadratic equation does not have real number solutions. (Recall, that the square root of a negative number is not defined with the real number system.)

• If the discriminant is positive, simplify using the rules for radicals. If the discriminant is a perfect square, the expression will simplify into two rational solutions. If the discriminant is not a perfect square, simplify the radicand and the fractional expression, and look for common factors to reduce.

• If the discriminant is zero, there is just one solution.

Example:

$$x = \frac{7 \pm \sqrt{49 - 24}}{4}$$

$$x = \frac{7 \pm \sqrt{25}}{4}$$

$$x = \frac{7 \pm 5}{4}$$

$$x = \frac{12}{4} \text{ or } \frac{2}{4}$$

$$x = 3 \text{ or } \frac{1}{2}$$

$$2x^2 - 7x + 3 = 0$$
$$2(3)^2 - 7(3) + 3 = 0$$
$$2(9) - 7(3) + 3 = 0$$
$$18 - 21 + 3 = 0$$
$$0 = 0 \quad \checkmark$$

$$2x^2 - 7x + 3 = 0$$
$$2\left(\frac{1}{2}\right)^2 - 7\left(\frac{1}{2}\right) + 3 = 0$$
$$2\left(\frac{1}{4}\right) - 7\left(\frac{1}{2}\right) + 3 = 0$$
$$\frac{1}{2} - \frac{7}{2} + 3 = 0$$
$$\frac{-6}{2} + 3 = 0$$
$$-3 + 3 = 0$$
$$0 = 0 \quad \checkmark$$

Solve the quadratic equation:

$x^2 - 5x + 9 = 5x$ Given the quadratic equation.

$x^2 - 10x + 9 = 0$ Rewrite given quadratic equation in standard form.

$x^2 - 10x + 9 = 0$ List the values for each of the
$a = 1, b = -10,$ *and* $c = 9$ letters in the quadratic formula.

$$x = \frac{-b \pm \sqrt{b^2 - 4ac}}{2a}$$

$$x = \frac{-(\ ) \pm \sqrt{(\ )^2 - 4(\ )(\ )}}{2(\ )}$$ Place empty parentheses for the letters in the quadratic formula.

$$x = \frac{-(-10) \pm \sqrt{(-10)^2 - 4(1)(9)}}{2(1)}$$ Place the values into the formula.

$$x = \frac{-(-10) \pm \sqrt{100 - 36}}{2(1)}$$ Work with the discriminant.

$$x = \frac{-(-10) \pm \sqrt{64}}{2(1)}$$ Simplify the radical.

$$x = \frac{10 \pm 8}{2} = \frac{18}{2} \ and \ \frac{2}{2}$$ Use the plus-minus sign to form two solutions.

$x = 9 \ and \ x = 1$ Simplified answers.

$(9)^2 - 5(9) + 9 = 5(9)$ Check each solution into the
$81 - 45 + 9 = 45$ original equation.
$0 = 0 \checkmark$

$(1)^2 - 5(1) + 9 = 5(1)$ Check each solution into the
$1 - 5 + 9 = 5$ original equation.
$0 = 0 \checkmark$

Additional practice in setting up and using the quadratic formula follows:

Equation to solve	Standard form	Identify letters (*a, b, c*)	Numbers inserted into formula
$x^2 - 12 = 4x$	$x^2 - 4x - 12 = 0$	$a = 1$ $b = -4$ $c = -12$	$x = \dfrac{-(-4) \pm \sqrt{(-4)^2 - 4(1)(-12)}}{2(1)}$
$2x^2 - 17x = -8$	$2x^2 - 17x + 8 = 0$	$a = 2$ $b = -17$ $c = 8$	$x = \dfrac{-(-17) \pm \sqrt{(-17)^2 - 4(2)(8)}}{2(2)}$
$x^2 - 16 = 0$	$x^2 - 16 = 0$	$a = 1$ $b = 0$ $c = -16$	$x = \dfrac{-(0) \pm \sqrt{(0)^2 - 4(1)(-16)}}{2(1)}$
$x^2 = 13x$	$x^2 - 13x = 0$	$a = 1$ $b = -13$ $c = 0$	$x = \dfrac{-(-13) \pm \sqrt{(-13)^2 - 4(1)(0)}}{2(1)}$

Trigonometry

Trigonometry uses ratios to find the measures of the sides and angles in a right triangle. The basic trigonometric ratios are defined from an acute angle of a right triangle. The definition of these ratios includes the terms *adjacent*, *opposite*, and *hypotenuse*. Let's define these terms using $\angle A$ as the acute angle.

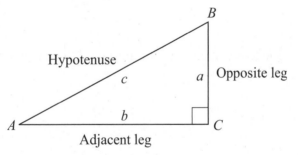

Notice $\overline{AC}$ is the leg adjacent to $\angle A$, and $\overline{BC}$ is the leg opposite $\angle A$.

It is common to label the angles using capital letters and the sides opposite the angles with their corresponding lowercase letter. The right angle of this triangle is labeled with a small box. The measure of this angle is 90°. The side opposite the right angle is called the hypotenuse. The other two angles are both acute (less than 90° in measure) and are said to be complementary angles (the sum of the two angles is 90°). The sides opposite the acute angles are sometimes referred to as legs.

There are three major trigonometric ratios: sine, cosine, and tangent. These are often abbreviated as sin, cos, and tan. These ratios have a wide range of applications. The trigonometric ratios are defined as follows:

- The sine of angle A is the ratio of the length of the opposite leg to the length of the hypotenuse.

$$\sin A = \frac{\text{Opposite leg}}{\text{Hypotenuse}} = \frac{a}{c}$$

- The cosine of angle A is the ratio of the length of the adjacent leg to the length of the hypotenuse.

$$\cos A = \frac{\text{Adjacent leg}}{\text{Hypotenuse}} = \frac{b}{c}$$

- The tangent of angle A is the ratio of the length of the opposite leg to the length of the adjacent leg.

$$\tan A = \frac{\text{Opposite leg}}{\text{Adjacent leg}} = \frac{a}{b}$$

Here are some practice problems.

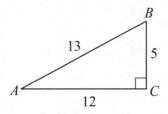

$$\sin A = \frac{5}{13} \qquad \sin B = \frac{12}{13}$$

$$\cos A = \frac{12}{13} \qquad \cos B = \frac{5}{13}$$

$$\tan A = \frac{5}{12} \qquad \tan B = \frac{12}{5}$$

The trigonometry ratios can be used to solve for missing values in a right triangle.

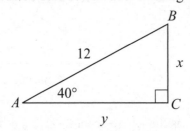

$$\sin 40 = \frac{x}{12} \qquad \cos 40 = \frac{y}{12}$$

On the AIMS test you will not be asked to solve for the values of x and y. This process requires the use of a calculator or a trigonometric table.

Matrices

A **matrix** is a rectangular arrangement of numbers in rows and columns. The plural for matrix is matrices. A matrix is used to organize related quantities. It is a chart that organizes numerical values in a way that can be easily read. Arithmetic operations can be performed on matrices.

Example: Suppose you owned a wood shop and had four types of wood: Ash, Walnut, Spruce, and Pine. Each type of wood was evaluated with a grade of clear, fair, and rough. The stock on hand of Ash in board foot is 100, 200, and 150 for each grade, respectively. The amount for Walnut is 75, 50, and 60. The amount for Spruce is 125, 100, and 75. The amount for Pine is 200, 250, and 225. This inventory information could be placed into a matrix. To organize the data you would choose to list one category as the rows and the other category as the columns. This matrix is arranged so that the rows are comprised of the types of wood and the columns are represented by the grade of the wood.

Inventory of Wood Available

	Clear	Fair	Rough
Ash	100	200	150
Walnut	75	50	60
Spruce	125	100	75
Pine	200	250	225

The matrix is said to have a dimension of four by three. You always state the row size first then followed by the column size. It is easy to confuse the rows with the columns. A method of remembering is to think that matrices are built out of ROWman COLumns. A matrix is given a name, usually a capital letter, like A or B. Sometimes as a reminder, the dimensions are written to the right of the letter, as in $A_{4\,3}$.

$$\begin{bmatrix} 100 & 200 & 150 \\ 75 & 50 & 60 \\ 125 & 100 & 75 \\ 200 & 250 & 225 \end{bmatrix}$$

The matrix from the wood inventory can be placed in standard matrix form as shown above. In this example the rows represented the type of wood and the columns represent the grade of the wood.

The values in a matrix are called elements of the matrix. The elements of a matrix have names, usually a lowercase letter the same as the matrix name, with the position of the element written as a subscript. The top row is row one. The leftmost column is column one.

From the wood matrix inventory, the value 200 representing the amount of Ash that is rated Fair is found in the first row and the second column. If the matrix is called matrix A then the 200 could be represented by the name of a_{12}. Note that if you switch the row number with the column number you are at the location of a_{21}. This value is 75 board feet and is describing the amount of Clear grade Walnut that you have on hand.

Two matrices can be equal to each other if they have the same dimensions and have the same values in the same positions. Consider the following examples:

$A = \begin{bmatrix} 1 & 2 \\ 3 & 4 \\ 5 & 6 \end{bmatrix}$ $B = \begin{bmatrix} 1 & 2 & 3 \\ 4 & 5 & 6 \end{bmatrix}$	The matrices contain the same numbers but the positions of these numbers are different. These are not equal matrices.
$A = \begin{bmatrix} 1 & 2 \\ 3 & 4 \\ 5 & 6 \end{bmatrix}$ $B = \begin{bmatrix} \dfrac{3}{3} & 6-4 \\ 3 & 2^2 \\ \dfrac{10}{2} & 2 \cdot 3 \end{bmatrix}$	The matrices have the same dimension and if you simplify the elements of matrix B, each entry equals the corresponding entry in matrix A. The matrices are equal.

There are several defined arithmetic operations for matrices. Matrices can be added and subtracted from each other, as well as scaled by multiplying by a number. This is called scalar matrix multiplication.

If two matrices have the same dimension (same number of rows and columns), then the matrix sum can be computed. The operation is performed by adding corresponding elements of the first matrix with the second matrix. The element a_{11} is combined with b_{11}, and so forth.

Let's go back to our wood shop. Suppose you ordered additional wood for your inventory.

Purchased Wood To Be Added to the Inventory

	Clear	Fair	Rough
Ash	50	25	100
Walnut	100	35	70
Spruce	100	50	25
Pine	225	150	125

To find the total available wood sorted by grade, we add the matrices together.

Current inventory

$$\begin{bmatrix} 100 & 200 & 150 \\ 75 & 50 & 60 \\ 125 & 100 & 75 \\ 200 & 250 & 225 \end{bmatrix}$$

New shipment

$$+ \begin{bmatrix} 50 & 25 & 100 \\ 100 & 35 & 70 \\ 100 & 50 & 25 \\ 225 & 150 & 125 \end{bmatrix}$$

Updated inventory

$$= \begin{bmatrix} 150 & 225 & 250 \\ 175 & 85 & 130 \\ 225 & 150 & 100 \\ 425 & 400 & 350 \end{bmatrix}$$

Subtraction with matrices is similar to addition of matrices. When the dimensions are the same, you subtract the corresponding elements. Just as with the real number system where subtraction is not commutative, subtraction of matrices is not commutative. Watch the order in which you subtract.

Suppose you had a project that required:

	Clear	Fair	Rough
Ash	10	30	40
Walnut	35	20	0
Spruce	0	0	0
Pine	100	75	200

By subtracting the amounts needed from your updated inventory, you will have the remaining amount in your new inventory.

Updated inventory   Needed for project   New inventory

$$
\begin{bmatrix} 150 & 225 & 250 \\ 175 & 85 & 130 \\ 225 & 150 & 100 \\ 425 & 400 & 350 \end{bmatrix} - \begin{bmatrix} 10 & 30 & 40 \\ 35 & 20 & 0 \\ 0 & 0 & 0 \\ 100 & 75 & 200 \end{bmatrix} = \begin{bmatrix} 140 & 195 & 210 \\ 140 & 65 & 130 \\ 225 & 150 & 100 \\ 325 & 325 & 150 \end{bmatrix}
$$

The last operation for matrices covered on the AIMS test is scalar multiplication. The word "scalar" is a term meaning a number. To multiply a matrix by a scalar, you multiply each element of the matrix by this number.

If we wanted to carry twice the inventory of wood, we would multiply our matrix representing our inventory by 2.

$$
2 \cdot \begin{bmatrix} 140 & 195 & 210 \\ 140 & 65 & 130 \\ 225 & 150 & 100 \\ 325 & 325 & 150 \end{bmatrix} = \begin{bmatrix} 280 & 390 & 420 \\ 280 & 130 & 260 \\ 450 & 300 & 200 \\ 650 & 650 & 300 \end{bmatrix}
$$

Mixed Practice with the following given matrices:

$$
A = \begin{bmatrix} 1 & 2 \\ 3 & 4 \end{bmatrix} \quad B = \begin{bmatrix} 9 & 8 \\ 7 & 6 \end{bmatrix} \quad C = \begin{bmatrix} 3 & 0 & -2 \\ 7 & 6 & 9 \end{bmatrix} \quad D = \begin{bmatrix} -5 & 1 & 4 \\ 3 & -3 & 5 \end{bmatrix}
$$

Example:

$$
A + B = \begin{bmatrix} 10 & 10 \\ 10 & 10 \end{bmatrix} \qquad A + C = \text{Cannot add—different dimensions}
$$

$$
B - A = \begin{bmatrix} 8 & 6 \\ 4 & 2 \end{bmatrix} \qquad 3D = \begin{bmatrix} -15 & 3 & 12 \\ 9 & -9 & 15 \end{bmatrix}
$$

Are you able to do each of the following?

❏ Identify patterns

❏ Recognize a function and determine its domain and range

❏ Describe, identify, and sketch graphs of real-world situations

❏ Express and interpret the relationship between data

❏ Determine if the graphs of two linear equations are
- Parallel
- Perpendicular
- Coincident
- Intersecting, but not perpendicular

❏ Maximum/minimum problems

❏ Mathematics expressions and equations
- Evaluate algebraic expressions
- Simplify algebraic expressions
  - Multiply and divide monomials
  - Add, subtract, and perform scalar multiplication with matrices
  - Calculate powers and roots of real numbers
- Translate
  - Words to algebra
  - Algebra to words
- Write a linear equation
  - Using a table of values
  - Using a data set that models a real-world situation
  - Given two points on the line
  - Given the slope and point on the line
  - Given the graph of the line
- Solve
  - Linear equations
  - Linear inequalities
  - Algebraic proportions
  - Square root radical equations
  - Systems of equations
  - Quadratic equations

❏ Identify the sine, cosine, and tangent ratios

❏ Determine slope, x-intercept, and y-intercept of a linear equation

❏ Solve formulas for specified variables

Practice Problems—Strand 3

1. Which of the following is a solution to the inequality:

$$4(3 - x) + 2x + 8 \geq 14$$

A. $x \leq 8$
B. $x \geq -3$
C. $x \geq -8$
D. $x \leq 3$

2. Find the solution to the following equation:

$$3\sqrt{2 - x} = 9$$

A. $x = -25$
B. $x = -7$
C. $x = 2$
D. $x = 5$

3. The formula for finding the volume of a right circular cone is given below. Solve the formula in terms of the variable h (height).

$$V = \frac{1}{3}\pi r^2 h$$

A. $V\frac{1}{3}\pi r^2 = h$

B. $3V \pi r^2 = h$

C. $\frac{3V\pi}{r^2} = h$

D. $\frac{3V}{\pi r^2} = h$

4. Solve the following quadratic:

$$x^2 - 7x + 6 = 0$$

A. $-6, -1$
B. $6, 1$
C. $6 \pm \sqrt{42}$
D. $1 \pm \sqrt{42}$

5. Evaluate the expression $-2x^2 + 11x + 9$ when $x = -2$.

A. $y = 22$
B. $y = -21$
C. $y = 21$
D. $y = -5$

6. Simplify $(3a^4b^5c)^4$.

A. $7a^{16}b^{20}c^4$
B. $3a^{16}b^{20}c^4$
C. $12a^8b^9c^5$
D. $81a^{16}b^{20}c^4$

7. Find the next three numbers in the pattern 25, 21, 17, 13, ___, ___, ___.

A. 12, 11, 10
B. 10, 7, 4
C. 17, 21, 25
D. 9, 5, 1

8. Which rule describes how to find the next number in the pattern $-5, -2, 1, 4$?

A. Divide by 4.
B. Subtract 3.
C. Add 3.
D. Add 4.

9. For safety reasons, the base angle of a 26-foot ladder should be no less than 65° between the ladder and the ground. Which trigonometric expression would be used to find how high a 26-foot ladder safely can reach?

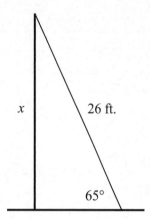

A. $\sin(65) = \dfrac{x}{26}$

B. $\cos(65) = \dfrac{x}{26}$

C. $\sin(65) = \dfrac{26}{x}$

D. $\cos(65) = \dfrac{26}{x}$

10. Solve the system of equations using substitution or elimination.

$$x + 2y = 6 \text{ and } x - 3y = -4$$

A. $(2, 2)$
B. $(-2, 2)$
C. $(2, -2)$
D. $(-2, -2)$

11. Tickets for a music concert were purchased at a rate of 1,000 tickets in 20 minutes. At this rate, how many tickets would have been purchased in 100 minutes?

A. 200
B. 500
C. 1,000
D. 5,000

12. Which equation is shown in the graph?

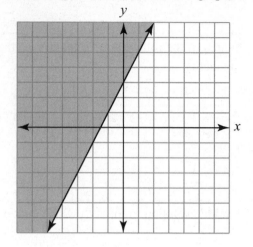

A. $y < 2x + 3$
B. $y \le 2x + 3$
C. $y > 2x + 3$
D. $y \ge 2x + 3$

13. Solve the system of equations. Which point on the graph is the solution to the system of linear equations?

$$x + y = 4$$
$$x - y = 2$$

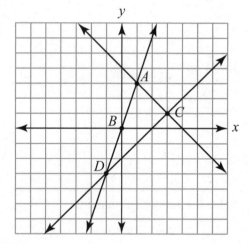

A. A
B. B
C. C
D. D

14. Which of these statements is true about the graphs of the equations below?

$$y = 3x + 5$$
$$3y = 9x - 5$$

A. The lines are parallel.
B. The lines are perpendicular.
C. The lines coincide.
D. The lines intersect, but are not perpendicular.

---

15. Which of the following is the sum $2A + 3B$?

$$A = \begin{bmatrix} -2 & 3 \\ 5 & -3 \end{bmatrix} \quad \text{and} \quad B = \begin{bmatrix} 2 & -3 \\ 7 & 0 \end{bmatrix}$$

A. $\begin{bmatrix} 2 & -3 \\ 31 & -6 \end{bmatrix}$ **B.** $\begin{bmatrix} -10 & -15 \\ 31 & -6 \end{bmatrix}$

C. $\begin{bmatrix} -2 & 3 \\ 21 & 0 \end{bmatrix}$ **D.** $\begin{bmatrix} 0 & 0 \\ 12 & -3 \end{bmatrix}$

---

16. Which of the following expressions is equivalent to the one shown below?

$$(3b^4 - 2b^3 - b + 7) - (3b^4 - 4b^3 + 5)$$

A. $2b^3 - b + 2$
B. $6b^4 + 2b^3 - b + 2$
C. $2b^3 - b + 12$
D. $6b^4 + 2b^3 - b + 12$

---

17. Bailey noticed that many of the students at her school had red hair. She randomly chose 25 of the students in her school and found that two of them had red hair. If Bailey's sample is representative, which of the following is closest to the number of the 2,000 students at her school who have red hair?

A. 40
B. 80
C. 160
D. 200

---

18. Maria was asked to find an expression that is not equivalent to 2^{12}. Which of the following is not equivalent to the given expression?

A. $(2^3)^4$
B. $(2^2 \cdot 2^2)^3$
C. $(2^6 \cdot 2^2)$
D. $\left(\dfrac{2^{15}}{2^3}\right)$

---

19. Solve the equation for the variable x:

$$9x - 2(4x + 5) = 2x - (4 - x) - 12$$

A. 16
B. 13
C. 9
D. 3

20. Which graph represents an ice cold glass of water that is left on the table on a typical Arizona day?

A.

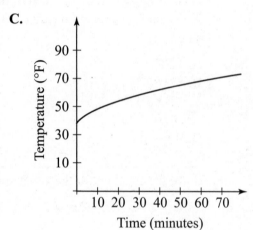

B.

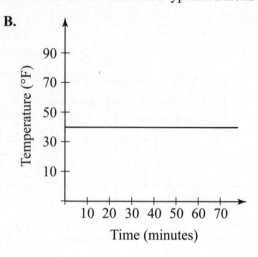

C.

D.

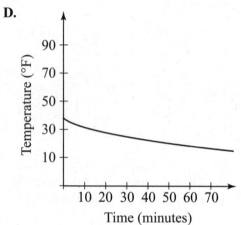

21. Which of the following graphs is not a graph of a function?

A.

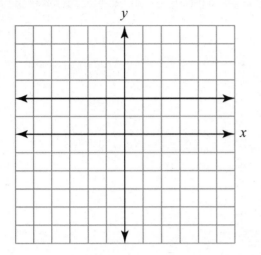

B.

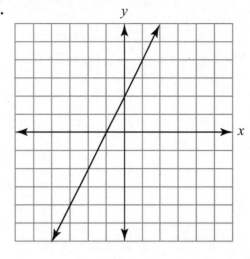

C.

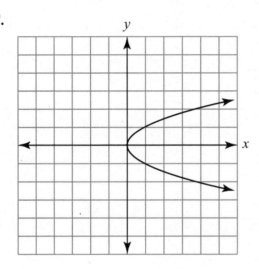

D.

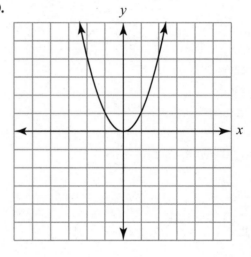

---

22. What is the domain and the range for the function $y = \sqrt{x+2}$?

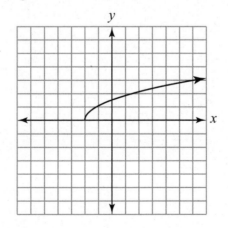

    A. Domain $x \geq 0$ and range $y \geq 0$
    B. Domain $x \geq 2$ and range $y \geq 0$
    C. Domain $x \geq -2$ and range $y \geq 0$
    D. Domain $x \geq 0$ and range $y \geq -2$

23. A backhoe is digging a trench. On its first sweep, the shovel follows a path modeled by the equation $y = x^2 - 3x - 4$, where x is the length in feet and y is the depth in feet of the hole. What is the deepest point of the hole?

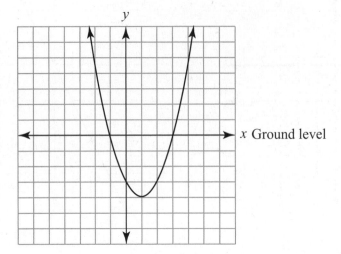

A. 1 foot

B. 2 feet

C. 3 feet

D. 4 feet

24. Simplify the following: $\dfrac{3x^5 y}{2xy^2} \cdot \dfrac{6xy^4}{4x^7}$

A. $\dfrac{9y^3}{4x^2}$

B. $\dfrac{9y^3}{8x^2}$

C. $\dfrac{9y^2}{4x}$

D. $\dfrac{9x^2 y^3}{4}$

25. Which of the following graphs indicates that one of the runners did NOT start 10 yards ahead of the other?

A.

Distance from starting point — Time (in seconds)

B.

Distance from starting point — Time (in seconds)

C.

Distance from starting point — Time (in seconds)

D.

Distance from starting point — Time (in seconds)

26. Which table accurately represents points from the given line graph?

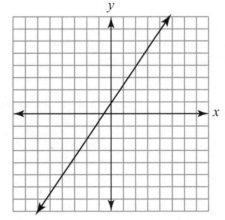

A.

x	y
0	−2
2	2
4	6

B.

x	y
−2	0
2	2
6	4

C.

x	y
−4	−2
−3	−4
−2	−6

D.

x	y
0	1
2	4
4	7

27. If the product of the slopes of two lines is −1, then what relationship exists between the lines?

    A. Parallel
    B. Horizontal
    C. Vertical
    D. Perpendicular

28. Write the sentence for the equation $11+\dfrac{x}{2}=16$.

    A. The product of 11 and x divided by 2 is equal to 16.
    B. The product of x and 2 increased by 11 is equivalent to 16.
    C. Eleven added to x divided by 2 is equal to 16.
    D. The quotient of x and 2 decreased by 11 is equivalent to 16.

29. Determine the slope of the line that passes through (2, 2) and (8, 5).

    A. $\dfrac{1}{2}$

    B. −2

    C. $\dfrac{10}{7}$

    D. 2

30. What is the equation of the graph that has a slope of 3 and passes through (−4, −3)?

    A. $y = 3x − 3$
    B. $y = 3x + 9$
    C. $y = 3x + 3$
    D. $y = 3x − 9$

31. What is the slope of all lines parallel to the line $4x − 5y = −1$?

    A. $\dfrac{-5}{4}$

    B. $\dfrac{-1}{5}$

    C. $\dfrac{4}{5}$

    D. $\dfrac{5}{4}$

32. What is the slope of all lines perpendicular to the line $3x + 4y = 17$?

    A. $\dfrac{-4}{3}$

    B. $\dfrac{-3}{4}$

    C. $\dfrac{3}{5}$

    D. $\dfrac{4}{3}$

33. Which best describes the lines
    $-x+3y=1 \ and \ y=\dfrac{1}{2}x-2$?

    A. Parallel
    B. Perpendicular
    C. Coincident
    D. Intersecting

34. To convert the temperature in Centigrade to Fahrenheit, you use the formula

    $F=\dfrac{9}{5}C+32$. What is 25°C when converted to Fahrenheit?

    A. 46°F
    B. 68°F
    C. 77°F
    D. 102°F

35. Simplify the expression $\dfrac{-24a^6b^3c}{6a^2bc}$.

    A. $-4a^4b^2$
    B. $-4a^3b^3$
    C. $-4a^4b^2c$
    D. $-4a^3b^3c$

Answer Key

1. **D**	7. **D**	13. **C**	19. **D**	25. **B**	31. **C**
2. **B**	8. **C**	14. **A**	20. **C**	26. **D**	32. **D**
3. **D**	9. **A**	15. **A**	21. **C**	27. **D**	33. **D**
4. **B**	10. **A**	16. **A**	22. **C**	28. **C**	34. **C**
5. **B**	11. **D**	17. **C**	23. **D**	29. **A**	35. **A**
6. **D**	12. **D**	18. **C**	24. **A**	30. **B**	

Answers Explained

1. **D** Work through each line shown here. Watch the reversal of the inequality sign in the last step when you divide both sides by a negative.

$$4(3 - x) + 2x + 8 \geq 14$$
$$12 - 4x + 2x + 8 \geq 14$$
$$20 - 2x \geq 14$$
$$-2x \geq -6$$
$$x \leq 3$$

2. **B** Notice that you can remove the coefficient of 3 by dividing both sides by 3. When you square both sides, you remove the radical.

$$3\sqrt{2 - x} = 9$$
$$\sqrt{2 - x} = 3$$
$$2 - x = 9$$
$$-x = 7$$
$$x = -7$$

3. **D** This solution is shown in three steps just for clarity.

$$V = \frac{1}{3}\pi r^2 h$$
$$3V = \pi r^2 h$$
$$\frac{3V}{\pi r^2} = h$$

4.   **B**   Work through each line shown here using the quadratic formula.

$$x = \frac{-b \pm \sqrt{b^2 - 4ac}}{2a}$$

$$x^2 - 7x + 6 = 0$$

$$x = \frac{-(\ ) \pm \sqrt{(\ )^2 - 4(\ )(\ )}}{2(\ )}$$

$$x = \frac{-(-7) \pm \sqrt{(-7)^2 - 4(1)(6)}}{2(1)}$$

$$x = \frac{7 \pm \sqrt{49 - 24}}{2}$$

$$x = \frac{7 \pm \sqrt{25}}{2}$$

$$x = \frac{7 \pm 5}{2}$$

$$x = \frac{12}{2}, \frac{2}{2}$$

$$x = 6, 1$$

5.   **B**   Replace the given value for x into the expression.

$$-2x^2 + 11x + 9$$
$$-2(-2)^2 + 11(-2) + 9$$
$$-2(4) + 11(-2) + 9$$
$$-8 - 22 + 9$$
$$-30 + 9$$
$$-21$$

6.   **D**   To simplify this expression with exponents, you begin by raising each factor in the parentheses to the power of 4. Remember that when raising a factor with an exponent to a higher power, you multiply the exponents.

$$(3a^4b^5c)^4$$
$$(3^4 a^{4 \cdot 4} b^{5 \cdot 4} c^{1 \cdot 4})$$
$$(81 a^{16} b^{20} c^4)$$

7.   **D**   Looking for a pattern, you will discover that each value is 4 less that the previous. Taking 4 from 13 will give you 9, taking 4 from 9 will give you 5, and, finally, taking 4 from 5 you will end with a 1.

8.   **C**   The numbers are increasing by 3. This pattern can be continued by the rule of adding 3 to the previous term to obtain the next term.

9.   **A**   The question is asking for the side opposite the given angle. You are given the hypotenuse length. The trigonometric relationship is sin.

$$\sin(A) = \frac{\text{Opposite}}{\text{Hypotenuse}}$$

$$\sin(65) = \frac{x}{26}$$

10.   **A**   To solve the system using substitution isolate a variable in one equation and place the results into the other equation. Simplify and solve the remaining equation.

$$1\text{st equation } x - 3y = -4$$
$$x = -4 + 3y$$
$$2\text{nd equation } x + 2y = 6$$
$$-4 + 3y + 2y = 6$$
$$-4 + 5y = 6$$
$$5y = 10$$
$$y = 2$$

Take your answer and substitute for y in the first equation.

$$x = -4 + 3y$$
$$x = -4 + 3(2)$$
$$x = -4 + 6$$
$$x = 2$$

11.   **D**   Set up a ratio for the given data and solve using the means-extremes product rule.

$$\frac{1{,}000 \text{ tickets}}{20 \text{ minutes}} = \frac{y \text{ tickets}}{100 \text{ minutes}}$$
$$(1{,}000)(100) = (20)(y)$$
$$100{,}000 = 20y$$
$$5{,}000 = y$$

12.   **D**   Notice the inequalities in each of the choices have the same slope and y-intercept. This question is asking you to determine whether you have a solid line or dashed line and which side of the line to shade. Remember the solid line from the graph means that you will choose the "equal to" sign. Choices A and C are eliminated. Notice that the shaded part of the graph includes point (–2, 2). This ordered pair will make a true statement for the inequality.

Place this point into choice B
$$y \le 2x + 3$$
$$2 \le 2(-2) + 3$$
$$2 \le -4 + 3$$
$$2 \le -1 \text{ false statement}$$

Place this point into choice D
$$y \ge 2x + 3$$
$$2 \ge 2(-2) + 3$$
$$2 \ge -4 + 3$$
$$2 \ge -1 \text{ true statement}$$

13.   **C**   To solve this system using the elimination method, combine the two equations and eliminate the y variable.

$$x + y = 4$$
$$\underline{x - y = 2}$$
$$2x = 6$$
$$x = 3$$

Replacing x with the value of 3 into either equation and solving for y, you will get $y = 1$. The ordered pair (3, 1) is the point labeled C.

14.   **A**   When linear equations are solved for y, the coefficient for the variable x represents the slope. The first equation is solved for y, but for the second equation, you need to divide through by 3.

$$3y = 9x - 5$$

$$y = 3x - \frac{5}{3}$$

The slopes for the lines are the same and the y-intercepts are different. These lines must be parallel. This is choice A.

15.   **A**   To determine this sum, you first multiply each entry in matrix A by the scalar 2 and matrix B by the scalar 3. Then add the corresponding entries of the first matrix to the second matrix.

$$2A = \begin{bmatrix} -4 & 6 \\ 10 & -6 \end{bmatrix} \quad \text{and} \quad 3B = \begin{bmatrix} 6 & -9 \\ 21 & 0 \end{bmatrix}$$

16.   **A**   Subtracting one polynomial from another, you distribute the minus sign with the second polynomial and then proceed to combine like terms.

$$\left(3b^4 - 2b^3 - b + 7\right) - \left(3b^4 - 4b^3 + 5\right)$$

$$3b^4 - 2b^3 - b + 7 - 3b^4 + 4b^3 - 5$$

$$\cancel{3b^4} - 2b^3 - b + 7 - \cancel{3b^4} + 4b^3 - 5$$

$$2b^3 - b + 2$$

17.   **C**   This problem is an example of experimental probability. From your sampling you determine that the ratio of students with red hair to all students is $\frac{2}{25}$. To compute the number of students with red hair in a larger population, set up a proportion. $\frac{2}{25} = \frac{x}{2,200}$. Using the means-extremes product rule the expression becomes

$$2 \cdot 2,200 = 25x$$

$$4,400 = 25x$$

$$160 = x$$

Choice C is the correct choice.

18.   **C**   In this example, you will need to simplify each choice. Choice A uses the power to power rule. Copy the base and multiply the exponents. $(2^3)^4 = 2^{12}$. Choice B uses a combination of power to power and multiplication rule with like bases. $(2^2 \cdot 2^2)^3 = (2^4)^3 = 2^{12}$. Choice C uses the multiplication rule with like bases $(2^6 \cdot 2^2) = 2^8$. Choice D uses the division rule $\left(\frac{2^{15}}{2^3}\right) = 2^{15-3} = 2^{12}$. Choice C answers the question.

19.   **D**   Work through each line shown here. Use the distributive property. Watch your signed numbers.

$$9x - 2(4x + 5) = 2x - (4 - x) - 12$$
$$9x - 8x - 10 = 2x - 4 + x - 12$$
$$1x - 10 = 3x - 16$$
$$-2x = -6$$
$$x = 3$$

20.   **C**   The graph that closely models this situation would be choice C. There is a gradual increase in the temperature over a period of time. Choice A depicts the water heating to over $90°$ in less than 40 minutes. Choice B shows no change in the temperature. Choice D shows the water getting colder over time.

21.   **C**   Graph C is not a function because it fails the vertical line test. Single values of x correspond to different y-values. In the study guide, this was described as "one to many."

22.   **C**   To answer the domain question, look at the graph and determine how far to the left and right the graph goes. In this graph, the graph touches $x = -2$ and continues to the right indefinitely. The domain is $x \geq -2$. The range can be answered by looking at how low and high the graph goes. The graph appears to touch the x-axis and slowly rise up. The range of the graph is $y \geq 0$.

23.   **D**   This question is looking for the minimum value as determined by the quadratic equation. From the graph the minimum depth is the lowest y-value, -4 feet.

24.   **A**   When given fractions in factored form, it may be easier to combine the factors of the numerators and combine the denominators. Remember, with the operation of multiplication of like bases, you are to add the exponents after copying the base.

$$\frac{3x^5 y}{2xy^2} \cdot \frac{6xy^4}{4x^7} = \frac{3 \cdot 6x^5 xyy^4}{2 \cdot 4xx^7 y^2} = \frac{3 \cdot 6x^{5+1} y^{4+1}}{2 \cdot 4x^{7+1} y^2} = \frac{3 \cdot 6x^6 y^5}{2 \cdot 4x^8 y^2}$$

Next, use the division rule for factors with like bases by subtracting the exponents and reducing coefficients.

$$\frac{3 \cdot 6x^6 y^5}{2 \cdot 4x^8 y^2} = \frac{3 \cdot \overset{3}{\cancel{6}} \, y^{5-2}}{\cancel{2} \cdot 4x^{8-6}} = \frac{9y^3}{4x^2}$$

25.   **B**   The y-intercept for each of the graph lines reveals where the runners start. Only choice B has the same y-intercept for both graphs. Both of the runners start 10 units from the starting point.

26.   **D**   Look at the graph where the line crosses a corner point of a grid square. Another great place to look would be either the x- or y-intercept. In this graph, the line crosses the y-axis at 1. The ordered pair $(0, 1)$ must be in the table of values. The line crosses the grid corner at the point $(2, 4)$. The same is true for the point $(4, 7)$. Choice D includes these points.

27.   **D**   Two lines are perpendicular when the product of their slopes is -1. It is sometimes stated that the slopes are negative reciprocals of each other. Likewise, two lines are said to be parallel when their slopes are the same and their y-intercepts are different. Horizontal slopes are equal to zero, and vertical lines have no slope.

28.   **C**   When translating an algebraic equation into words, watch for key words describing the operations, and look for the word *is* which is represented by the equal sign. In this problem: Eleven added to x divided by 2 is equal to 16.

29.   **A**   To compute the slope using the formula from the reference sheet:

$$m = \frac{y_2 - y_1}{x_2 - x_1} = \frac{(5) - (2)}{(8) - (2)} = \frac{3}{6} = \frac{1}{2}.$$ The correct choice is A.

30.   **B**   Using the slope-intercept formula, place the given slope and point into the equation. Next, solve the equation for the remaining variable b.

$$y = mx + b$$
$$-3 = 3(-4) + b$$
$$-3 = -12 + b$$
$$9 = b$$

Replacing the slope and the new found b the equation becomes $y = 3x + 9$.

31.   **C**   The slope value for all lines parallel to a given line is the same. First, find the slope of the given line by solving for y.

$$4x - 5y = -1$$
$$-5y = -4x - 1$$
$$y = \frac{-4x}{-5} - \frac{1}{-5}$$
$$y = \frac{4x}{5} + \frac{1}{5}$$

The slope is $\dfrac{4}{5}$ for all lines parallel to the given line.

32.   **D**   The slope value for a line perpendicular to a given line is the negative reciprocal of the slope of the given line. First, find the slope of the given line by solving for y.

$$3x + 4y = 17$$
$$4y = -3x + 17$$
$$y = \frac{-3x}{4} + \frac{17}{4}$$

The negative reciprocal for $\dfrac{-3}{4}$ is $\dfrac{4}{3}$. A nice check is the fact that the product of the slopes is equal to -1.

$$\frac{4}{3} \cdot \frac{-3}{4} = -1$$

33.   **D**   The solution can be found by looking at the slopes of each line. The first equation can be placed into slope-intercept form by solving for y.

$$-x + 3y = 1$$
$$3y = x + 1$$
$$y = \frac{1}{3}x + 1$$

Comparing the slopes, they are different. The lines intersect. The lines are not perpendicular because their slopes are not negative reciprocals.

34.   **C**   Replacing the variable C with the given value, 25°, and simplifying the equation the temperature converts to 77°F.

$$F = \frac{9}{5}C + 32$$

$$F = \frac{9}{5}(25) + 32$$

$$F = \frac{9}{\cancel{5}}\left(\overset{5}{\cancel{25}}\right) + 32$$

$$F = 45 + 32 = 77°$$

35.   **A**   Simplifying this problem requires use of the rule for division of like bases with exponents.

$$\frac{-24a^6b^3c}{6a^2bc} = -4a^{6-2}b^{3-1}c^{1-1} = -4a^4b^2c^0 = -4a^4b^2$$

Remember, raising to the zero power is equal to 1.

Geometry and Measurement

When you complete your study of Strand 4, you will be able to:

- Identify the attributes of special triangles
  - Isosceles triangles
  - Equilateral triangles
  - Right triangles
- Identify the quadrilaterals
  - Quadrilateral
  - Parallelogram
  - Rectangle
  - Rhombus
  - Square
  - Trapezoid
- Recognize, draw, and label three-dimensional figures and their nets
- Solve problems related to a variety of geometric concepts
  - Complementary, supplementary, and congruent angles
  - Circles and the arcs, angles, and segments related to them
  - Triangle inequality property
  - Special-case right triangles
  - Contextual situations related to geometry
- Determine when triangles are congruent or similar
- Apply spatial reasoning to create transformations
- Graph equations and inequalities and determine the effect of changes in constants and coefficients
  - Linear equation
  - Linear inequality
  - Quadratic equation
- Determine the midpoint and the distance between points
- Determine the solution to a system of equations
- Calculate the area and volume of geometric shapes
- Solve for missing measures in a pyramid
- Find the length of an arc and the area of a sector of a circle
- Find the sum of the interior and exterior angles of a polygon
- Solve problems involving scale factor and similar triangles

The fourth strand introduces the topic of geometry. Some students look at geometry as a refreshing change from algebra. Students often say that they can visualize geometric concepts, unlike algebra, which, for some, is quite abstract.

Geometry helps explain the world around you. Try an experiment. Look up from this book and see how many two-dimensional and three-dimensional objects you can identify. Look at the different shapes and sizes. Colors and textures enhance what you are seeing. Try to focus on basic shapes and how these shapes interact with each other. We live in a three-dimensional world so you probably see many basic figures like cylinders, spheres, prisms, and pyramids. You don't necessarily use those names when you talk about the objects. Instead, you might use more common words such as can, ball, and box.

Geometry can also be related to algebra, which sometimes leads to a better understanding of algebra. Geometry and some other areas of mathematics are used to help enhance your critical thinking abilities, your understanding of logic and structure.

Like any other topic, part of learning geometry deals with the vocabulary. Although the new terminology may be cumbersome at first, the end result leads to a well-defined subject.

Geometry Basics

We begin our study of geometry with the discussion of three very basic geometric figures—point, angle, and line. Using these words (and many others), we develop definitions and concepts about the geometric figures of the world in which we live.

A **point**, represented by a dot and named with a capital letter, has no dimension.

Point A $A \cdot$

Angle Concepts

Angles
• Acute
• Right
• Obtuse
• Straight

We named angles in triangles using one capital letter when we discussed trigonometry. Another method is to use three letters. This is quite common when the angle is not necessarily part of a triangle.

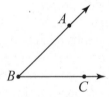

This angle is written as $\angle ABC$ or $\angle CBA$. Notice that the letter B is in the middle. It is called the vertex of the angle. An m before the angle symbol means the measure of the angle. The common unit of measure for angles is degrees.

First let's review some angle vocabulary.

- **Acute angle**—an angle whose measure is greater than $0°$ and less than $90°$.

- **Obtuse angle**—an angle whose measure is greater than $90°$ and less than $180°$.

- **Right angle**—an angle whose measure is $90°$.

- **Straight angle**—an angle whose measure is $180°$.

Line Concepts

Basic linear notation
• Line $\overleftrightarrow{AB}$
• Line segment $\overline{AB}$
• Ray $\overrightarrow{AB}$
• Measure of segment AB

A **line** is named with two capital letters that identify points on the line. It is straight and continues forever in both directions as shown by the arrows.

$\overrightarrow{AC}$ or $\overleftrightarrow{CA}$

A line can also be named with a single italic lowercase letter.

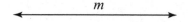

line *m*

Because a line is continuous, we do not measure its length.

A **line segment** is part of a line. Unlike the line, it has two endpoints, and we can find its length. We name the segment with a segment bar over the top of the two capital letter endpoint names.

$\overline{MN}$ or $\overline{NM}$

When we talk about the measure of the segment, the bar is omitted. The measure of the segment is a number, not a set of points.

MN = 12 feet

A **ray** is also part of a line. It extends infinitely from an endpoint in only one direction. It also does not have a length. We name rays starting with the endpoint and include one other point on the ray.

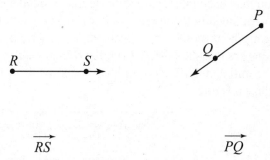

$\overrightarrow{RS}$ $\overrightarrow{PQ}$

When two rays share the same endpoint, they form an angle. The endpoint is the vertex of the angle. If the rays go in opposite directions from the shared endpoint, the angle is a **straight angle**.

$\overrightarrow{RT}$ and $\overrightarrow{RS}$ form $\overleftrightarrow{TS}$ or $\overleftrightarrow{ST}$

Parallel ||
Perpendicular ⊥

Parallel lines are lines on the same surface that do not intersect. Notice the symbol representing parallel in the figure below.

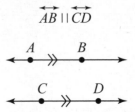

Two lines on the same surface can intersect. The intersecting lines form four angles. When the four angles are right angles, the lines are called **perpendicular lines.** The symbol for perpendicular is ⊥.

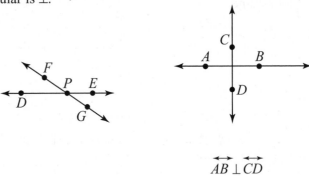

Notice that, in the figure on the left, there are two acute and two obtuse angles. ∠DPF and ∠GPE are acute angles. ∠FPE and ∠DPG are obtuse angles.

Polygons

Figures that have two dimensions (length and height), such as triangles, squares, and octagons, are also named with capital letters according to the number of vertices they have. The vertex of a polygon is the point where two consecutive sides meet. The word "vertices" is the plural form of the word "vertex."

You may encounter some of these many-sided shapes, like the traffic stop sign, which is an eight-sided polygon (an octagon). If you study bees, you will discover the six-sided (hexagon) honeycomb in a beehive. As a group, these figures are called polygons. **Polygons** are two-dimensional figures of many sides. The base word "poly" means many.

The initial classification of shapes comes from the number of sides used to produce an enclosed figure.

Number of sides	Name
3	Triangle
4	Quadrilateral
5	Pentagon
6	Hexagon
7	Heptagon
8	Octagon

This table provides a sample of the common names given to polygons. There are many more. Mathematicians even have a name for the shape with unknown number of sides. If you have n number of sides, the shape is called an n-gon.

This AIMS preparation text will concentrate on three- and four-sided figures, but you should be aware of the existence of polygons with more sides.

Triangles

A triangle has three vertices and three sides. The triangle below could be named $\triangle ABC$. The three letters in the name of the triangle can be written in any order.

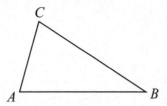

We classify the various triangles by the types of angles and the length of sides. We use these angle names to classify triangles by their angles.

Triangle name	Definition	Diagram
Right triangle	A triangle with exactly one right angle	
Acute triangle	A triangle with three acute angles	
Obtuse triangle	A triangle with exactly one obtuse angle	
Equiangular	A triangle with three equal angles	

We also classify triangles by the length of their sides.

Triangle name	Definition	Diagram
Scalene triangle	A triangle with no two equal sides	
Isosceles triangle	A triangle with two equal sides (The angles opposite the equal sides are also equal.)	
Equilateral triangle	A triangle with three equal sides	

Did you notice the markings on the triangle sides and angles? The markings indicate congruent angles and congruent sides. An equilateral triangle is also equiangular and vice versa. So the equilateral/equiangular triangle could be marked showing all sides equal as well as angles equal.

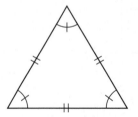

Angle and side classifications combined		
Acute triangles	Scalene	
	Isosceles	
	Equilateral	
Obtuse triangles	Scalene	
	Isosceles	
Right triangles	Scalene	
	Isosceles	

Triangle Relationships

Angles

Recall that the sum of the measures of the angles of a triangle is 180°.

> The sum of the measures of the angles of a triangle is 180°.

$$m\angle A + m\angle B + m\angle C = 180°$$

Also, every triangle has at least two acute angles. Both of these concepts can lead us into application questions.

Examples: What is the measure of the angles in the triangle below?

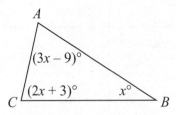

$$(2x + 3) + (x) + (3x - 9) = 180°$$
$$6x - 6 = 180°$$
$$6x = 186°$$
$$x = 31°$$

$$m \angle A = 84, m \angle B = 31, m \angle C = 65$$

Given triangle ABC as shown:

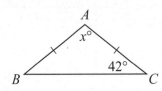

Find the measure of angle A. Recognize that the triangle is isosceles and the base angles are equal. Therefore, m $\angle B = 42$.

$$42° + 42° + x = 180°$$
$$84° + x = 180°$$
$$x = 96°$$

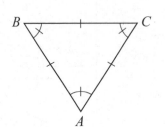

Find all the measures of the angles of the equilateral triangle. Remember, an equilateral triangle is equiangular.

$$x + x + x = 180°$$
$$3x = 180°$$
$$x = 60°$$

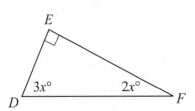

Find the measures of the acute angles of the right triangle DEF.

$$3x + 2x + 90° = 180°$$
$$5x = 90°$$
$$x = 18°$$
$$2x = 36°, 3x = 54°$$

Sides

> The sum of any two sides of a triangle is greater than the third side.

The sum of any two sides of a triangle will always be greater than the third side. This is called the **triangle inequality property**.

Do the following lengths form a triangle?

	Sides	Calculations	Result
A	3, 4, 5	$3 + 4 > 5$; $4 + 5 > 3$; $3 + 5 > 4$	Yes
B	2, 3, 3	$2 + 3 > 3$; $3 + 3 > 2$; $3 + 2 > 3$	Yes
C	4, 8, 16	$4 + 8 > 16$—Not true	No
D	5, 5, 5	$5 + 5 > 5$	Yes
E	6, 8, 10	$6 + 8 > 10$; $8 + 10 > 6$; $6 + 10 > 8$	Yes
F	4, 9, 13	$4 + 9 > 13$—Not true	No

What happens when the triangle inequality property is not true for all pairs of sides? Here is an attempt to draw example C:

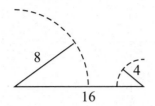

Notice the sum of sides with lengths of 4 cm and 8 cm is not greater than the third side with a length of 16 cm. So, a triangle cannot be formed.

Special Cases for Right Triangles

The Pythagorean theorem, a unique relationship of the sides of a right triangle, has many applications. In a right triangle, the sum of the squares of the legs equals the square of the hypotenuse. The converse of this is also true. You can show that you have a right triangle by finding the square of the two shorter sides, finding the sum of those squares, and comparing that to the square of the longest side. If the sum equals the square of that longest side, then the triangle is a right triangle.

$$a^2 + b^2 = c^2$$

Using algebra, given any two sides of a right triangle, the third side can be calculated with the use of the Pythagorean theorem.

Given side $a = 6$ and $c = 10$
What is the measure of side b?

$$a^2 + b^2 = c^2$$
$$(6)^2 + b^2 = (10)^2$$
$$36 + b^2 = 100$$
$$b^2 = 64$$
$$b = 8$$

In the trigonometry review from Strand 3, the ratio between two sides of a right triangle were defined with three trigonometric ratios: *sine, cosine, and tangent.*

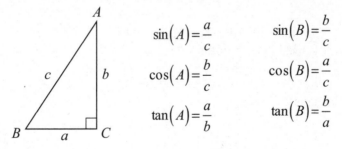

$$\sin(A) = \frac{a}{c} \qquad \sin(B) = \frac{b}{c}$$
$$\cos(A) = \frac{b}{c} \qquad \cos(B) = \frac{a}{c}$$
$$\tan(A) = \frac{a}{b} \qquad \tan(B) = \frac{b}{a}$$

Two special right triangles occur frequently in geometry, as well as in practical applications.

30°-60°-90° *Triangle Relationship*

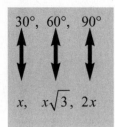

A right triangle whose acute angles measure 30° and 60° is generally called a 30°-60°-90° triangle.

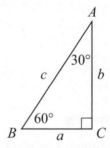

Recall from the trigonometry section that the leg opposite angle *A* is named using the lowercase *a*? The same is true for each of the other side angle combinations.

In a 30°-60°-90° triangle the length of side *c*, the hypotenuse, is twice as long as the length of side *a*. Side *b* is the product of side *a* and the square root of three.

Example: Consider a 30°-60°-90° triangle with the leg opposite the 30° angle equal to 5. The hypotenuse is 10, or twice the short leg, and the leg opposite the 60° angle is $5\sqrt{3}$.

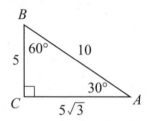

$$a = 5, b = 5\sqrt{3}, \text{ and } c = 10$$

In general terms the measure of the angles and the sides opposite them correspond according to this diagram.

30°, 60°, 90°

$$x, \quad x\sqrt{3}, \quad 2x$$

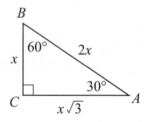

The value of *x* (the short leg) is used in each of the side lengths.

Examples:

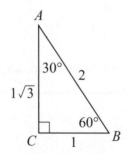

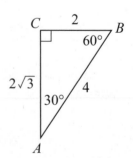

Did you observe that the smallest side was always opposite the smallest angle (30°) and the largest side opposite the biggest angle (90°)?

45°-45°-90° Triangle Relationship

A right triangle whose acute angles both measure 45° is generally called a 45°-45°-90° triangle.

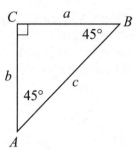

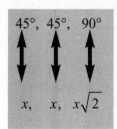

When you are given the length of one leg, you are really given both leg lengths. Remember, if the base angles are equal, the triangle is isosceles and the sides opposite the equal angles are equal in length. The length of the hypotenuse is the leg length times radical two.

Example: Consider a 45°-45°-90° triangle with a

side = 5. The hypotenuse is $5\sqrt{2}$.

$a = 5$, $b = 5$, and $c = 5\sqrt{2}$

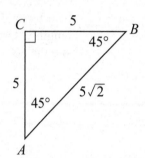

In general terms, the measure of the angles and the sides opposite them correspond according to this diagram.

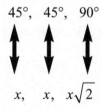

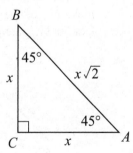

Examples:

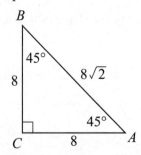

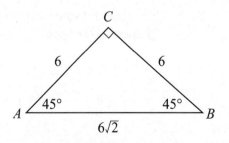

Are you curious about why these relationships are true?

To show and explain why the 30°-60°-90° and 45°-45°-90° relationships are true is beyond the AIMS test requirements. However, with a little algebra and some geometric concepts, you can show the relationships. Ask your tutor or math teacher if you need help with these steps.

Start with an equilateral triangle of any size. For example, in the triangle below the sides are all equal to $2x$. Remember, the angles are equal and therefore each equals 60°. Draw a perpendicular bisector to one side. Notice the right angle, and that one of the 60° angles is divided into two 30° angles.

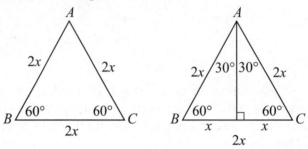

Two 30°-60°-90° triangles are formed. Note the side opposite the original 60° angle is bisected and is half of the hypotenuse of the newly formed right triangle. Its measure is length x. Use algebra and the Pythagorean theorem to show the length of the side opposite the 60° angle.

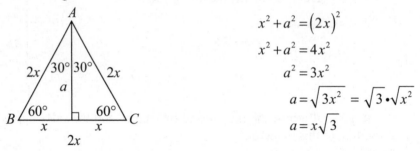

$$x^2 + a^2 = (2x)^2$$
$$x^2 + a^2 = 4x^2$$
$$a^2 = 3x^2$$
$$a = \sqrt{3x^2} = \sqrt{3} \cdot \sqrt{x^2}$$
$$a = x\sqrt{3}$$

The steps for showing side relationships in a 45°-45°-90° triangle are similar. We make use of the fact that the triangle is isosceles and the legs are equal in length. Next, label the hypotenuse to be h, and use the Pythagorean theorem to solve for h.

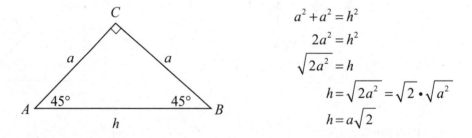

$$a^2 + a^2 = h^2$$
$$2a^2 = h^2$$
$$\sqrt{2a^2} = h$$
$$h = \sqrt{2a^2} = \sqrt{2} \cdot \sqrt{a^2}$$
$$h = a\sqrt{2}$$

Here are some sample problems to check your understanding of the side length relationships with special triangles.

Example: You want to hang a shelf on the wall and you want to use a brace to support the weight. If the shelf is 12 inches wide, how long must the brace be if you want the angle between the wall and brace to be 30°?

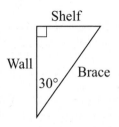

The shelf is 12 inches wide. With a 30° angle, the brace will be twice the shelf length—24 inches.

Example: Suppose you want a 45° angle between the wall and the brace. How long will this brace be?

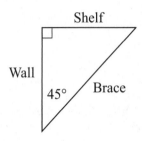

The shelf is 12 inches wide. With a 45° angle, the brace will be $12\sqrt{2} \approx 17$ inches.

Did you notice the "wavy" equal sign? This is the mathematical symbol for approximately equal to.

Example: Suppose you want a 60° angle between the wall and the brace. How long will this brace be?

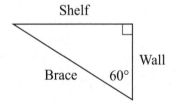

The shelf is 12 inches wide. If the brace makes a 60° angle with the wall, the angle opposite the wall is 30°. You must first find the length of the segment on the wall side of the triangle.

$$x\sqrt{3} = 12$$

$$x = \frac{12}{\sqrt{3}}$$

$$x = \frac{12}{\sqrt{3}} \cdot \frac{\sqrt{3}}{\sqrt{3}}$$

$$x = \frac{12\sqrt{3}}{3}$$

$$x = 4\sqrt{3}$$

$$2x = 8\sqrt{3}$$

The shelf is opposite the 60° angle. Its length is equal to $x\sqrt{3}$. The segment in the triangle opposite the 30° angle is $4\sqrt{3}$. The length of the brace is $8\sqrt{3}$, which is approximately 14 inches.

Check:

$$\left(\text{Shelf}\right)^2+\left(\text{Wall}\right)^2=\left(\text{Brace}\right)^2$$

$$\left(12\right)^2+\left(4\sqrt{3}\right)^2=\left(8\sqrt{3}\right)^2$$

$$144+\left(16\cdot3\right)=\left(64\cdot3\right)$$

$$144+48=192$$

$$192=192$$

Example: Suppose that you have a 16 foot ladder to lean against a wall. You are told that the base angle between the ladder and ground must be at least 60°. What are the possible heights that the top of the ladder will touch the wall? (Note: It is generally assumed that the wall is perpendicular to the ground.)

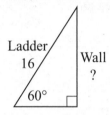

If you place the ladder flat against the wall, the top of the ladder would touch the wall 16 feet from the ground (not very safe or practical!).

Next, compute the minimum height the ladder will touch, keeping in mind the requirement of a 60° safety angle between the ladder and the ground.

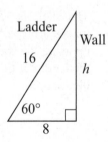

The triangle formed is 30°-60°-90°. The distance to the base of the wall and foot of the ladder is one-half the (hypotenuse) ladder. This is 8 feet. The ladder reaches a height of $8\sqrt{3}\approx14$ feet .

So the range of heights that the ladder could reach is 14 feet ≤ heights ≤ 16 feet.

Extra for Experts

Example: To compute any other base angle for the foot of the ladder you would use trigonometry and either a calculator or a trigonometric table. Side h is opposite the base angle, B. The ladder is the hypotenuse, which is 16 feet.

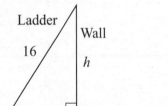

$$\sin\left(B\right)=\frac{h}{16}$$

$$16\sin\left(B\right)=h$$

Assign any value for angle B and use a calculator or trigonometric table.

Example: Suppose you have a 13 foot ladder and the top must touch 12 feet up the wall. How far from the base must you place the foot of the ladder?

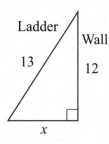

Using the Pythagorean theorem,

$$x^2 + 12^2 = 13^2$$
$$x^2 + 144 = 169$$
$$x^2 = 25$$
$$x = 5$$

The base of the ladder will be 5 feet from the wall.

Example: Suppose you have a rectangular opening 3 meters wide by 4 meters tall. What is the tallest square piece of cardboard that you can fit through this opening? Since the height is 4 meters, some students would guess that as the answer. If you pass the cardboard through the opening on the diagonal, you can have a taller piece. How tall? The diagonal is the hypotenuse for a right triangle with sides of 3 meters and 4 meters.

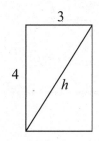

$$3^2 + 4^2 = h^2$$
$$9 + 16 = h^2$$
$$25 = h^2$$
$$h = 5$$

The tallest piece of cardboard to fit through opening would be a bit smaller than 5 meters.

Quadrilaterals

The prefix "quad" means four and "lateral" means sides. A **quadrilateral** is a four-sided polygon. The various classifications of quadrilaterals are also based on lengths of sides and angles, but quadrilaterals may have parallel sides. Whether the opposite sides are parallel or not provides additional characteristics.

Quadrilaterals
• Quadrilateral
• Parallelogram
• Rectangle
• Rhombus
• Square
• Trapezoid

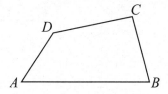

Quadrilateral *ABCD*

A quick check of the glossary states that parallel lines are lines (in the same surface) that never intersect and are always the same distance apart.

Since a quadrilateral has just four sides, it makes sense to talk about sides across from (opposite) each other. A quadrilateral can have no sides parallel, one pair of opposite sides parallel, or both pairs of opposite sides parallel.

A quadrilateral with one pair of parallel sides is called a **trapezoid**. It might be helpful to picture in your mind a triangle with its top cut off.

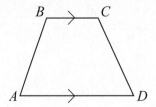

Trapezoid *ABCD*

The use of the arrowheads in the diagram shows the lines that are parallel.

A **parallelogram** is a quadrilateral with both pairs of sides parallel.

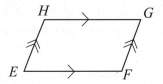

Parallelogram *EFGH*

If you add the characteristic of a right angle to a parallelogram, the familiar shape formed is a **rectangle**.

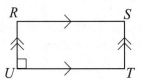

Rectangle *RSTU*

If you don't add the characteristic of a right angle to a parallelogram, but instead the characteristic that all four sides are equal, you have defined a **rhombus**.

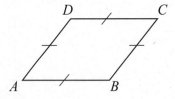

Rhombus *ABCD*

The term "congruent" is used when talking about segments or angles that are equal. Therefore, we could say that the four sides of a rhombus are congruent.

$$\overline{AB} \cong \overline{BC} \cong \overline{CD} \cong \overline{AD}$$

When a right angle and equal sides are both in a parallelogram, the quadrilateral is a **square**.

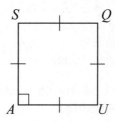

Square *SQUA*

A **kite** is a quadrilateral with a special characteristic that two distinct pairs of consecutive sides are congruent. The word kite is an appropriate name. A kite in geometry looks just like the kite you fly in the sky.

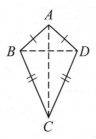

Kite *ABCD*

All of the figures mentioned here are members of the family of quadrilaterals. There are characteristics that distinguish one member from another; there are characteristics that define their relationship to each other. Study the figure below, generally called the quadrilateral hierarchy, to see the upward relationship among the shapes.

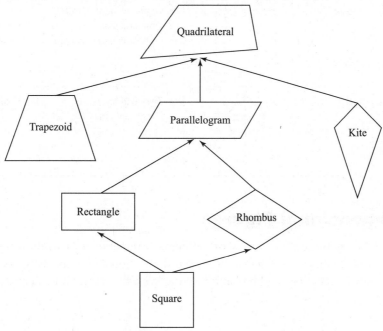

Quadrilateral Family

Notice that a square is both a rectangle and a rhombus—see how the arrowheads go up. It is also a parallelogram and a quadrilateral. A rectangle is a parallelogram, which is a quadrilateral. Again, you follow the arrowheads up. It would not be true to say that a rhombus is a square. The square has right angles as well as congruent sides; a rhombus has congruent sides. To say that a rectangle is a square is incorrect because all of the sides of a rectangle don't have to be the same size.

Summary Chart of Some Distinguishing Characteristics of the Quadrilateral Families

Shape	Picture	Characteristics	
Trapezoid		• Only two parallel sides	
Parallelogram		• Opposite sides parallel • Opposite sides congruent	• Opposite angles congruent • Diagonals bisect each other
Rectangle		• Opposite sides parallel • Opposite sides congruent • Opposite angles congruent	• Diagonals bisect each other • Diagonals congruent • Four right angles
Rhombus		• Opposite sides parallel • Opposite sides congruent • Opposite angles congruent	• Diagonals are perpendicular • Four sides congruent • Diagonals bisect each other
Square		• Opposite sides parallel • Opposite sides congruent • Opposite angles congruent	• Diagonals bisect each other • Four right angles • Four sides congruent
Kite		• Two distinct pair of congruent sides	• Diagonals are perpendicular

Three-Dimensional Figures

Just as there are a variety of mathematical names for two-dimensional polygons, there are also mathematical names for three-dimensional objects, as well as some common names. Study the chart and become familiar with the names and shapes.

Math Term	Common Name	Diagram	Net
Cube	Square block		
Rectangular prism	Box		
Triangular prism	Light refracting prism		
Cylinder	Can		
Cone	Cone		
Pyramid	Pyramid		
Sphere	Ball		None

Notice the last column includes the net of the object. A **net** is the two-dimensional surface of a three-dimensional object. If you fold the net along the dotted lines, the result will be the three-dimensional object.

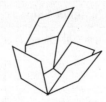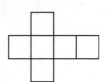

As you "fold" the object in your mind, ask yourself: "Do the sides come together?" Then try to reverse the process and lay the object out flat in your mind. Later, we will review the volume and surface area formulas for these three-dimensional figures. Being able to mentally visualize these nets will be helpful in understanding the usage of the formulas.

Angle Concepts

In this section we will continue to define geometric properties and incorporate our algebra skills as we use the properties.

Complementary Angles

Two angles are said to be complementary if the sum of their measures is 90°.

Complementary—sum of 90°
Supplementary—sum of 180°

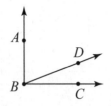

∠ABC is a right angle.

m ∠ABD = 63 and m ∠DBC = 27

$$63° + 27° = 90°$$

63° is the complement of 27°
27° is the complement of 63°

Example: If Q is the measure of an angle, what is the measure of its complement?

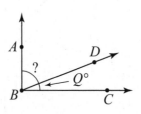

∠ABC is a right angle.

If Q is the measure of one angle, the complement angle measure is $90 - Q$.

$$Q + (90 - Q) = 90$$
$$90 = 90$$

Example: The measure of one angle is four times the measure of its complement. What is the measure of the larger angle?

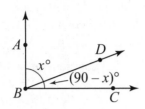

∠ABC is a right angle.

$$x = 4(90 - x)$$
$$x = 360 - 4x$$
$$5x = 360$$
$$x = 72°$$

Complement of x is $90 - x$

$$90 - 72 = 18$$

Angles measure 72° and 18°

Answer: measure of larger ∠ = 72°

Supplementary Angles

Two angles are said to be supplementary if the sum of their measures is 180°.

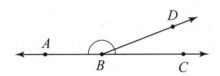

∠ABC is a straight angle.

∠ABD = 150° and ∠DBC = 30°

$$150° + 30° = 180°$$

150° is the supplement of 30°
30° is the supplement of 150°

Example: If P is the measure of an angle, what is the measure of its supplement?

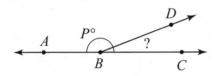

∠ABC is a straight angle.

If P is the measure of one angle, the supplement angle measure is $180 - P$.

$$P + (180 - P) = 180$$
$$180 = 180$$

Example: The measure of one angle is nine times the measure of its supplement. What is the measure of the smaller angle?

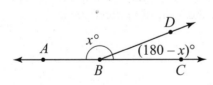

∠ABC is a straight angle.

$$x = 9(180 - x)$$
$$x = 1,620 - 9x$$
$$10x = 162$$
$$x = 162°$$

Supplement of x is $180 - x$

$$180 - 162 = 18$$

Angles measure 162° and 18°

Answer: measure of smaller ∠ = 18°

Example: The supplement of an angle is 4 times the measure of the complement of the same angle. What is the measure of all of the angles?

Let x = measure of one angle. The measure of the supplement angle is $180 - x$. The measure of the complement of the given angle is $90 - x$.

$$180 - x = 4(90 - x)$$
$$180 - x = 360 - 4x$$
$$3x = 180$$
$$x = 60°$$
$$180 - x = 120°$$
$$90 - x = 30°$$

Check:
Supplement = 4 • (complement)
$$120 = 4(30)$$
$$120 = 120$$

Congruent Angles

Two angles are said to be congruent if their measures are equal.

Example:

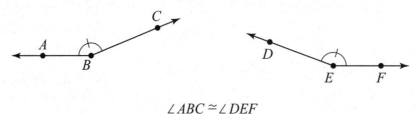

$$\angle ABC \cong \angle DEF$$

Find the measure of two congruent angles that are also complementary angles.

Let x = measure of one angle. Because the angles are said to be congruent, the other angle also has a measure of x. The fact that they are complementary angles reveals that their sum is 90°.

$$x + x = 90$$
$$2x = 90$$
$$x = 45°$$

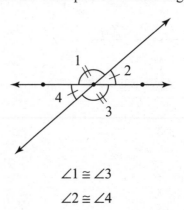

Vertical Angles

The opposite angles formed when two lines intersect are **vertical angles**. The intersecting lines always form two pair of vertical angles. In the diagram below, angles 1 and 3 are vertical angles. Angles 2 and 4 are the other pair of vertical angles.

$$\angle 1 \cong \angle 3$$
$$\angle 2 \cong \angle 4$$

Vertical angles are congruent.

Example:

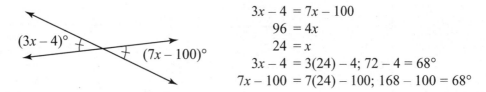

$$3x - 4 = 7x - 100$$
$$96 = 4x$$
$$24 = x$$
$$3x - 4 = 3(24) - 4; \ 72 - 4 = 68°$$
$$7x - 100 = 7(24) - 100; \ 168 - 100 = 68°$$

Parallel lines were mentioned in the discussion of quadrilaterals. Parallel lines also generate questions about congruent angles.

Angles Around Transversals

When a line intersects two parallel lines, eight angles are formed. The line that intersects the two parallel lines is called a **transversal**. Various angles share common characteristics and special names.

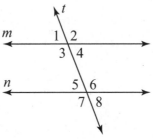

$m \parallel n$

Types of angles	Angle from diagram	Description
Exterior angles	$\angle 1$, $\angle 2$, $\angle 7$, and $\angle 8$	Exterior angles are outside the parallel lines.
Interior angles	$\angle 3$, $\angle 4$, $\angle 5$, and $\angle 6$	Interior angles are between the parallel lines.
Alternate interior angles	$\angle 3$ and $\angle 6$ $\angle 4$ and $\angle 5$	Alternate interior angles are on opposite sides of the transversal, one at each vertex, between the parallel lines.
Alternate exterior angles	$\angle 1$ and $\angle 8$ $\angle 2$ and $\angle 7$	Alternate exterior angles are on opposite sides of the transversal, one at each vertex, outside the parallel lines.

Types of angles	Angle from diagram	Description
Corresponding angles	∠2 and ∠6 ∠4 and ∠8 ∠1 and ∠5 ∠3 and ∠7	Corresponding angles are on the same side of the transversal, one at each vertex, one between the parallel lines, and one outside the parallel lines. Another way of saying this is that they are in the same position in the circle of four angles at each vertex (e.g., both in the upper right).

Some of these pairs of angles are congruent. Study the chart below.

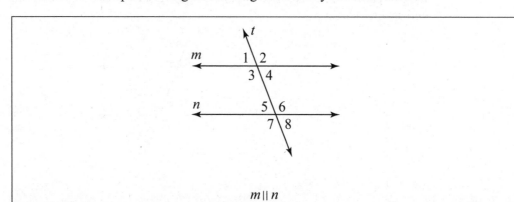

$m \| n$

Pairs of angles	Congruent pairs
Alternate exterior angles	∠1 ≅ ∠8 and ∠2 ≅ ∠7
Alternate interior angles	∠3 ≅ ∠6 and ∠4 ≅ ∠5
Corresponding angles	∠1 ≅ ∠5, ∠3 ≅ ∠7, ∠2 ≅ ∠6, ∠4 ≅ ∠8
Vertical angles	∠1 ≅ ∠4, ∠2 ≅ ∠3, ∠5 ≅ ∠8, ∠6 ≅ ∠7

Examples:

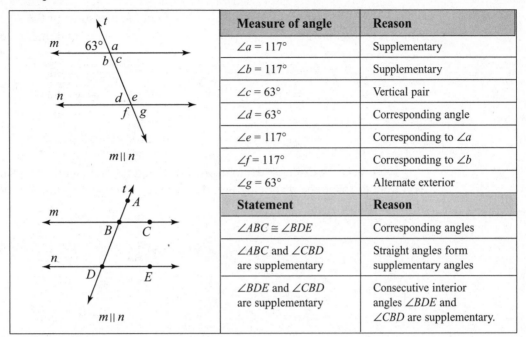

Measure of angle	Reason
∠a = 117°	Supplementary
∠b = 117°	Supplementary
∠c = 63°	Vertical pair
∠d = 63°	Corresponding angle
∠e = 117°	Corresponding to ∠a
∠f = 117°	Corresponding to ∠b
∠g = 63°	Alternate exterior

Statement	Reason
∠ABC ≅ ∠BDE	Corresponding angles
∠ABC and ∠CBD are supplementary	Straight angles form supplementary angles
∠BDE and ∠CBD are supplementary	Consecutive interior angles ∠BDE and ∠CBD are supplementary.

Consecutive Interior Angles

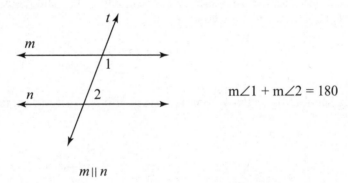

$$m\angle 1 + m\angle 2 = 180$$

In the discussion about parallelograms, do you remember that the opposite sides of a parallelogram are parallel? In the diagram below, side *AD* intersects sides *AB* and *DC*.

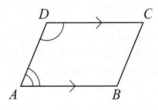

▱ *ABCD*

Side *AD* can be considered as a transversal to the parallel lines. Do you see that the angles formed (angle *A* and angle *D*) are consecutive angles? Therefore, applying the concept from above, angle *A* and angle *D* are supplementary.

Example:

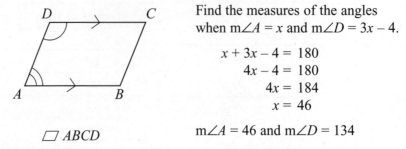

Find the measures of the angles when $m\angle A = x$ and $m\angle D = 3x - 4$.

$$x + 3x - 4 = 180$$
$$4x - 4 = 180$$
$$4x = 184$$
$$x = 46$$

$\square$ ABCD

$m\angle A = 46$ and $m\angle D = 134$

The preceding example demonstrated that angle A and angle D are supplementary. If you were to rotate the parallelogram, the previous statement would also be true for angle A and angle B.

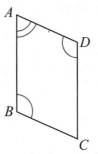

Since angle A and angle D are supplementary, and angle A and angle B are supplementary, their sums are equal.

$$m\angle A + m\angle D = m\angle A + m\angle B$$
$$m\angle D = m\angle B$$

By subtracting angle A from both sides of this equation, angle D is equal to angle B.

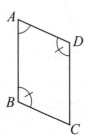

Angles such as B and D in a parallelogram are called opposite angles. The other pair of opposite angles in this parallelogram are angles A and C. Every parallelogram has two pair of congruent opposite angles.

Example:

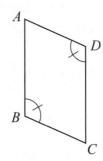

Find the measure of angle B.
$m\angle B = 2x - 14$
$m\angle D = x + 47$

$$2x - 14 = x + 47$$
$$x = 61$$
$$m\angle B = 108$$

Another instance in which we deal with congruent angles is in isosceles triangles. Remember, the angles opposite the congruent sides are themselves congruent.

Example:

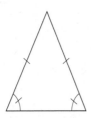 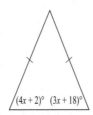

$$4x + 2 = 3x + 18$$
$$x = 16$$
$$4(16) + 2 = 66°$$
$$3(16) + 18 = 66°$$

Interior and Exterior Angles

There are special relationships among the angles of a polygon. Polygons have both interior and exterior angles. Seeing the interior angles of polygons is straightforward. They are simply the angles on the inside of the polygon. The **exterior angles** are formed by extending a side at each vertex. The angle formed is supplementary to its adjacent interior angle. In the diagram below, angle 4 is an exterior angle.

Sum of the interior angles $= (n - 2)180°$
Sum of the exterior angles $= 360°$

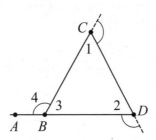

1. $m\angle 3 + m\angle 4 = 180$ 1. Straight angles equal 180°.
2. $m\angle 3 + m\angle 2 + m\angle 1 = 180$ 2. Sum of measures of triangle equal 180°.
3. $m\angle 3 + m\angle 4 = m\angle 3 + m\angle 2 + m\angle 1$ 3. Substitution
4. $m\angle 4 = m\angle 2 + m\angle 1$ 4. Subtraction

On the AIMS test you will be asked to find the sum of the measures of the interior and exterior angles of any polygon. The AIMS test will have a reference sheet that contains these formulas. Let's see how we might verify the formula for the sum of the interior angles.

We know that the sum of the interior angles of a triangle is 180°. A quadrilateral can be divided into two triangles. Therefore, the sum of its interior angles is two times 180° or 360°. Look at the diagram to see how this applies to polygons with more sides.

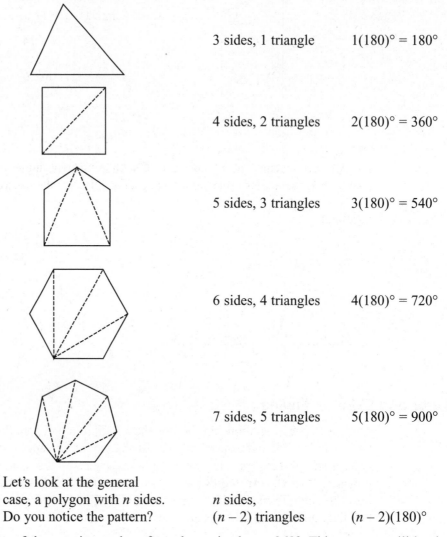

3 sides, 1 triangle	$1(180)° = 180°$
4 sides, 2 triangles	$2(180)° = 360°$
5 sides, 3 triangles	$3(180)° = 540°$
6 sides, 4 triangles	$4(180)° = 720°$
7 sides, 5 triangles	$5(180)° = 900°$

Let's look at the general case, a polygon with n sides. Do you notice the pattern?

n sides, $(n - 2)$ triangles $(n - 2)(180)°$

The sum of the exterior angles of a polygon is always 360°. This concept will be developed using induction in Strand 5.

Sum of the exterior angles of all polygons = 360°

Example: Suppose that you had a question on your AIMS test asking you to find the sum of the interior and exterior angles of an octagon.

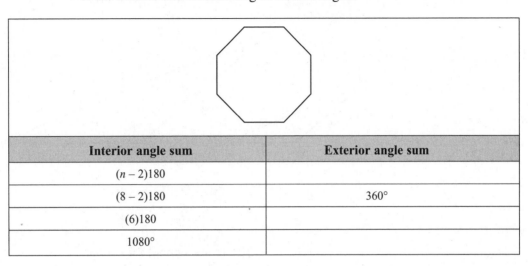

Interior angle sum	Exterior angle sum
$(n - 2)180$	
$(8 - 2)180$	$360°$
$(6)180$	
$1080°$	

Congruent and Similar Triangles

We now need to investigate congruency and similarity of triangles. Let's begin, as usual, by adding these two words to the ever-growing vocabulary list.

> Congruent triangles
> • SSS
> • SAS
> • ASA
> • AAS

Congruent triangles—Triangles that are the same shape and size. That means their corresponding angles are equal in measure and their corresponding sides are equal in length.

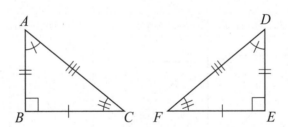

Congruent Triangles

$\triangle ABC \cong \triangle DEF$

Note the symbol for congruency.
Note the order of the vertices.

Similar triangles—Triangles that have corresponding (matching) angles that are equal and corresponding sides that are proportional.

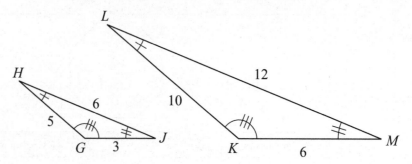

Similar Triangles

$$\triangle GHJ \sim \triangle KLM$$

Note the symbol for similar.
Note the order of the vertices.

Let's start by discussing congruency. Objects that are congruent are a perfect match in size and shape—identical in every way. You may need to turn, flip or slide one object to position it on top of the other, but the end result is that the objects coincide. (We will look at the turning, flipping, and sliding in the next section. This is called transformation of shapes.)

Showing that two triangles are congruent involves a few basic concepts. There are four different options to show that two triangles are congruent.

Congruency for Triangles

SSS	SAS	ASA	AAS
Side-Side-Side	Side-Angle-Side	Angle-Side-Angle	Angle-Angle-Side

Some of the congruent parts needed to prove two triangles congruent are not always obvious.

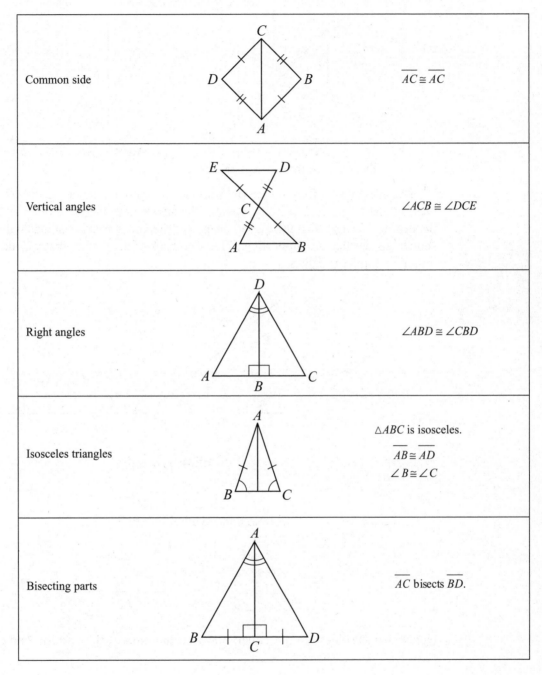

Common side		$\overline{AC} \cong \overline{AC}$
Vertical angles		$\angle ACB \cong \angle DCE$
Right angles		$\angle ABD \cong \angle CBD$
Isosceles triangles		$\triangle ABC$ is isosceles. $\overline{AB} \cong \overline{AD}$ $\angle B \cong \angle C$
Bisecting parts		$\overline{AC}$ bisects $\overline{BD}$.

Examples: Which of the following could be used to determine congruency?

SSS, SAS, ASA, AAS

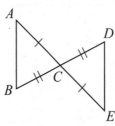
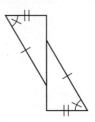

Answer: SAS   Answer: SAS
Did you see the vertical angles?

Next, let's move from congruent triangles to similar triangles. Like congruent triangles, similar triangles have equal angles. The angles have the same number of degrees. If the degrees change, then the overall shape is different. Similar triangles must have corresponding sides that are proportional. Let's go back to the picture that was next to the definition of similar triangles.

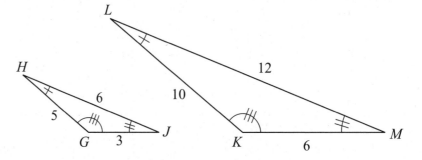

Similar Triangles

$$\triangle GHJ \sim \triangle KLM$$
Note the symbol for similar.

Angles	Sides
$\angle G \cong \angle K$	
$\angle H \cong \angle L$	$\dfrac{GH}{KL} = \dfrac{GJ}{KM} = \dfrac{HJ}{LM}$
$\angle J \cong \angle M$	

Notice that the sides are not equal; however, the ratio of the corresponding sides is the same.

$$\frac{GH}{KL} = \frac{GJ}{KM} = \frac{HJ}{LM} \quad \longleftrightarrow \quad \frac{5}{10} = \frac{3}{6} = \frac{6}{12}$$

The pairs of corresponding sides have been written as ratios. Notice that the ratios are equal. This ratio is called the scale factor for the similar triangles. In our example, that scale factor is ½.

Similar triangles all have a scale factor. That scale factor can be used to find missing sides. In the diagram on the next page, the value of x is missing. Use the fact that the triangles are given as similar (and therefore their corresponding pairs of sides are proportional) to find the missing value.

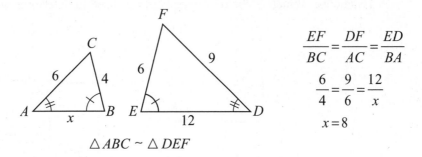

$$\frac{EF}{BC} = \frac{DF}{AC} = \frac{ED}{BA}$$

$$\frac{6}{4} = \frac{9}{6} = \frac{12}{x}$$

$$x = 8$$

$$\triangle ABC \sim \triangle DEF$$

Table of Similar Triangles

AA	SAS	SSS
Angle–Angle	Side–Angle–Side	Side–Side–Side

You might wonder why only two angles are required for similar triangles when a triangle has three angles. Consider triangles *ABC* and *DEF* as shown.

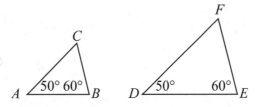

We know that the measures of the angles of a triangle add up to 180°.

$$50° + 60° + m\angle C = 180°$$
$$110° + m\angle C = 180°$$
$$m\angle C = 70°$$

The same can be shown for triangle *DEF*. So, each corresponding angle of triangle *ABC* is congruent to the corresponding angles in triangle *DEF*.

Examples: Which of the following could be used to determine similarity?

SSS, SAS, AA

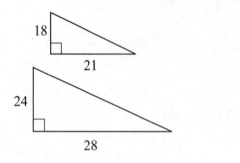
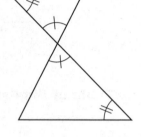

Answer: SAS   Answer: AA

Circles, Arcs, Angles, and Related Segments

Another shape that is a foundational part of geometric properties is the circle. In this section, we will continue to add vocabulary terms and look at the mathematical relationships between the lines, segments, arcs, and angles of circles.

A **circle** is the set of all points that are a given distance from a given point. The given point is the center, and the given distance is the radius of the circle.

Circle Facts

Circles are named using their center. The circle below is called circle P, sometimes written $\odot P$.

The **radius** of a circle is a line segment from the center of the circle to a point on the circle. The **diameter** of the circle is a line segment through the center of the circle whose endpoints are on the circle. The length of the diameter is twice the length of the radius.

The **area** of a circle is the region within the circle. The formula for finding the area of a circle is shown.

$$A = \pi r^2$$

The **circumference** is the distance around the circle. Suppose the circle was made by a string. If you cut the string and lay it out as a straight line, the length would be the same as the circumference. The formula for circumference is shown.

$$C = \pi d = 2\pi r$$

Pi (π) is used to determine both the area and the circumference of a circle. **Pi** is an irrational number and its numerical value is an approximation. Traditionally, there is a decimal or fractional approximation for π.

$$\pi \approx 3.14 \ \text{ or } \ \pi \approx \frac{22}{7}$$

Examples:

Given a circle with a radius of 7 centimeters	
Area of the circle	Circumference of the circle
$A = \pi r^2$	$C = 2\pi r$
$A = \pi(7)^2$	$C = 2\pi(7)$
$A = 49\pi$	$C = 14\pi$
$A \approx 49(3.14)$	$C \approx 14\left(\dfrac{22}{7}\right)$
$A \approx 154 \text{ cm}^2$	$c \approx 44 \text{ cm}$

In addition to diameters and radii (plural for radius) there are more definitions related to circles.

A **chord** of a circle is a line segment whose endpoints are on the circle. A diameter is a special type of chord—a chord that contains the center of the circle.

A **secant** of a circle is a line that contains two points of a circle. Every secant to a circle contains a chord of the circle.

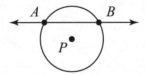

A **tangent** is a line in the same surface as a circle that touches the circle at one point. The tangent line is perpendicular to the radius at the point of tangent.

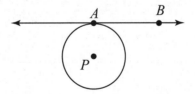

Even though secants and tangents are defined as lines, we sometimes work with parts of those lines.

A **secant segment** is a part of a secant line that joins a point (C) outside the circle to the chord $\left(\overline{AB}\right)$ on the secant line.

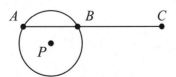

A **tangent segment** is a part of a tangent line between the point of tangency (*A*) and a point outside the circle (*B*).

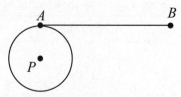

Summary Chart of Formulas—Segment Lengths

Name	Diagram	Formula
Chord–Chord		$\dfrac{a}{b}=\dfrac{c}{d}$ or $a \cdot d = b \cdot c$
Secant–Secant		$\dfrac{a}{c}=\dfrac{d}{b}$ or $a \cdot b = d \cdot c$
Tangent–Secant		$\dfrac{c}{a}=\dfrac{a}{b}$ or $a^2 = b \cdot c$
Tangent–Tangent		$a = b$

Examples:

Name	Diagram	Formula
Chord–Chord	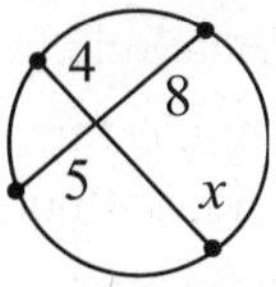	$\dfrac{4}{5} = \dfrac{8}{x}$ $4x = 40$ $x = 10$
Secant–Secant	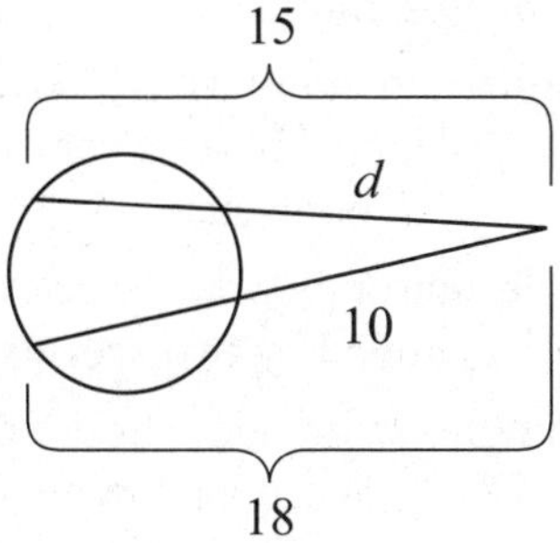	$\dfrac{a}{c} = \dfrac{d}{b}$ $\dfrac{18}{15} = \dfrac{d}{10}$ $180 = 15d$ $d = 12$
Tangent–Secant	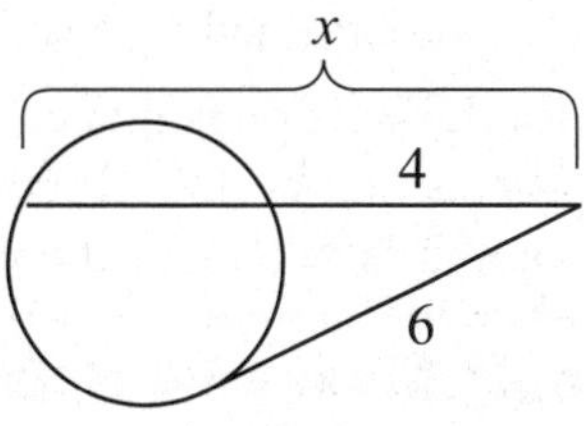	$\dfrac{c}{a} = \dfrac{a}{b}$ $\dfrac{x}{6} = \dfrac{6}{4}$ $4x = 36$ $x = 9$
Tangent–Tangent	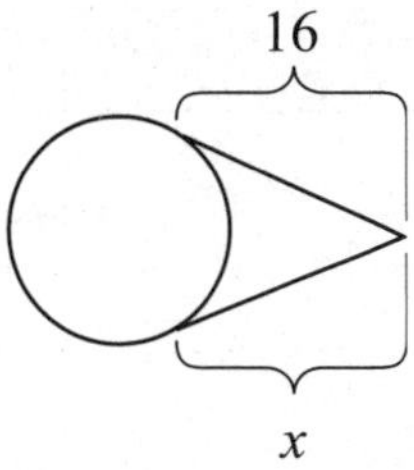	$a = b$ $x = 16$

Circle, Angles, and Intercepted Arcs

Before beginning with the formulas, let's add a few words to our vocabulary.

An **arc** of a circle is a part of the circumference of the circle. The measure of an arc is given in degrees because it is associated with an angle. We name arcs using the endpoints of the arc. In the chart below the arcs are identified by name and kind of arc.

Diagram	List arcs	Definitions
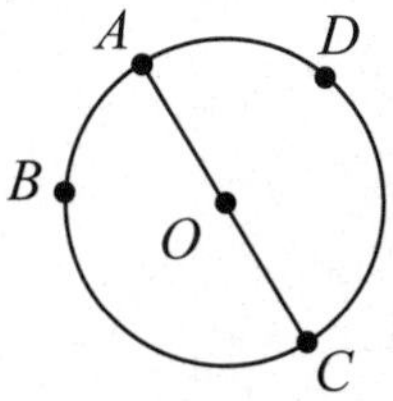	$\overarc{AB}, \overarc{AD}, \overarc{BC}, \overarc{CD}, \overarc{BD}$	Minor arcs are less than half a circle (180°) and are named with two letters.
	$\overarc{ABC}, \overarc{ADC}$	Semicircles are half a circle (180°) and must be named with three letters.
	$\overarc{ABD}, \overarc{BCD}, \overarc{CDB}, \overarc{DBC}$	Major arcs are more than half a circle (180°) and must be named with three letters.

We have discussed finding measures of segments in and about a circle. Now let's take a look at some angles related to circles. Solving problems with circles, angles, and intercepted arcs is dependent on the placement of the vertex of the angle on the circle.

Central angle—an angle whose vertex is at the center of the circle and whose sides are radii of the circle.

Chord-chord angle—an angle formed by two chords that intersect at a point inside the circle, but not at the circle's center.

Inscribed angle—an angle whose vertex is on the circle and whose sides are determined by two chords.

Tangent-chord angle—an angle whose vertex is on the circle and whose sides are determined by a tangent and a chord that intersect at the tangent's point of contact.

Secant-secant angle—an angle whose vertex is outside a circle and whose sides are determined by two secants.

Tangent-tangent angle—an angle whose vertex is outside a circle and whose sides are determined by two tangents.

Secant-tangent angle—an angle whose vertex is outside a circle and whose sides are determined by a secant and a tangent.

The good news for this section is that each of the formulas you need to find the measure of the angles and arcs can be found on the AIMS reference sheet provided with the test. The challenge for you is to choose the proper formula for the given conditions.

Formulas for Angle Measures Related to Circles

Name	Diagram	Formula
Vertex in the circle		
Central angle		$m\angle X = m\,\overarc{BC}$
Chord–chord angle		$m\angle X = \dfrac{1}{2}\left(m\,\overarc{AB} + m\,\overarc{CD}\right)$

Name	Diagram	Formula
Vertex on the circle		
Inscribed angle		$m\angle X = \frac{1}{2}\left(m\widehat{BC}\right)$
Tangent–chord angle		$m\angle X = \frac{1}{2}\left(m\widehat{BC}\right)$
Vertex outside of the circle		
Secant-secant angle		$m\angle X = \frac{1}{2}\left(m\widehat{CD} - m\widehat{AB}\right)$
Tangent-tangent angle		$m\angle X = \frac{1}{2}\left(m\widehat{BAC} - m\widehat{BC}\right)$
Secant-tangent angle		$m\angle X = \frac{1}{2}\left(m\widehat{BCD} - m\widehat{AB}\right)$

Examples:

Name	Diagram	Formula
Vertex in the circle		
Central angle		Given $m\widehat{BC}=62°$ $m\angle X = m\widehat{BC}$ $m\angle X = 62°$
Chord–chord angle		$m\angle X = \dfrac{1}{2}\left(m\widehat{AB}+m\widehat{CD}\right)$ $m\angle X = \dfrac{1}{2}\left(88+26\right)$ $m\angle X = \dfrac{1}{2}\left(114\right)$ $m\angle X = 57°$
Vertex on the circle		
Inscribed angle		$m\angle X = \dfrac{1}{2}\left(m\widehat{BC}\right)$ $m\angle X = \dfrac{1}{2}\left(130\right)$ $m\angle X = 65°$
Tangent-chord angle		$m\angle X = \dfrac{1}{2}\left(m\widehat{BC}\right)$ $m\angle X = \dfrac{1}{2}\left(100\right)$ $m\angle X = 50°$

Name	Diagram	Formula
	Vertex outside of the circle	
Secant-secant angle		$m\angle X = \dfrac{1}{2}\left(m\overset{\frown}{CD} - m\overset{\frown}{AB}\right)$ $m\angle X = \dfrac{1}{2}\left(57-31\right)$ $m\angle X = \dfrac{1}{2}\left(26\right)$ $m\angle X = 13°$
Tangent-tangent angle		$m\angle X = \dfrac{1}{2}\left(m\overset{\frown}{BAC} - m\overset{\frown}{BC}\right)$ $m\angle X = \dfrac{1}{2}\left(233-127\right)$ $m\angle X = \dfrac{1}{2}\left(106\right)$ $m\angle X = 53°$
Secant-tangent angle		$m\angle X = \dfrac{1}{2}\left(m\overset{\frown}{BCD} - m\overset{\frown}{AB}\right)$ $m\angle X = \dfrac{1}{2}\left(200-70\right)$ $m\angle X = \dfrac{1}{2}\left(130\right)$ $m\angle X = 65°$

Arc Length and Sector Area

In working with circles, we use the formulas for circumference and area. These formulas enable us to look at part of the distance around a circle, called arc length, and a part of the area of a circle called a sector.

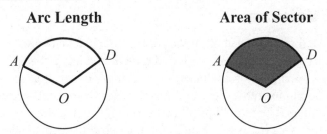

Arc Length   **Area of Sector**

Remember, the measure of an arc is given in degrees. However, the length of an arc is a distance and uses the same units as the radius. The measure of the length of an arc is a fractional part of the circumference of a circle. Recall that circumference is the distance around a circle. The formula for computing circumference, given the diameter, is $C = \pi d$. The formula for computing circumference, given the radius, is $C = 2\pi r$.

The arc length formula incorporates both the circumference and the ratio of arc measure in degrees to 360 (one revolution of a circle).

Arc Length

$$\text{Length of } \overset{\frown}{AB} = \frac{m\overset{\frown}{AB}}{360°}\left(2\pi r\right)$$

The arc length formula is found on the AIMS reference sheet.

Just as the arc length was a fractional part of circumference, a sector is the fractional part of the area of a circle. The sector of a circle is like a wedge of pizza.

Area of a Sector

$$A = \pi r^2 \left(\frac{\text{degrees in corresponding arc}}{360°} \right)$$

The units used when measuring arc length are the same as the units of the radius. However, the units used in measuring the area of a sector must be square units. If the radius of a circle is measured in inches, then the arc length is measured in inches, and the sector area is measured in inches squared. In a word problem, you may need to replace π with its decimal or fractional approximations, 3.14 or 22/7.

Example: Given that the radius of a circle is 6 inches and the measure of an arc is 60°, what is the measure of the arc length and area of the sector?

Arc length	Diagram	Area of sector
$\text{Length of } \overset{\frown}{AB} = \frac{m\overset{\frown}{AB}}{360°}\left(2\pi r\right)$ $\text{Length of } \overset{\frown}{AB} = \frac{60}{360°}\left(2\pi 6\right)$ $\text{Length of } \overset{\frown}{AB} = 2\pi \approx 6.28 \text{ in.}$	6, 60°	$A = \pi r^2 \left(\frac{\text{degrees in arc}}{360°} \right)$ $A = \pi 6^2 \left(\frac{60}{360°} \right)$ $A = 6\pi \approx 18.84 \text{ in.}^2$

Area and Volume

Before looking at all of the formulas to calculate area and volume on the reference sheet, let's review the units of measurement and basic shapes.

Length is sometimes referred to as linear measure. It is the measurement from one point to another. Typical units of measure are centimeters, inches, feet, and miles, among others. The term "linear" measure implies one-dimensional measure.

Area refers to the measure of a two-dimensional shape. The units of measure are written as square units. Examples include centimeters squared (cm^2) or inches squared (in.^{2}). It doesn't matter if you refer to area as units squared or square units. However, when abbreviated, you will see the unit followed by the superscript (raised) 2.

Volume refers to the measure of capacity of a three-dimensional object. The units for volume are stated in cubic terms such as centimeters cubed (cm^3) or cubic inches (in.^{3}). Again, notice that the exponent is 3 and is a superscript written after the unit.

Area

For the purpose of the AIMS test, it is helpful to look at figures as combinations of some basic shapes. Take for instance, the stop sign (an octagon). It is an eight-sided regular polygon. (The term "regular" means all of the sides and angles are congruent.) With two lines you can picture the octagon as a rectangle and two trapezoids.

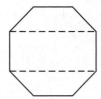

With two more lines, you create a square, four rectangles, and four triangles.

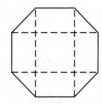

Example: Suppose your track coach wanted you to mow the grass on the infield of the school's track. You could approximate the area by looking at the shape as a rectangle and two semicircles (or one complete circle).

The straightaway is 100 meters; the radius of the curve is 30 meters.

A = Rectangle + Circle
$A = \ell \cdot w + \pi r^2$
$A = 100 \cdot 60 + \pi 30^2$
$A = 6{,}000 + 900\pi$ m^2

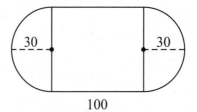

Example: You are planning to put grass around your pool in your backyard. It is necessary to calculate the area in which the grass will be planted. To do this you would subtract the area of the rectangular pool from the area of the rectangular backyard.

The dimensions of the backyard are 25 feet by 50 feet and the pool is 15 feet by 35 feet.

A = Yard − Pool
$A = 25 \cdot 50 - 15 \cdot 35$
$A = 725$ ft.^{2}

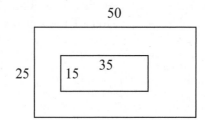

Example: You are planning to cover the side of some steps with decorative stone. One way to find the area of that surface is to divide the region into smaller known shapes. This region can be divided in several ways. Another way would be to enclose the region with a larger rectangle and subtract the area of the smaller rectangle from the larger rectangle.

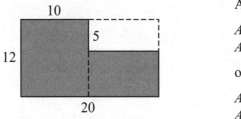

Area of inside rectangles:

$A = 12(10) + 10(7)$
$A = 190 \text{ ft.}^2$

or

$A = 20(12) - 5(10)$
$A = 240 - 50 = 190 \text{ ft.}^2$

Three-Dimensional Figures

Vocabulary for Three-Dimensional Figures

Term	Letter	Description
Volume	V	Measure of cubic units in the interior of a three-dimensional figure
Surface area	T	The sum of the measure of the areas of the faces of a three-dimensional figure
Area of base	B	Area of the parallel and congruent faces of a cylinder or prism
Radius	r	Distance from the center of a circle to the circle
Height (altitude)	h	Perpendicular segment from the vertex to the center of the base.
Lateral face	none	Triangular face of a pyramid
Slant height	ℓ	The height of the lateral face of a pyramid
Perimeter	P	The distance around the base when finding surface area

Here is a table of the common three-dimensional objects. The chart includes volume and surface area. Remember to label volume using cubic units and surface area using square units. The surface area formulas come from the area of the faces.

Volume of Three-Dimensional Figures

Figure names	Diagram	Example problems
Prism		$V = Bh$ $\quad = (l \cdot w)h$ $\quad = (15 \cdot 10)5$ $\quad = 150 \cdot 5$ $\quad = 750$ cu units
Square pyramid		$V = \dfrac{1}{3}Bh$ $\quad = \dfrac{1}{3}\left(s^2\right)h$ $\quad = \dfrac{1}{3}\left(10^2\right)12$ $\quad = \dfrac{1}{3} \cdot 100 \cdot 12$ $\quad = 400$ cu units
Cylinder		$V = Bh$ $\quad = (\pi r^2)h$ $\quad = (\pi 3^2)12$ $\quad = 9\pi \cdot 12$ $\quad = 108\pi$ cu units
Cone		$V = \dfrac{1}{3}Bh$ $\quad = \dfrac{1}{3}\left(\pi r^2\right)h$ $\quad = \dfrac{1}{3}\left(\pi 4^2\right)10$ $\quad = \dfrac{1}{3}\left(16\pi\right)10$ $\quad = \dfrac{1}{3}\left(160\pi\right)$ $\quad = \dfrac{160\pi}{3}$ cu units
Sphere		$V = \dfrac{4}{3}\left(\pi r^3\right)$ $\quad = \dfrac{4}{3}\left(\pi 3^3\right)$ $\quad = \dfrac{4}{3}\left(27\pi\right)$ $\quad = 36\pi$ cu units

Surface Area of Three-Dimensional Figures

Figure names	Diagram	Example problems
Prism		$TA = 2B + Ph$ $= 2\,(l \cdot w) + (2l + 2w)h$ $= 2(15 \cdot 10) + (2 \cdot 15 + 2 \cdot 10)5$ $= 2(150) + (30 + 20)5$ $= 300 + 250$ $= 550 \text{ units}^2$
Square pyramid		$TA = B + \dfrac{1}{2}P\ell$ $= s^2 + \dfrac{1}{2}\big(4s\big)\ell$ $= 10^2 + \dfrac{1}{2}\big(4 \cdot 10\big)13$ $= 100 + \dfrac{1}{2}\big(40\big)13$ $= 100 + 260$ $= 360 \text{ units}^2$
Cylinder		$TA = 2\pi rh + 2\pi r^2$ $= 2\pi \cdot 3 \cdot 12 + 2\pi \cdot 3^2$ $= 72\pi + 18\pi$ $= 90\pi \text{ units}^2$
Cone		$TA = \pi r\ell + \pi r^2$ $= \pi \cdot 4 \cdot 2\sqrt{29} + \pi 4^2$ $= 8\pi\sqrt{29} + 16\pi \text{ units}^2$ Compute the slant height using Pythagorean theorem.
Sphere		$TA = 4\pi r^2$ $= 4\pi \cdot 3^2$ $= 4\pi \cdot 9$ $= 36\pi \text{ units}^2$

Occasionally, a question related to pyramids gives some information that doesn't directly match the variables from the formula. You may need to look for relationships between the given information and formula variables.

$\overline{PY}$ is a slant height.

$\overline{PY}$ is the hypotenuse of $\triangle PXY$.

$\overline{PX}$ is the altitude of the pyramid.

$\overline{PX}$ is the leg of the right $\triangle PXY$.

$\overline{XY}$ is the leg of the right $\triangle PXY$.

XY is one-half of AB.

$\overline{AB}$ is a side of the square base of pyramid.

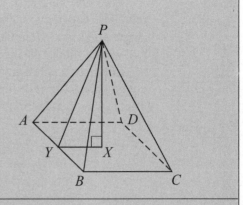

When finding the slant height and altitude of a pyramid, there are two basic concepts that are required. Look at right triangle PXY in the diagram above.

Because X is at the center of square $ABCD$	Using the Pythagorean theorem
$XY = \dfrac{1}{2} BC$	$(PX)^2 + (XY)^2 = (PY)^2$

Example:

Find the slant height of the pyramid with a square base of 12 inches and an altitude of 8 inches.

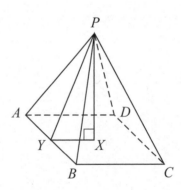

From above, $XY = \dfrac{1}{2} BC$.

$$XY = \frac{1}{2}(12)$$

$$XY = 6$$

From above, $(PX)^2 + (XY)^2 = (PY)^2$.

$$(8)^2 + (6)^2 = (PY)^2$$
$$64 + 36 = (PY)^2$$
$$100 = (PY)^2$$
$$PY = 10 \text{ inches}$$

Coordinate Geometry

This section on coordinate geometry takes a closer look at the equation of lines and parabolas. The equation of a line is a linear equation and the equation of a parabola is a quadratic equation. Linear and quadratic functions were explained in earlier sections.

In this section, the principles and properties of linear and quadratic equations are investigated in relationship to the coordinate axis.

The coordinate axis, x-y axis, helps to visualize the behavior of the functions and other relations. The horizontal axis is called the x-axis. The vertical axis is called the y-axis. A single point plotted on the x-y axis is listed as an ordered pair (x, y). The ordering of the two values simply means that the x value is listed first, followed by the y value.

In function notation, the variable x represents the input values called the domain. The output values from the function, called the range values, are represented by the y variable. This is why the function notation f(x) is often replaced by the variable y.

$f(x) = 2x$ The function called f doubles the input values.

$y = 2x$ The linear equation displays a graph showing for each input value of x, the y value is twice the size.

Review the graph below, noting the plotting of points and the labeling of the quadrants. The quadrants are labeled I, II, III, and IV, counterclockwise, starting in the upper right corner. Quadrants represent regions whose boundaries are the x-y axis. Occasionally, it is convenient to refer to the regions of the coordinate plane. Take note of the positive and negative values of x and y and their placement from the center of the x-y axis, as you plot the points.

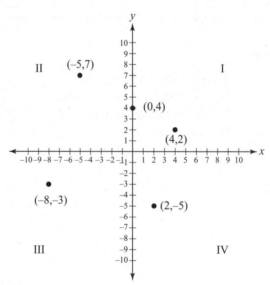

Graphing Lines

The general equation for the equation of a line is below.

$$Ax + By = C$$

The letters A, B, and C represent real numbers. (The only exception is that A and B cannot both be zero.) A and B are called the coefficients of the variables x and y. Inasmuch as there is not a variable with the letter C, it is called a constant.

A line is made up of points. The graph of a line is represented by the points. All of the ordered pairs that make the linear equation true are called solutions of the equation and, when plotted, form a straight line.

Let's look at each step needed to graph the line with coefficients of 6 and 2 and a constant of 12.

$$6x + 2y = 12$$

Table of Values Method

One method for graphing the line is to determine the x-y pairs that satisfy the equation. You can choose a value for x or y and place the value into the equation. With one variable remaining, you can solve for the other value of the coordinate pair.

It is a good practice to find four points and plot them. If the points do not begin to form a pattern of a line, you should check your calculations.

Example:

Let $x = 0$

Let $x = 1$

Let $x = -1$

Let $y = 0$

x	$6x + 2y = 12$	y	Ordered Pairs
0	$6(0) + 2(y) = 12$	6	$(0,6)$
1	$6(1) + 2(y) = 12$	3	$(1,3)$
-1	$6(-1) + 2(y) = 12$	9	$(-1,9)$
2	$6(x) + 2(0) = 12$	0	$(2,0)$

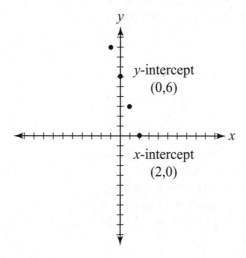

Look at the first and the fourth points from the example above. The first point is the y-intercept. The y-intercept is the point where the graph touches the y-axis. The x value of the y-intercept is always zero. The fourth point is the x-intercept. The y value for this point is always zero and the point is on the x-axis.

$$6x + 2y = 12$$

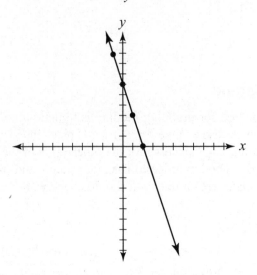

Slope-intercept form of a line
$y = mx + b$

Slope

$$m = \frac{y_2 - y_1}{x_2 - x_1}$$

In this book's algebra section (Strand 3), the method of solving for the variable y was shown. Recall that this method changes the equation into slope-intercept form.

By plotting the y-intercept first, we have a starting point to begin to draw the graph. Using the definition of slope, in fraction form, we can plot the vertical change from the numerator value and the horizontal change from the denominator value.

Example:

$$6x + 2y = 12$$

$$2y = -6x + 12$$

$$y = -3x + 6$$

Start by plotting the y-intercept (0, 6).
From the y-intercept apply the slope of $-3/1$ (down 3 and right 1).

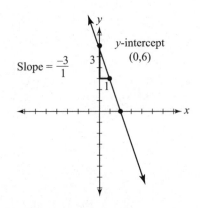

Direction Summary

Direction	Sign of number
Up	Positive
Down	Negative
Right	Positive
Left	Negative

Let's take a closer look at the coefficients and constant for a graph of a line.

Examples: If we change the value of A, there is a change in the slope.

$$4x + 2y = 12$$
$$2y = -4x + 12$$
$$y = -2x + 6$$

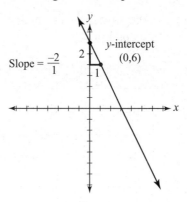

If we change the value of B, there is a change in the slope and intercept.

$$6x + 6y = 12$$
$$6y = -6x + 12$$
$$y = -x + 2$$

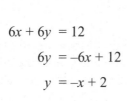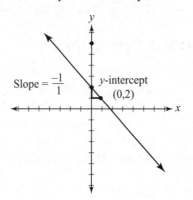

If we change the value of C, there is a change in the intercept.

$$6x + 2y = 6$$
$$2y = -6x + 6$$
$$y = -3x + 3$$

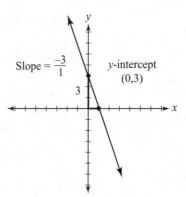

Summary of the coefficients	
$Ax + By = C$ $By = -Ax + C$ $y = \dfrac{-A}{B}x + \dfrac{C}{B}$	Slope $= \dfrac{-A}{B}$ y-intercept $\left(0, \dfrac{C}{B}\right)$

Graphing Inequalities

We can extend the procedure of graphing a line to graphing a linear inequality. Start by treating the inequality as an equality. Graph this equation using the method with which you are most comfortable. Then just two additional steps are needed.

Select Solid or Dotted Line

The first step is to look at the given inequality to determine if the problem has the equal to condition.

Symbol	Meaning	Graph line
<	Less than	Dotted line
≤	Less than or equal to	Solid line
>	Greater than	Dotted line
≥	Greater than or equal to	Solid line

Notice that it is a single horizontal bar that represents "equal to" in addition to the inequality symbol. With the equal to condition, draw a solid line for the graph. If there is no horizontal bar, then draw a dotted line.

Decide Which Side To Shade

The second additional step is to identify all of the ordered pairs that make the inequality a true statement. The solutions are found on one side or the other of the line you just have drawn. Shading is the way to show which values make the inequality true.

Pick a point, any point, *not* on your line. Place this *x-y* value into the inequality and simplify. If the resulting inequality statement is true, shade the side where your point is located. If the inequality statement is not true, shade the other side.

Examples:

$$4x + 4y < 8$$

Line Preparation

Table of values for $4x + 4y = 8$		Solve for y
$\begin{array}{c\|c} x & y \\ \hline 0 & 2 \\ 2 & 0 \\ 1 & 1 \\ -1 & 3 \end{array}$	Plot points of grid.	$y < -x + 2$ Plot the y-intercept $(0, 2)$. Draw slope $m = \dfrac{-1}{1}$.
		Choose a dotted line. Pick a point $(0, 0)$. $4(0) + 4(0) < 8$ $0 < 8$ True Shade below the line.

$$y \geq 2x - 1$$

Line Preparation

Table of Values for $y = 2x - 1$		Solve for y
$\begin{array}{c\|c} x & y \\ \hline 0 & -1 \\ \frac{1}{2} & 0 \\ 2 & 3 \\ -2 & -5 \end{array}$	Plot points of grid.	$y \geq 2x - 1$ Plot the y-intercept $(0, -1)$. Draw slope $m = \dfrac{2}{1}$.
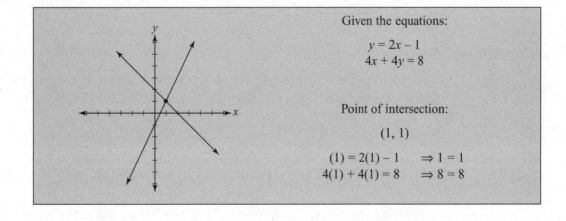		Choose a solid line. Pick a point $(0, 0)$. $\quad (0) \geq 2(0) - 1$ $\quad\quad 0 \geq -1$ True Shade above the line.

Systems of Linear Equations

Graphs are also very helpful in visualizing the solutions to a system of linear equations. If you are good at graphing lines, you will find the solution of linear systems easy. Identify where the two graphed lines intersect. The ordered pair for this intersection point is the only x-y pair that makes both equations true at the same time.

Solution of a linear system
- 1—Intersecting lines
- 0—Parallel lines
- Infinite—Coinciding lines

Identify the Solution

Given the equations:

$$y = 2x - 1$$
$$4x + 4y = 8$$

Point of intersection:

$$(1, 1)$$

$$(1) = 2(1) - 1 \quad \Rightarrow 1 = 1$$
$$4(1) + 4(1) = 8 \quad \Rightarrow 8 = 8$$

Examples: What happens if the graphs of the two lines have the same slope and do not cross? Geometrically, we say that the lines are parallel. As for the solution to this system, you would state that there is no solution.

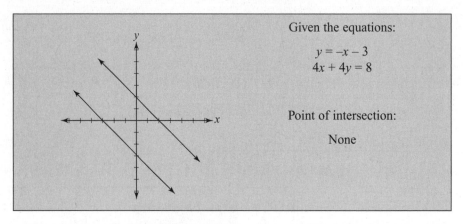

Given the equations:

$$y = -x - 3$$
$$4x + 4y = 8$$

Point of intersection:

None

What happens if after graphing one line, the next line to be drawn is on top of the first line drawn? Geometrically, we say the lines coincide (are coincident). We do not have a unique solution, but rather every ordered pair that makes one linear equation true makes the other linear equation true. Because a line contains an infinite number of points, we say the solution to the system has an infinite number of solutions. Even though there are infinitely many ordered pair solutions, they all must be on the line for the equation to be a solution to the system.

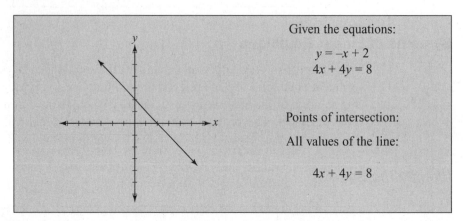

Given the equations:

$$y = -x + 2$$
$$4x + 4y = 8$$

Points of intersection:

All values of the line:

$$4x + 4y = 8$$

Midpoint and Distance

If you are given the endpoints of a line segment, the length of the segment as well as the midpoint can be found using two formulas that are on the AIMS reference sheet.

Midpoint   Distance

$$\text{Midpoint of } \overline{AB} = \left(\frac{x_2 + x_1}{2}, \frac{y_2 + y_1}{2} \right) \qquad AB = \sqrt{(x_2 - x_1)^2 + (y_2 - y_1)^2}$$

Note that the answer to a question asking you for a midpoint is an ordered pair. The midpoint is a point. It is the average of the given x-coordinates followed by the average of the y-coordinates. Think of averaging just as you would average two numbers. Add the two numbers together and divide by two. Choose your first endpoint and label the coordinates as x_1 and y_1; label the second endpoint as x_2 and y_2.

Example: Find the midpoint given the endpoints (2, 1) and (6, 9).

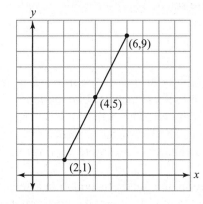

$$\text{Midpoint} = \left(\frac{x_2 + x_1}{2}, \frac{y_2 + y_1}{2} \right)$$

$$\text{Midpoint} = \left(\frac{(\ ) + (\ )}{2}, \frac{(\ ) + (\ )}{2} \right)$$

$$\left(\frac{6+2}{2}, \frac{9+1}{2} \right) = (4, 5)$$

To calculate the distance between any two points, it is helpful to label each endpoint x-y coordinates with subscripts. As you did when you found a midpoint, choose your first point and label the coordinates as x_1 and y_1; label and the second endpoint as x_2 and y_2. Place the values into the formula and simplify.

$$\text{distance} = \sqrt{\left(x_2 - x_1 \right)^2 + \left(y_2 - y_1 \right)^2}$$

Example: Line segment AB has endpoints of (1, − 2) and (7, 6). Find the distance. For the point (1, –2), let $x_1 = 1$ and $x_2 = -2$, and for the second point (7, 6), let $y_1 = 7$ and $y_2 = 6$.

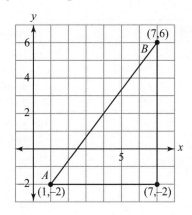

$$\text{Distance} = \sqrt{\left(1 - 7 \right)^2 + \left((-2) - 6 \right)^2}$$

$$\text{Distance} = \sqrt{\left(-6 \right)^2 + \left(-8 \right)^2}$$

$$\text{Distance} = \sqrt{36 + 64} = \sqrt{100} = 10$$

Extra for Experts

For some students knowing how a formula is developed helps them remember the formula and make sense of the solution.

The graph of the problem above is shown. Notice two additional lines are shown. There is a horizontal line through the bottom point and a vertical line through the top point. These lines, along with the line segment $\overline{AB}$, form a right triangle. When you subtract the x-coordinates in the distance formula, you are finding the length of the horizontal leg. When you subtract the y-coordinates, you are finding the length of the vertical leg. (The subtractions may yield negative numbers, and distance is not a negative value.) Did you see that in the distance formula you had to square the differences? When you square a negative number, the answer is always a positive number. That will take care of any negative difference and set up the next step in which you will use the Pythagorean theorem. In a right triangle, the sum of the square of the legs is equal to the square of the hypotenuse. Finally, take the square root to find the length of the hypotenuse (the length of $\overline{AB}$).

Quadratic Equation

The last equation whose graph is on the x-y coordinate system that we will study is the quadratic equation. Its graph looks somewhat like the letter "U." The quadratic graph is a model for many everyday activities. The path of a home run ball hit off a bat follows the upside down quadratic graph. The fireworks launched at a Fourth of July celebration follow a quadratic graph. Skateboarders are quite familiar with this curve. They call this shape a half pipe.

As with the linear equation, the quadratic equation also has a general form. Note however, that unlike the linear equation, the quadratic equation is already solved for y in terms of x.

$$y = ax^2 + bx + c$$
$$a \neq 0$$

The values of a, b, and c can be any real numbers, and only a cannot be zero. Notice the general form is written in **descending order** for the exponent. The term "descending order" means the terms are arranged so the exponents on the variable go from highest to lowest.

The general form shows three terms. Each term has a name and contributes to the graph of the quadratic equation.

Term	Name	Influence
ax^2	Quadratic term	Main contribution—The direction of the opening of the graph. If a is positive, the graph opens up. If a is negative, the graph opens down.
bx	Linear term	This value will shift the graph to the left or right of the y-axis.
c	Constant term	This value is the y-intercept. The value of c becomes the y-coordinate in the ordered pair $(0, c)$.

A line can be drawn with just two points. Often the x- and y-intercepts are used because they are quick to calculate and plot. Graphing a quadratic equation is a little more involved.

Let's begin by looking at two major features of the quadratic equation, the vertex and axis of symmetry.

Vertex

The quadratic equation has an extreme value located at the vertex of the graph. This vertex point is either the highest point on the graph of the quadratic equation or the lowest point. It is quite common in solving applications with quadratic equations to ask for the minimum or maximum value. Finding the vertex gives that value.

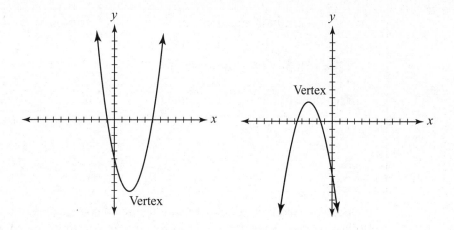

Axis of Symmetry

The second major feature helps to draw the graph. It is called the axis of symmetry. The axis of symmetry is the dividing center line of the graph. If you were to fold the curve on the axis of symmetry as the center line, the left half of the graph would match exactly with the right half.

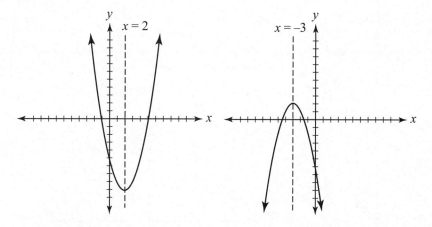

The vertex is the only point of the quadratic that is on the axis of symmetry. The axis of symmetry is always a vertical line in a quadratic graph and its equation is written in the form of $x =$ some number.

Let's take a look at an example quadratic and find the vertex, y-intercept, and axis of symmetry.

$$\text{Quadratic equation } y = x^2 - 4x - 5$$
$$a = 1, b = -4, \text{ and } c = -5$$

When you have the quadratic equation in standard form, the following equation is used to find the axis of symmetry.

$$x = \frac{-b}{2a}$$

Quadratic equation	Axis of symmetry
$y = x^2 - 4x - 5$ $a = 1, b = -4,$ and $c = -5$ The value of c is the y-coordinate in the y-intercept $(0, -5)$.	$x = \dfrac{-b}{2a}$ $x = \dfrac{-(-4)}{2(1)}$ $x = \dfrac{4}{2} = 2$

Once you find this value of x using the given coefficients of a and b, you have the x-coordinate of the vertex. Place this x value into the quadratic equation for each occurrence of x and evaluate the expression for y. This y value is the y-coordinate of the vertex.

Quadratic equation	Axis of symmetry	Vertex
$y = x^2 - 4x - 5$ $a = 1, \ b = -4,$ and $c = -5$	$x = \dfrac{-b}{2a}$ $x = \dfrac{-(-4)}{2(1)}$ $x = \dfrac{4}{2} = 2$	$y = x^2 - 4x - 5$ $y = (2)^2 - 4(2) - 5$ $y = 4 - 8 - 5$ $y = -9$ $(2, -9)$

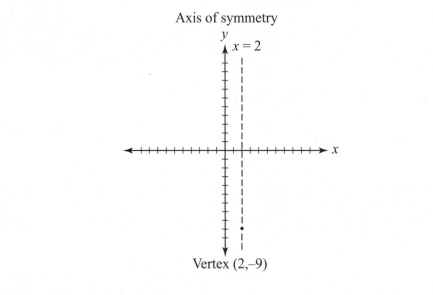

Axis of symmetry

Vertex (2, -9)

Finally, choose two or more values of x from either side of the axis of symmetry. Find the corresponding y values again by placing these x values into the quadratic equation.

Pick values larger or smaller than axis value of 2.

x	$x^2 - 4x - 5$	y
1	$(1)^2 - 4(1) - 5$	-8
0	$(0)^2 - 4(0) - 5$	-5
-1	$(-1)^2 - 4(-1) - 5$	0

Ordered Pairs

$(1, -8)$

$(0, -5)$

$(-1, 0)$

Use the feature of the axis of symmetry to find the reflected points across this axis of symmetry. The graph of the quadratic should begin to take shape. The process of picking x values on either side of the axis of symmetry can be repeated as necessary to find additional points to plot.

Table points and reflected points	Final graph
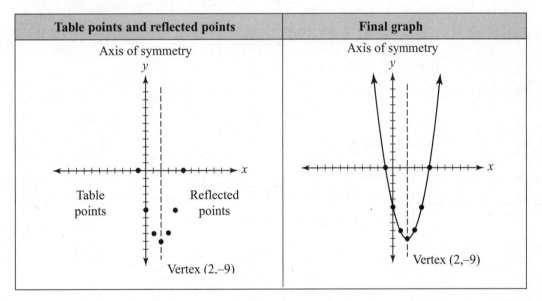	

Graphing the Quadratic

Example:

Equation $y = -x^2 - 6x - 7$		
1. $a = -1$, $b = -6$, and $c = -7$		
2. Let $x = 0$ (to find the y-intercept)	$y = -(0)^2 - 6(0) - 7$ $y = -7$	$(0, -7)$
3. Axis of symmetry	$x = \dfrac{-B}{2A}$ $x = \dfrac{-(-6)}{2(-1)}$ $x = \dfrac{6}{-2} = -3$	$x = -3$

4. Vertex	$y = -x^2 - 6x - 7$ $y = -(-3)^2 - 6(-3) - 7$ $y = -9 + 18 - 7$ $y = 2$	$(-3, 2)$

5. Table values	 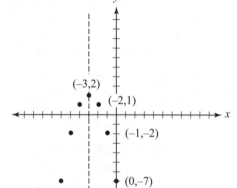	Points $(-2, 1)$ $(-1, -2)$ $(0, -7)$

6. Plot points and reflected points	
$(-3, 2)$ vertex $(-2, 1)$ reflects to $(-4, 1)$ $(-1, -2)$ reflects to $(-5, -2)$ $(0, -7)$ reflects to $(-6, -7)$	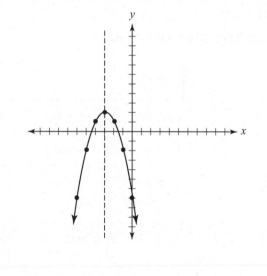

7. Sketch curve	

Extra for Experts

When students are asked what comes to mind with the word quadratic, some mention the "U" shape as just presented. Some mention the exponent of 2 for the variable, x. Some recall the quadratic formula.

So, just how does the quadratic formula relate to the graph of the quadratic?

$$y = ax^2 + bx + c \qquad x = \frac{-b \pm \sqrt{b^2 - 4ac}}{2a}$$

The solutions for the quadratic formula are the x-intercepts for the graph. When solving a quadratic, the goal is to find the values of x that make the quadratic true. These same values help to identify where the graph of the quadratic intersects the x-axis. Write the values in ordered pairs with the y-coordinate as zero.

Quadratic	x-Intercepts	Graph
$y = x^2 - 6x + 5$	(1, 0) and (5, 0)	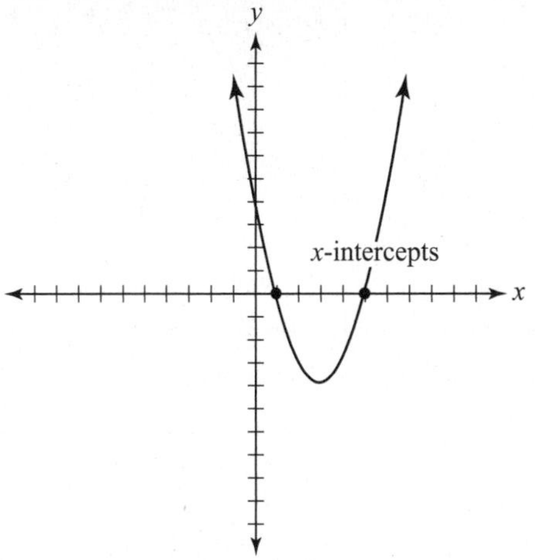

Do you recall looking at the discriminant portion of the quadratic formula? The discriminant revealed quick information about the solutions. That same information can easily be matched up with the types and number of intercepts. Study the chart below.

Discriminant	Number of roots (solutions to the related quadratic equation)	Number of intercepts	Example graph
$b^2 - 4ac > 0$	Two roots	Two intercepts	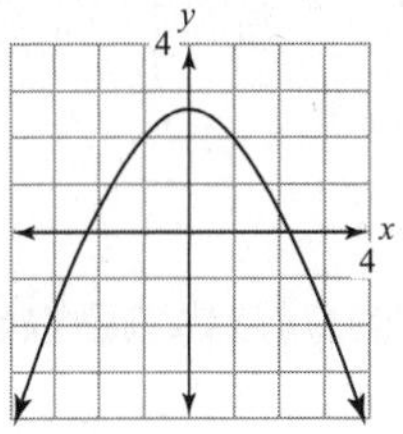
$b^2 - 4ac = 0$	One root (sometimes called a double root)	One intercept (said to touch the graph—a point of tangency)	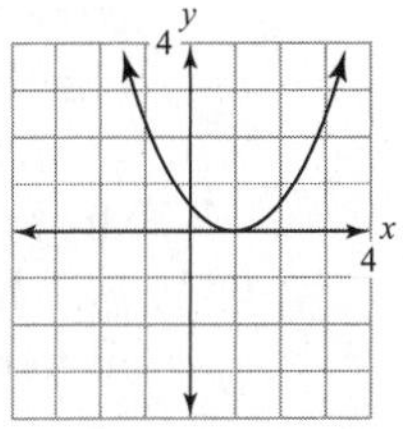

Discriminant	Number of roots (solutions to the related quadratic equation)	Number of intercepts	Example graph
$b^2 - 4ac < 0$	No real root	No x-intercept	

Quadratic	Discriminant	x-Intercept(s)	Graph
$y = x^2 - 2x - 3$	$(-2)^2 - 4(1)(-3)$ $4 \qquad + 12$ $16 > 0$	$(-1, 0)$ and $(3, 0)$	
$y = x^2 - 4x + 4$	$(-4)^2 - 4(1)(4)$ $16 \quad - 16$ $0 = 0$	$(2, 0)$	
$y = x^2 - 4x + 6$	$(-4)^2 - 4(1)(6)$ $16 \quad - 24$ $-8 < 0$	No x-intercept	

Transformations

For many students the word "transformation" usually brings images of change. When the temperature is below freezing, water is transformed into ice, and sufficient heat causes ice to melt into water.

Transformation in geometry involves a change in position or a change in size or both. The transformation of a square will still result in a square.

The vocabulary of transformation includes the following words. The words in the definitions, "slide," "flip," and "turn," may be the words you used in elementary or middle school to define transformations. Those words are generally not used on the AIMS test. It is important that you use the words "translate," "reflect," and "rotate." An additional type of transformation that will be on the AIMS test is dilation.

The initial figure in a transformation is called the **pre-image**. A pre-image is defined as a picture or object before it undergoes a transformation. The result of a transformation is called the **image**.

Transformation Words

Transformations
• Translation
• Reflection
• Rotation
• Dilation

A **translation** (slide) is a transformation that slides each point of a figure the same distance in the same direction. The distance and direction are given in the problem. The image is always congruent (same size) to its pre-image.

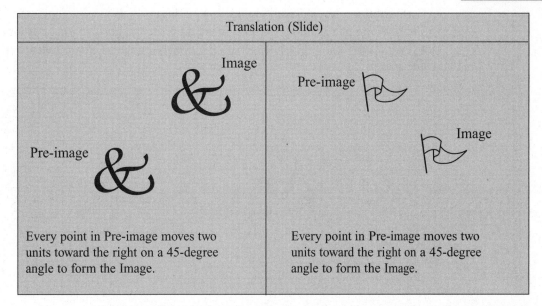

A **reflection** (flip) is a transformation that creates a mirror image across a line of reflection. The line of reflection is given. The image is always congruent to its pre-image.

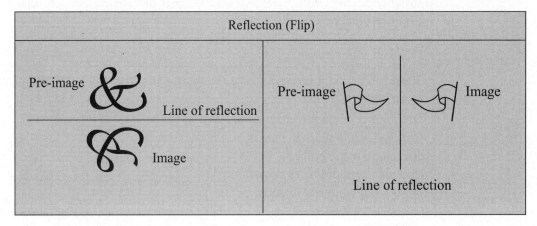

A **rotation** (turn) is a transformation in which a figure is turned about a fixed point called the center of rotation. The center and direction of the rotation are given. The image is always congruent to its pre-image. (Diagram of a rotation is on the next page.)

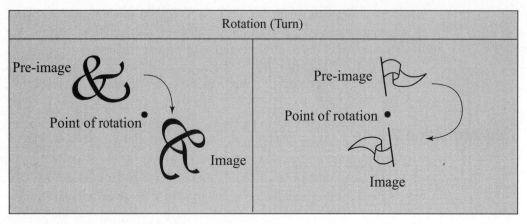

Dilation is a transformation that either enlarges or reduces a geometric figure proportionally. There is a scale factor used in creating the image. The image is always similar to its pre-image. Remember this means they have the same shape but not necessarily the same size.

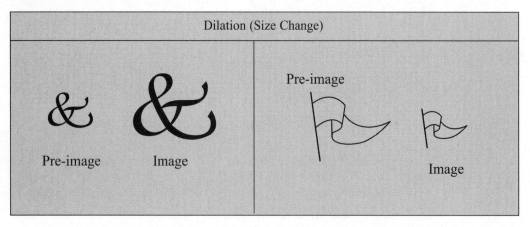

Transformed geometric shapes are labeled in a special manner. If point A is the pre-image, then A' is the image. A' is read A prime. Notice in the example below $\triangle ABC$ is transformed into $\triangle A'B'C'$. Sometimes a point is its own image; you do not use the prime notation for that point.

Example:

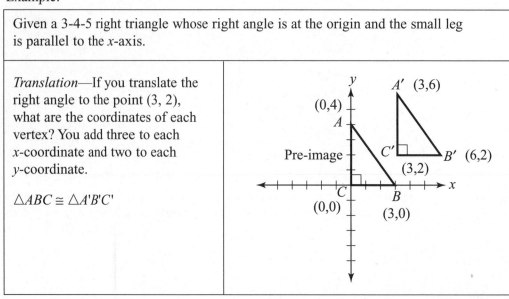

Given a 3-4-5 right triangle whose right angle is at the origin and the small leg is parallel to the x-axis.

Translation—If you translate the right angle to the point (3, 2), what are the coordinates of each vertex? You add three to each x-coordinate and two to each y-coordinate.

$\triangle ABC \cong \triangle A'B'C'$

Reflection—With the *y*-axis as the line of reflection, what are the coordinates of each vertex? Notice that points *A* and *C* are their own image. △*ABC* ≅ △*AB'C*	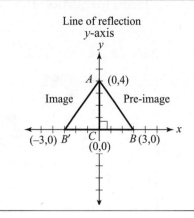
Rotation—If you rotate the right triangle about the point (3, 0) 90° clockwise, what are the coordinates of each vertex as a result of the rotation? Notice that point *B* is its own image. △*ABC* ≅ △*A'BC'*	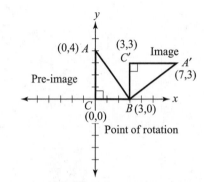
Dilation—If you dilate the right triangle by a scale factor of two, what are the coordinates of each vertex? Notice that point *C* is its own image. △*ABC* ~ △*A'B'C* △*ABC* ≇ △*A'B'C*	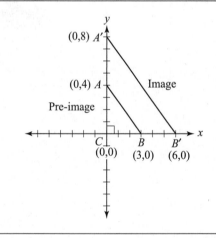

Transformations

Translate	Reflect
Image Pre-image	Line of reflection Pre-image Image
Pre-image Image	Line of reflection Image Pre-image
Pre-image Image	Pre-image Line of reflection Image

Rotate	Dilate

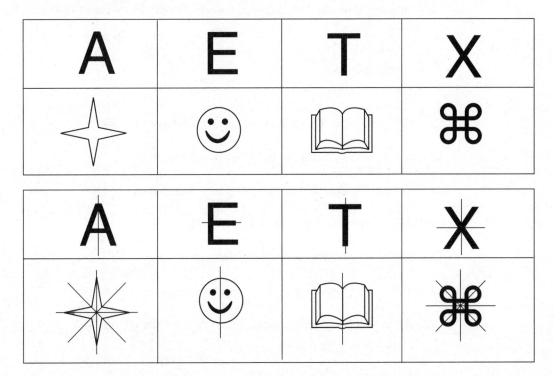

Extra for Experts

The line of symmetry plays an important role in mathematics. Can you find the line(s) of symmetry in the various shapes and pictures?

Are you able to do each of the following?

❑ Identify the attributes of special triangles
- Isosceles triangles
- Equilateral triangles
- Right triangles

❑ Identify the quadrilaterals
- Quadrilateral
- Parallelogram
- Rectangle
- Rhombus
- Square
- Trapezoid

❑ Recognize, draw, and label three-dimensional figures and their nets

❑ Solve problems related to a variety of geometric concepts
- Complementary, supplementary, and congruent angles
- Circles and the arcs, angles, and segments related to them
- Triangle inequality property
- Special-case right triangles
- Contextual situations related to geometry

❑ Determine when triangles are congruent or similar

❑ Apply spatial reasoning to create transformations

❑ Graph equations and inequalities and determine the effect of changes in constants and coefficients
- Linear equation
- Linear inequality
- Quadratic equation

❑ Determine the midpoint and the distance between points

❑ Determine the solution to a system of equations

❑ Calculate the area and volume of geometric shapes

❑ Solve for missing measures in a pyramid

❑ Find the length of an arc and the area of a sector of a circle

❑ Find the sum of the interior and exterior angles of a polygon

❑ Solve problems involving scale factor and similar triangles

Practice Problems—Strand 4

1. What is the sum of the interior angles of the regular polygon shown below?

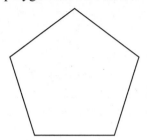

   A. 360°
   B. 540°
   C. 720°
   D. 900°

2. Mama's pizza divides the pizza into six equal pieces that have a 60° central angle. Find the area of a slice of your favorite pizza with a radius of 12 inches.

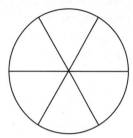

   A. 2π square inches
   B. 6π square inches
   C. 12π square inches
   D. 24π square inches

3. Which of the following graphs represents the linear equation:

$$2x - 3y = 6$$

A.

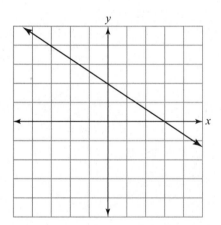

B.

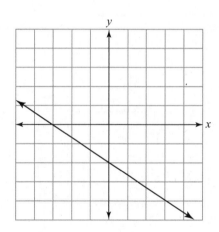

C.

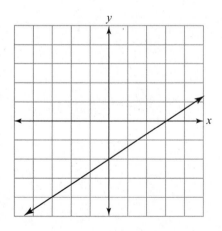

D.

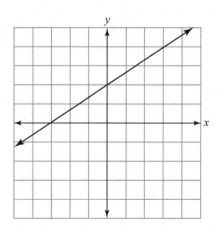

4. In the diagram below, congruent parts are marked. Which of the methods below would you use to prove the triangles congruent?

A. AAS (Angle–Angle–Side)
B. SAS (Side–Angle–Side)
C. ASA (Angle–Side–Angle)
D. SSS (Side–Side–Side)

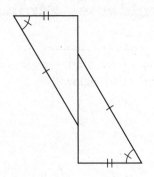

5. The diagram below is a net of the surface area of what three-dimensional object?

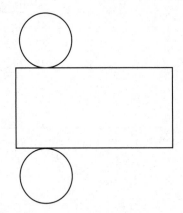

A.

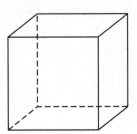

B.

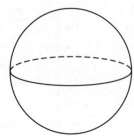

C.

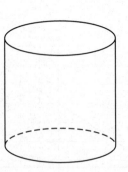

D.

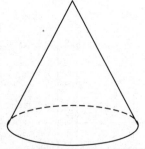

6. Given a circle with the radius of 9 centimeters. What is the length of $\overarc{AB}$ formed by a central angle of 120°?

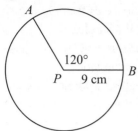

A. 3π cm

B. 6π cm

C. 9π cm

D. 27π cm

7. Which statement is true about squares and parallelograms?

A. Each has opposite sides parallel and four right angles, but the parallelogram has congruent sides while the square does not.

B. Each has opposite sides congruent and parallel but a square has four right angles while a parallelogram does not necessarily have four right angles.

C. Each has opposite sides congruent and four right angles.

D. Each has opposite sides congruent and parallel, but a parallelogram has four right angles while a square does not.

8. The scale in a scale drawing Tracey made of her backyard is 1 inch = 12 feet. A line $3\frac{1}{2}$ inches long represents how many feet?

A. 36 ft

B. 40 ft

C. 42 ft

D. 48 ft

9. Find the surface area of a triangular prism whose bases are right triangles with legs 3 feet and 4 feet and hypotenuse is 5 feet and whose height is 2 feet.

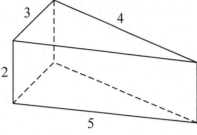

A. 14 square feet

B. 30 square feet

C. 36 square feet

D. 120 square feet

10. Find the volume of the prism.

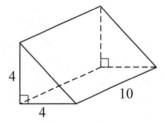

A. 18 in.³

B. 26 in.³

C. 80 in.³

D. 160 in.³

11. What reason would you use to state that $\triangle ABC \cong \triangle ADC$?

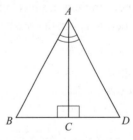

A. SSS (Side-Side-Side)

B. SAS (Side-Angle-Side)

C. ASA (Angle-Side-Angle)

D. AAS (Angle-Angle-Side)

12. Which two angles when added together always equals the measure of ∡1?

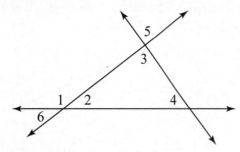

A. ∠2 + ∠6
B. ∠4 + ∠5
C. ∠3 + ∠4
D. ∠2 + ∠3

13. Given the following similar triangles △ABC ~ △DEF, what is the length of $\overline{DE}$?

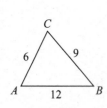 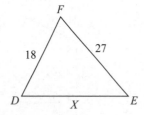

A. 12
B. 18
C. 24
D. 36

14. Mandi wants to wrap a present that is in a box with the dimensions of 4 × 5 × 6 inches. What is the minimum amount of wrapping paper needed to cover the box?

A. 64 square inches
B. 120 square inches
C. 148 square inches
D. 250 square inches

15. Which of the following is always TRUE?

A. A quadrilateral is a trapezoid.
B. A rectangle is a square.
C. A square is a rhombus.
D. A parallelogram is a kite.

16. Which of the following statements is TRUE?

A. A scalene triangle has no congruent sides.
B. An isosceles triangle is also an equilateral triangle.
C. An equilateral triangle has exactly two congruent sides.
D. A scalene triangle has exactly two congruent sides.

17. Points A, B, and C are on the circle O. What is the measure of $\overarc{ACB}$?

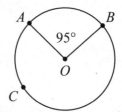

A. 95°
B. 190°
C. 265°
D. 275°

18. If a polygon is translated, which of the following characteristics of the polygon are the same?

I. Side lengths
II. Area
III. The coordinates of the vertices

A. I and II
B. I and III
C. II and III
D. I, II, and III

19. What is the distance between points $M(-3, -1)$ and $N(2, 3)$ on the graph below?

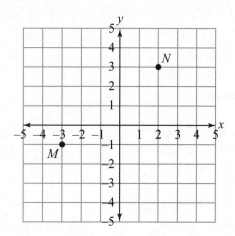

A. $\sqrt{5}$

B. $\sqrt{17}$

C. $\sqrt{41}$

D. $\sqrt{45}$

---

20. Which of the following represents the graph of the equation below?

$$y = -x^2 + 20$$

A.

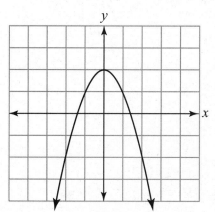

B.

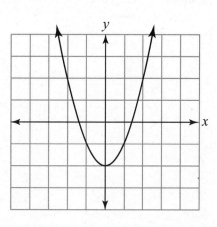

C.

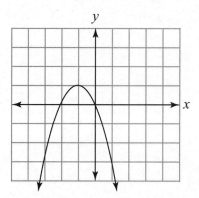

D.

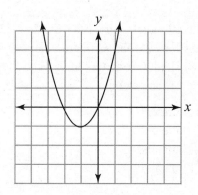

21. Which of the following transformations pro-duces a figure congruent to the original one?

   A. A transformation that changes the ordered pair for each vertex by adding 2 to the *x*-coordinate and dividing the *y*-coordinate by 2.
   B. A transformation that changes the ordered pair for each vertex by multiply-ing the *x*- and *y*-coordinates by 2.
   C. A transformation that changes the ordered pair for each vertex by adding 2 to the *x*-coordinate and multiplying the *y*-coordinate by 2.
   D. A transformation that changes the ordered pair for each vertex by adding ½ to the *x*-coordinate and subtracting ½ from the *y*-coordinate.

22. The points *A*, *B*, and *C* lie on circle *Q* below, in which $\overline{BC}$ is a diameter. In circle *Q*, what is the measure of angle *CAB*, in degrees?

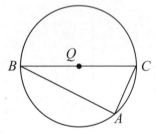

   A. 60°
   B. 90°
   C. 180°
   D. 360°

23. A tree 10 meters tall casts a shadow 25 meters long. At the same time, a person casts a shadow 5 meters long. Find the height of the person.

   A. 1.5 meters
   B. 2.0 meters
   C. 2.5 meters
   D. 2.7 meters

24. Find the volume of the cone with a radius of 5 inches and an altitude of 12 inches.

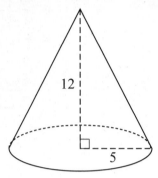

   A. 20π in.^{3}
   B. 40π in.^{3}
   C. 100π in.^{3}
   D. 300π in.^{3}

25. What is the slant height of the cone with a radius of 5 inches and an altitude of 12 inches?

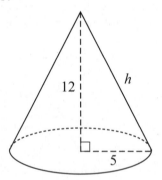

   A. 10 in.
   B. 12 in.
   C. 13 in.
   D. 26 in.

26. Find the area of the side of this house with the dimensions given as shown below.

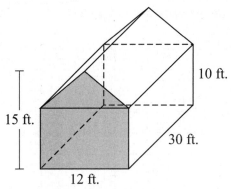

A. 100 square feet
B. 120 square feet
C. 150 square feet
D. 180 square feet

---

27. Given a right triangle with an acute angle of 45 degrees and a leg of length 26 units. Find the other leg.

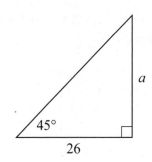

A. 18.4
B. 26
C. 36.8
D. 52

28. A support wire is connected to the top of a telephone pole. If the angle the support wire makes with the ground is 60° and the length of the support wire is 25 feet, how far is the base of the telephone from the support wire?

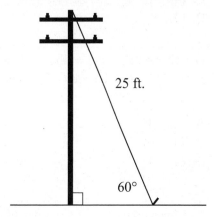

A. 5 feet
B. 10 feet
C. 12.5 feet
D. 15.5 feet

---

29. A telephone pole breaks and falls down as shown. What was the original height of the telephone pole?

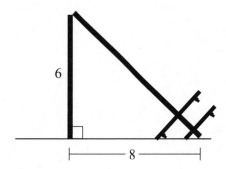

A. 60 feet
B. 24 feet
C. 16 feet
D. 14 feet

---

30. The diameter of a circle has endpoints $(2, -7)$ and $(-4, 1)$. What is the length of the diameter?

A. $6\sqrt{2}$

B. $\sqrt{10}$

C. $2\sqrt{10}$

D. 10

31. Find the midpoint of a segment whose endpoints are (2, – 6) and (10, 4).

    A. (6, – 1)
    B. (6, 1)
    C. (4, – 1)
    D. (4, 1)

32. Determine the center of the circle if the endpoints of a diameter are (– 2, 4) and (6, – 8).

    A. (2, – 12)
    B. (2, – 2)
    C. (4, – 6)
    D. (4, – 4)

33. Two lines contain opposite sides of a parallelogram. The equation for one line is $y = \frac{2}{3}x - 3$, and the second line is vertically shifted up 6 units. What is the equation of the second line?

    A. $y = \frac{2}{3}x + 3$

    B. $y = \frac{-2}{3}x + 3$

    C. $y = \frac{-2}{3}x - 3$

    D. $y = \frac{2}{3}x - 9$

34. $\triangle A'B'C'$ is the image of $\triangle ABC$ after a translation of 5 units left and 3 units down. What are the vertices for $\triangle A'B'C'$?

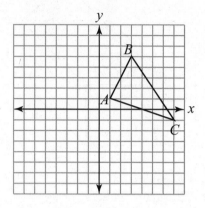

    A. $A'(1, 1)$, $B'(3, 5)$, and $C'(7, –1)$
    B. $A'(-4, -2)$, $B'(-2, 2)$, and $C'(2, -4)$
    C. $A'(-2, -4)$, $B'(2, -2)$, and $C'(-4, 2)$
    D. $A'(6, 4)$, $B'(8, 8)$, and $C'(12, 2)$

35. If the triangle ABC is reflected about the vertical line $x = 2$, what would be the new coordinates for A'?

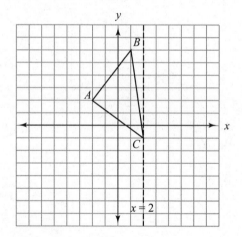

    A. (2, – 2)
    B. (6, 2)
    C. (2, 2)
    D. (6, – 2)

Answer Key

1. **B**	8. **C**	15. **C**	22. **B**	29. **C**
2. **D**	9. **C**	16. **A**	23. **B**	30. **D**
3. **C**	10. **C**	17. **C**	24. **C**	31. **A**
4. **B**	11. **C**	18. **A**	25. **C**	32. **B**
5. **C**	12. **C**	19. **C**	26. **C**	33. **A**
6. **B**	13. **D**	20. **A**	27. **B**	34. **B**
7. **B**	14. **C**	21. **D**	28. **C**	35. **B**

Answers Explained

1. **B** The formula from the reference sheet for the sum of the interior angles is $S = (n - 2)180°$, where S is the sum and n is the number of sides.
$$S = (n - 2)180°$$
$$S = (5 - 2)180°$$
$$S = (3)180°$$
$$S = 540°$$

2. **D** Using the formula for the area of a sector for the reference sheet.
$$A = \pi r^2 \left(\frac{\text{Degrees in corresponding arc}}{360°} \right)$$
$$A = \pi (12)^2 \left(\frac{60°}{360°} \right)$$
$$A = 144\pi \left(\frac{\overset{1}{\cancel{60°}}}{\underset{6}{\cancel{360°}}} \right)$$
$$A = 24\pi \text{ sq in.}$$

3. **C** Transforming the linear equation, $2x - 3y = 6$, into $y = mx + b$ is one method for graphing.
$$2x - 3y = 6$$
$$-3y = -2x + 6$$
$$y = \frac{2}{3}x - 2$$

    First, locate the y-intercept, -2. Apply the slope of $\frac{2}{3}$ by a vertical change of 2 and a horizontal change of 3. Choices B and C have the correct y-intercept. Choices C and D have the correct slope. The correct choice is C.

4. **B** Looking at the marks for congruency, the problem has side-angle-side congruency (SAS).

5. **C** The net having two circular parts helps to rule out all choices but the cylinder.

6. **B** The reference sheet has the formula for arc length.

Length of $\overarc{AB} = \dfrac{m\overarc{AB}}{360°}(2\pi r)$. Replace the given information for the measure of the arc, 120°, and the radius, 9 centimeters into the formula and simplify.

$$\text{Length of } \overarc{AB} = \frac{m\overarc{AB}}{360°}(2\pi r)$$

$$\text{Length of } \overarc{AB} = \frac{120}{360°}\left(2\pi(9)\right)$$

$$\text{Length of } \overarc{AB} = \frac{\overset{1}{\cancel{120}}}{\underset{3}{\cancel{360}}}\left(18\pi\right)$$

$$\text{Length of } \overarc{AB} = 6\pi \text{ cm}$$

7. **B** Parallelograms and squares do share the attributes of opposite sides parallel and congruent. Only squares must have four right angles.

8. **C** Since each inch represents 12 feet, 3 inches would represent 36 feet. A half inch would represent 6 feet. Add 6 feet to the 36 feet; the line represents 42 feet.

9. **C** Using the surface area of the triangular prism formula from the reference sheet: $T = 2B + Ph$. The B represents the area of the base. P represents the perimeter of base, and the h represents the height. In this triangle with sides of 3, 4, and 5 the top (and bottom) is a right triangle. The area of the triangle is $\frac{1}{2}(3)(4) = 6$.

The perimeter is 3 + 4 + 5 = 12. If we plug into the equation, $T = 2(6) + (12)(2) = 36$ sq ft.

10. **C** The formula for volume of a triangular prism:
$V = Bh$, where B is the area of a base and h is the height. The area of the triangle base is 8 and the height is 10 units. The volume is 80 cubic units.

11. **C** Two sets of congruent angles are labeled. The third part for congruency is not directly apparent. The two triangles share the side AC. Congruency can be stated because of angle-side-angle (ASA).

12. **C** Angle 1 is an exterior angle to the triangle and is a supplementary angle to angle 2. The interior angles, $\angle 2, \angle 3,$ and $\angle 4$, add to 180°.

$\angle 1 + \angle 2 = 180°$ and $\angle 2 + \angle 3 + \angle 4 = 180°$

$\angle 1 + \angle 2 = \angle 2 + \angle 3 + \angle 4$

$\angle 1 = \angle 3 + \angle 4$

13. **D** In similar triangles corresponding sides are proportional.

$$\frac{6}{18} = \frac{12}{x}$$

Solve the proportion using the means-extremes product rule.

$$\frac{\overset{1}{\cancel{6}}}{\underset{3}{\cancel{18}}} = \frac{12}{x}$$

$$x = 36$$

Notice that reducing the fraction made the calculations easier.

14. **C** The question is asking for the total surface area. From the reference sheet the formula is $T = 2B + Ph$. The B represents the area of the base. P represents the perimeter of base, and the h represents the height. Drawing a picture of this box and labeling the sides with the dimensions may be helpful. The formula uses the perimeter and area of the base. If you assign the base dimensions to be 4×5, the perimeter is 18 and the area is 20. The height will take on the last dimension of 6. Placing these values into the formula:

$T = 2B + Ph$
$T = 2(20) + (18)(6)$
$T = 40 + 108$
$T = 148$ sq in.

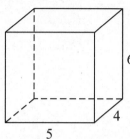

15. **C** Choice C is correct because each condition for a rhombus is satisfied by a square. Choice A—A quadrilateral is not always a trapezoid. Choice B—A rectangle is not always a square. Choice D—A parallelogram is a kite only when its adjacent sides are congruent.

16. **A** The definition of a scalene triangle is no congruent sides. An isosceles triangle is not required to have all sides equal (equilateral), just two or more sides congruent. Choice C is false because the equilateral triangle must have all sides equal.

17. **C** $\angle AOB$ is a central angle and the measure of arc AB ($m\,\overset{\frown}{AB}$) is equal to the measure of the central angle ($95°$). The entire circle contains $360°$. The measure of $\overset{\frown}{ACB}$ can be computed by subtracting $95°$ from $360°$.

18. **A** Translating a polygon moves the image. It does not change the length of the sides or the area.

19. **C** Using the distance formula from the reference sheet:

$$d = \sqrt{\left(x_2 - x_1\right)^2 + \left(y_2 - y_1\right)^2}$$
$$d = \sqrt{\left(2 - (-3)\right)^2 + \left(3 - (-1)\right)^2}$$
$$d = \sqrt{\left(2 + 3\right)^2 + \left(3 + 1\right)^2}$$
$$d = \sqrt{\left(5\right)^2 + \left(4\right)^2}$$
$$d = \sqrt{25 + 16}$$
$$d = \sqrt{41}$$

20. **A** In the quadratic, $y = -x^2 + 2$, the coefficient to the quadratic term is negative causing the graph to open down. The constant is 2 and this represents the y-intercept. Choice A is the only graph with a y-intercept of 2 and that opens down.

21. **D** To maintain congruency in transforming figures, you may only add or subtract from the vertices. Multiplying or dividing will enlarge or shrink a figure. Adding and subtracting will slide the figure from its current position to a new position but not change it dimensions.

22. **B** $\angle BAC$ is an inscribed angle and the measure of an inscribed angle is one-half the measure of its intercepted arc. Since $\overline{BC}$ goes through the center, it is a diameter and $m\overset{\frown}{BC}$ is 180 degrees. $m\angle BAC = \frac{1}{2} \cdot 180 = 90°$.

23. **B** This situation can be pictured as two similar triangles. In similar triangles the sides are proportional and can be solved with the means-extremes product rule. (Remember to reduce first.)

$$\frac{Shadow}{Height} = \frac{25}{10} = \frac{5}{x}$$

$$\frac{\overset{5}{\cancel{25}}}{\underset{2}{\cancel{10}}} = \frac{5}{x}$$

$$5x = 10$$

$$x = 2 \text{ meters}$$

24. **C** The volume formula for a cone is found on the reference sheet. The variable r represents the radius, and the height is represented by h.

$$V = \frac{1}{3}\pi r^2 h$$

$$V = \frac{1}{3}\pi (5)^2 (12)$$

$$V = \frac{1}{3}\pi 25 (12)$$

$$V = \frac{1}{\cancel{3}}\pi 25 \left(\overset{4}{\cancel{12}}\right)$$

$$V = 100\pi \text{ in.}^3$$

25. **C** The slant height, h, of the cone is the hypotenuse of the right triangle formed by the cone. This length can be calculated by using the Pythagorean theorem. The variable c represents the hypotenuse in this theorem.

$$
\begin{aligned}
a^2 + b^2 &= c^2 \\
(5)^2 + (12)^2 &= c^2 \\
25 + 144 &= c^2 \\
169 &= c^2 \\
13 &= c
\end{aligned}
$$

So, the slant height is 13 inches.

26. **C** This side of the house can be viewed as two basic geometric shapes, the rectangle and the triangle. The area of the rectangle is $\ell \times w = 12 \times 10 = 120 \text{ ft.}^2$. The triangle area is $\frac{1}{2}bh$. Note the overall height is given as well as the height of the rectangular side. Subtracting these two dimensions will yield the height of the triangle.

$$\frac{1}{2}bh = \frac{1}{2}(12)(5) = 30 \text{ ft.}^2$$

Together the area is 150 square feet.

27.   **B**   In a 45°-45°-90° triangle, the legs are equal.

28.   **C**   In a 30°-60°-90° triangle the smallest leg (opposite the 30° angle) is half the hypotenuse. This hypotenuse is 25 feet. The desired length is 12.5 feet.

29.   **C**   The toppled pole forms a right triangle. Using Pythagorean theorem to calculate the upper part of the pole,

$$a^2 + b^2 = c^2$$
$$(6)^2 + (8)^2 = c^2$$
$$36 + 64 = c^2$$
$$100 = c^2$$
$$10 = c$$

The question wants the original height. Add the toppled part with the part still standing (6 + 10 = 16 feet).

30.   **D**   Using the distance formula from the reference sheet,

$$d = \sqrt{\left(x_2 - x_1\right)^2 + \left(y_2 - y_1\right)^2}$$
$$d = \sqrt{\left(-4 - (2)\right)^2 + \left(1 - (-7)\right)^2}$$
$$d = \sqrt{\left(-6\right)^2 + \left(1 + 7\right)^2}$$
$$d = \sqrt{\left(-6\right)^2 + \left(8\right)^2}$$
$$d = \sqrt{36 + 64}$$
$$d = \sqrt{100} = 10$$

31.   **A**   Using the midpoint formula from the reference sheet:

$$\left(\frac{\left(x_2 + x_1\right)}{2}, \frac{\left(y_2 + y_1\right)}{2} \right)$$
$$\left(\frac{(10 + 2)}{2}, \frac{\left(4 + (-6)\right)}{2} \right)$$
$$\left(\frac{12}{2}, \frac{-2}{2} \right)$$
$$\left(6, -1 \right)$$

32.   **B**   The center of a circle is the midpoint of the diameter. Using the midpoint formula from the reference sheet,

$$\left(\frac{\left(x_2 + x_1\right)}{2}, \frac{\left(y_2 + y_1\right)}{2} \right)$$
$$\left(\frac{\left(6 + (-2)\right)}{2}, \frac{(-8 + 4)}{2} \right)$$
$$\left(\frac{4}{2}, \frac{-4}{2} \right)$$
$$\left(2, -2 \right)$$

33. **A** The opposite sides of a parallelogram are parallel to each other. The slope of parallel lines are same. Only choices A and D have the same slope as the given line. A vertical shift of 6 units up increases the y-intercept by 6 units. Choices A and B have an increase in their y-intercepts of 6 units. Only choice A meets both requirements.

34. **B** A translation of 5 units left subtracts 5 from each x-coordinate. A translation of 3 units down subtracts 3 from each y-coordinate. Choice B matches these requirements.

35. **B** A reflection about a vertical line places the image on the other side of the line. Only the x-coordinate changes. The vertical position does not change. In this problem the point A is 4 units from the line of reflection. This point will be 4 units on the opposite side. Point A starts at $(-2, 2)$ and reflects to $(6, 2)$, choice B.

Structure and Logic

When you complete your study of Strand 5 you will be able to:

- Determine whether a given procedure is valid
  - Simplifying expressions
  - Solving equations
  - Solving inequalities
- Work with algorithms
- Work with if-then statements
- Write, identify, and analyze valid conjectures
- Create and critique valid inductive and deductive arguments
- Identify or construct counterexamples
- State the inverse, converse, or contrapositive of a given statement and determine its truth
- Construct a simple formal or informal proof
- Verify characteristics of a given geometric figure

In Strands 1 through 4, we reviewed and/or developed concepts related to mathematics. But it isn't enough just to have concepts and skills—we have to draw on our knowledge of those concepts and skills to solve new and unfamiliar problems.

The title of Strand 5 is Structure and Logic. It is an attempt to join a step-by-step problem-solving approach together with the "what-if" attitude. Problem solving, critical thinking, and mathematical justification are important skills to develop. They appear in all areas of mathematics as well as other areas of study. They should not be reserved for special activities or special times, but should be a part of classroom discussions no matter what the topic.

Logic Principles

Think about how you think when answering questions. This is not simply confined to doing math problems, but problem solving in general.

- Are you a facts and figures type of person? Do you like to be given the necessary details and formulas so as to answer the posed question and then move to the next question?

- Do you answer a question with a question? When you are given a formula or a statement, do you like to analyze the solution as well as the problem?

The first set of questions above describes a doer; the second set of questions more accurately describes a thinker. For some students logical thinking and drawing conclusions is difficult, but for others it can be quite rewarding.

It's important that answers make sense and that they are reasonable. Your answers have to fit into real life. For example, if you work a problem asking for the amount of water needed to fill a swimming pool and your answer is 10 quarts of water, would that be reasonable? You probably couldn't take a bath in 10 quarts of water let alone fill a swimming pool.

To help determine whether an answer is reasonable or not, you might ask yourself these questions:

- Does this answer make sense?

- Could this really happen?

- Is the answer found in the choices given?

To those ends, Strand 5 contains problems that ask you to use reasoning and logic to solve mathematical problems in real situations. It contains problems that ask you to evaluate the steps used in finding your solution. The procedures used to solve problems need to be carefully followed so that the solutions are mathematically correct.

Terms Unique to Strand 5

A **valid argument** is an argument that is correctly inferred or deduced from a given statement. An **invalid argument** reaches a conclusion by arguing incorrectly from a given statement.

VALID:

If Janelle gets a job, then she can afford to buy a car. Janelle gets a job; therefore, she can afford to buy a car.

These statements represent a valid argument. If the initial statement is true, the second statement logically follows.

INVALID:

If Janelle gets a job, then she can afford to buy a car. Janelle can afford to buy a car; therefore, she got a job.

These statements represent an invalid argument. Even if the initial statement is true, the second statement is not necessarily true. Janelle's parents may have given her the money to buy the car. The conclusion of buying a car does not mean she got a job.

Notice that the topic of the if clause became the topic of the conclusion. Usually this is an indication that there may be a problem, and you need to examine the resulting statement very carefully. Ask yourself, is the second statement a direct result of the first?

A **premise** is a given statement that is accepted as true. The initial sentence, "If Janelle gets a job, she can afford to buy a car," is an example of a premise. If the premise is initially a false statement, it would undermine all the subsequent conclusions. It is possible to begin with a false statement and still develop true logical conclusions. Below is an example of a false statement leading to a true conclusion:

I looked out the window and saw that the ground was wet. I concluded that it rained. Therefore, I do not have to water my lawn.

My premise that it rained last night may be false, but my conclusion that I do not need to water the lawn is still true. The premise is based upon the ground being wet, which may or may not have been caused by rain.

As you begin to solve a problem, you have to decide what you are given and what you want to find. The given information and what you want to find can be written in the form of a conditional.

A **conditional** is a logical compound statement that has two parts: a hypothesis and a conclusion. The **hypothesis** is the first part of the compound statement that sets the stage for reaching the conclusion. Often a conditional is written in if-then form where the if clause is called the hypothesis and the then clause is called the conclusion. The **conclusion** is a result of the hypothesis.

> If—Hypothesis
> Then—Conclusion

Examples:

- If <u>ice is floating in the lake</u>, then <u>the water is cold</u>.
            hypothesis   conclusion

- If <u>I want to go to a four year college</u>, then <u>I need to take four years of mathematics</u>.
            hypothesis   conclusion

- If <u>two angles are supplementary</u>, then <u>the sum of their angles is 180 degrees</u>.
            hypothesis   conclusion

- <u>You will do well on the test</u> if <u>you eat a hearty breakfast</u>.
            conclusion   hypothesis

Note that, in the last example, the conclusion was stated before the hypothesis. The hypothesis is related to the if part of the conditional, even if it comes near the end of the sentence.

So, how do we verify a conditional? There are times when you can immediately say a conditional is false.

Examples:

Conditional false statements	Reason
If an angle is 45°, then it is obtuse.	Contradicts the definition of obtuse.
If Mary received an A on her math test, then she failed the test.	An A is not typically a failing grade.
If two angles are congruent, then they are right angles.	Congruent angles are not required to be right angles; instead, they have the same measure.

To show that a conditional is true, we can use some statements that are related to the conditional. Sometimes proving that a variation of the conditional is true will prove that the given conditional is true. The variations of the given conditional are called its inverse, converse, and contrapositive. Before we define these words, we need to define the word "negation."

The **negation** of a statement is the opposite of the original statement. If the statement is true, then the negation is false; if the statement is false, the negation is true. To negate the if part of a conditional is to replace the clause after the if with a new clause that is not true. Adding the word "not" can negate the clause.

Statement: If *A*, then *B*.
Inverse: If not *A*, then not *B*.
Converse: If *B*, then *A*.
Contrapositve: If not *B*, then not *A*.

Statement	Negation
The cat is black.	The cat is not black.
The ball is not red.	The ball is red.
Angle A is a right angle.	Angle A is not a right angle.

The **inverse** of a conditional is a new sentence obtained by negating both the hypothesis and the conclusion of the given conditional.

GIVEN CONDITIONAL	If two lines are perpendicular, then they form right angles.
INVERSE	If two lines are NOT perpendicular, then they do NOT form right angles.

The **converse** of a conditional is a new sentence obtained by exchanging the hypothesis and the conclusion of the given conditional.

GIVEN CONDITIONAL	If two lines are perpendicular, then they form right angles.
CONVERSE	If two lines form right angles, then they are perpendicular.

Notice that it is important to maintain the subject of the conditional sentence. The subject, "two lines," stays at the beginning of each of the sentences. The parts that reverse are "form right angles" and "are perpendicular."

If **two lines** form right angles, then they are perpendicular.

The **contrapositive** of a conditional is a new sentence obtained by exchanging the negation of the conclusion with the negation of the hypothesis of the given conditional.

GIVEN CONDITIONAL	If two lines are perpendicular, then they form right angles.
CONTRAPOSITIVE	If two lines do NOT form right angles, then they are NOT perpendicular.

Here is a chart to help learn these new logical conditionals.

Statement	Action	Example
Conditional	Given	If *A*, then *B*
Inverse	Both NOT	If NOT *A*, then NOT *B*
Converse	Switch order	If *B*, then *A*
Contrapositive	Both NOT and switch order	If NOT *B*, then NOT *A*

Example:

Type	Sentence
Given conditional	If ice is floating in the lake, then the water is cold.
Inverse	If ice is NOT floating in the lake, then the water is NOT cold.
Converse	If the water is cold, then ice is floating in the lake.
Contrapositive	If the water is NOT cold, then ice is NOT floating in the lake.

The given conditional appears logical and can be true. But, ice alone floating doesn't logically imply cold water. Swimming in cold water doesn't mean that ice will float by you.

Example:

Type	Sentence
Given conditional	If two angles are straight angles, then they are congruent.
Inverse	If two angles are not straight angles, then they are not congruent.
Converse	If two angles are congruent, then they are straight angles.
Contrapositive	If two angles are not congruent, then they are not straight angles.

In the previous examples, not all of the sentences are true. Just because a conditional sentence is true, it does not mean its inverse and converse are true. In the inverse and the converse examples above, the two congruent angles can be angles other than straight angles. Any two angles that have the same measure are congruent to each other.

Example:

Type	Sentence
Given conditional	If $m\angle A = 99°$, then $\angle A$ is obtuse.
Inverse	If $m\angle A \neq 99°$, then $\angle A$ is NOT obtuse.
Converse	If $\angle A$ is obtuse, then $m\angle A = 99°$.
Contrapositive	If $\angle A$ is NOT obtuse, then $m\angle A \neq 99°$.

Again, just because the given conditional was true, the inverse and converse are not necessarily true. In the inverse, just because $m\angle A$ is not 99° does not mean it is not obtuse. Looking at the converse, if $m\angle A$ is obtuse, it does not have to be 99°.

Example:

Type	Sentence
Given conditional	If the sides of a triangle are equal, then the triangle is equilateral.
Inverse	If the sides of a triangle are NOT equal, then the triangle is NOT equilateral.
Converse	If a triangle is equilateral, then the sides of the triangle are equal.
Contrapositive	If a triangle is NOT equilateral, then the sides of the triangle are NOT equal.

In this example, each statement is true. Do you recognize this relationship as the definition of equilateral triangles?

Example:

Type	Sentence
Given conditional	If two angles are supplementary, then the sum of their angles is 180°.
Inverse	If two angles are NOT supplementary, then the sum of their angles is NOT 180°.
Converse	If the sum of two angles is 180°, then they are supplementary.
Contrapositive	If the sum of two angles is NOT 180°, then they are NOT supplementary.

In the preceding example, the given conditional, inverse, converse, and contrapositive are all true. All four statements can be used in a logical argument. It makes logical sense to associate supplemental angles with the measure of 180 degrees. Good definitions are written in a form such that all four statements are true.

When starting with a premise, we assume truth. What if we are uncertain? What if we are curious? We satisfy our curiosity and questions in the form of a proof or argument. A proof (deductive reasoning) is a series of statements that lead to a conclusion. Inductive reasoning is an argument that is based on observations.

Deduction and Induction

Deduction—Logical steps

Deduction is the process of reasoning with a series of logical steps in which a conclusion is drawn directly from a statement or a set of statements that are assumed to be true. Below is an example of a simple deductive argument.

The rule at your school is that you must attend all of your classes in order to participate in sports after school. Jim went to all of his classes on Monday. Therefore, Jim was able to play in a soccer game after school on Monday.

The series of steps used in the deductive process are generally written in if-then format. They must be arranged in a certain order. This order must follow a pattern where the conclusion of one statement is the hypothesis of the next statement. Below are examples that demonstrate the connection between each conclusion and its matching hypothesis.

Examples: If A, then B.
If B, then C.
If C, then D.
If D, then E.

If $\angle A$ and $\angle B$ are right angles, then m$\angle A = 90°$ and m$\angle B = 90°$.
If m$\angle A = 90°$ and m$\angle B = 90°$, then m$\angle A$ = m$\angle B$.
If m$\angle A$ = m$\angle B$, then $\angle A \cong \angle B$.

If $ABCD$ is a square, then $ABCD$ is a rectangle.
If $ABCD$ is a rectangle, then $ABCD$ is a parallelogram.
If $ABCD$ is a parallelogram, then $AB = DC$ and $AD = BC$.

On the AIMS test, you may be asked to put a group of related if-then statements in logical order. This ordering will allow the statements to flow from one statement to another.

First, look at each statement for a phrase that appears only in a hypothesis. This statement is the first statement in the logical ordering of the sentences. The remaining statements follow a logical order. Each hypothesis connects to the preceding conclusion.

Example: Out of order:
If car, then dog.
If dog, then egg.
If apple, then boy.
If boy, then car.

In order: If apple, then boy. | Apple appears only in a hypothesis. This should be first.

In order: If apple, then boy. / If boy, then car. | Now look at the conclusion of the first statement. This conclusion is the hypothesis of the next statement.

In order: If apple, then boy. / If boy, then car. / If car, then dog. | Now look at the conclusion of the second statement. This conclusion is the hypothesis of the next statement.

In order: If apple, then boy. / If boy, then car. / If car, then dog. / If dog, then egg. | The hypothesis of the remaining statement is the conclusion of the third statement.

Below are two additional examples of putting conditionals into logical order. One involves colors and the other is about geometry.

Examples: Out of order:
If pink, then green.
If red, then blue.
If blue, then pink.
If green, then purple.

In order:
If red, then blue.
If blue, then pink.
If pink, then green.
If green, then purple.

Out of order:
If $\angle CDA$ is a right angle, then m$\angle CDA = 90°$.
If $\overline{CD}$ is an altitude drawn to $\overline{AB}$, then $\overline{CD} \perp \overline{AB}$.
If $\overline{CD} \perp \overline{AB}$, then $\angle CDA$ is a right angle.

In order:
If $\overline{CD}$ is an altitude drawn to $\overline{AB}$, then $\overline{CD} \perp \overline{AB}$.
If $\overline{CD} \perp \overline{AB}$, then $\angle CDA$ is a right angle.
If $\angle CDA$ is a right angle, then m$\angle CDA = 90°$.

A second type of argument used in reaching conclusions is induction. It doesn't necessarily follow the if-then format found in deductive reasoning. Instead, the emphasis is on observations.

Induction—Observations

Induction is the process of making a generalization based on observation of specific cases or patterns. Below is an example of a simple inductive argument.

Example: Describe the pattern in the numbers

$$-7, -21, -63, -189, \ldots$$

Notice that each number in the pattern is three times the previous number.

$$-7 \quad -21 \quad -63 \quad -189$$
$$\cdot 3 \quad \cdot 3 \quad \cdot 3$$

By observing the pattern, a generalization can be made that the sequence of numbers can be continued by multiplying the last term by three to obtain the next term.

Conjecture—Unproven statement

A **conjecture** is an unproven statement based on observations or an anticipated result. It is a statement that says you think something is true—you haven't proved it yet; you just think it's true. A proof begins with a premise and a proposed conclusion related to the premise. This conclusion is called a conjecture.

What would you say to the statement that the square of a number is always larger than the original number? It seems like an obviously true statement. The square of 3 is 9; 9 is bigger than 3. So, is our conjecture always true?

The key in the last statement is the word "always." If you found five values that hold true, are you convinced that your conjecture is "always" true? How about 50 values or 100 values that make the conditional true? Exhaustion or apathy is sure to set in. What if you were to find just one number that makes the statement false? You would have to change the word "always" because you now know of an example that doesn't support the initial conjecture. That one example is called a counterexample.

Counterexample—A single example to show the conjecture is false.

A **counterexample** is an example that shows that a conjecture is not always true. We need to find only one false example and we can declare the conjecture to be false.

Is $(0.5)^2 > 0.5$ true?

$0.25 > 0.5$ is false.

Therefore, 0.5 is a counterexample.

In this case, the square of the number is not larger than the original number. So, even though the conjecture is true for some numbers, we found one counterexample and can declare the conjecture is not always a true statement.

Let's try another. Wouldn't it seem reasonable to say that the sum of two numbers is larger than either number? So, our conjecture would be "If I add two numbers together, then the sum is larger than either of the two original numbers."

$$x + y \geq x \text{ and } x + y \geq y$$

If I want to prove that this is not always true, I need to find a counterexample. The tendency is to choose counting numbers first, such as $3 + 7 = 10$.

$$3 + 7 \geq 3 \text{ and } 3 + 7 \geq 7$$

It seems like every group of counting numbers we pick will generate a true statement. Let's consider some other numbers in our number system. Suppose our numbers are –2 and 5.

$$-2 + 5 \geq -2 \quad \text{and} \quad -2 + 5 \geq 5$$
$$3 \geq -2 \quad \text{and} \quad 3 \geq 5$$

Three is bigger than –2, but it is not bigger than 5. So, our initial conjecture is false.

All the examples of conjectures so far have been related to numbers. Let's consider a conjecture related to the angles formed when two lines intersect. Remember, in pairs these angles are called vertical angles.

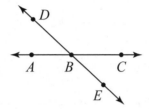

$\angle ABD$ and $\angle EBC$ are vertical angles.
$\angle DBC$ and $\angle ABE$ are vertical angles.

Are the pairs of vertical angles congruent? That sounds like a conjecture. Using a protractor on the diagram, we could measure the pairs of vertical angles to show they have the same value. However, they may not have exactly the same value due to lack of accuracy. Even if they did have the same measures, this is not enough to prove it true for all of the possible pairs of vertical angles that could be formed by two intersecting lines.

Does this mean we can't come up with a counterexample? How do we prove or disprove our conjecture that vertical angles are equal? In mathematics, we can prove or disprove conjectures using deductive reasoning. Earlier, we defined deductive reasoning; **deductive reasoning** is the process of reasoning from a set of statements that are assumed to be true.

Let's try to prove the conjecture that when two lines intersect, the vertical angles are equal.

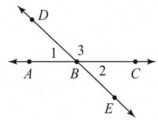

	Statement		Justification
1.	$\overleftrightarrow{AC}$ and $\overleftrightarrow{DE}$ intersect at B forming vertical angles 1 and 2 as shown in the diagram.	1.	Given condition
2.	$\angle ABC$ and $\angle DBE$ are straight angles.	2.	Straight lines form straight angles. (Some textbooks call them linear pairs.)
3.	$m\angle ABC = m\angle DBE$	3.	If two angles are straight angles, then they are equal.
4.	$m\angle 1 + m\angle 3 = m\angle 3 + m\angle 2$	4.	Substitution
5.	$m\angle 1 + m\angle 2$	5.	Subtraction

Using deductive reasoning, we have proved that pairs of vertical angles are equal. Our conjecture is true.

The conjecture about vertical angles was already something that we wondered about. Sometimes we arrive at conjectures by experimenting and looking at patterns. Earlier, this was defined as induction. Look at the chart below showing the number of segments that can be drawn between three, four, five, and six noncollinear points. What conjecture could we make about the number of segments to connect seven points?

Number	3	4	5	6	7
Picture					
Number of segments	3	6	10	15	?

Do you recognize a pattern? The number of segments increases every time, but it does not increase the same amount each time. However, there is a pattern in how the number of segments increases. The difference in the number of segments increases by one each time.

Number of segments   3   6   10   15   ?

Difference in terms   3   4   5   ?

So our conjecture is that you can connect seven noncollinear points with 21 different segments.

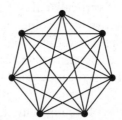

What would be your conjecture for eight noncollinear points?

In conclusion, a valid deductive argument is one that is concluded using previously accepted statements. An inductive argument is based on observations.

To develop our ideas about how an inductive argument can lead to a deductive proof, let's consider the perimeter and area of a square. Is there ever a time when the area of a square is the same numeric measure as the perimeter of the same square?

Let's look at this inductively. The following table lists the side, perimeter, and area of several squares.

Length of side of square	Perimeter	Area
1	4	1
2	8	4
3	12	9
4	16	16
5	20	25
6	24	36
⋮	⋮	⋮
s	$4s$	s^2

We can conclude from the table that when the length of the side is 4, the perimeter and the area are numerically equal. Remember the units on the area and the perimeter are different. If the side has a length of 4 inches, the perimeter is 16 inches, but the area is 16 inches squared.

We found one instance when the perimeter and the area are the same. Is that the only time? In looking at the table, notice that as the length of the side increases, the area increases at a faster rate than the perimeter. This would suggest that the side length of 4 is the only time the perimeter and area have the same numeric measure.

Reaching a conclusion inductively is not sufficient to conclude that a statement is always true. It is necessary to complete a deductive proof using a series of logical steps before you can be certain that your conjecture is true.

The formulas for the perimeter and area of a square are in the last line of the table. Our original conjecture asked if there was ever a time when the perimeter and area were equal. Now our new conjecture asks, "Is this the only time they are equal?" Let's create an algebraic equation stating that they are equal and see if there is a solution for this equation.

$$s^2 = 4s$$
$$s^2 - 4s = 0$$

This is a quadratic equation and can be solved using the quadratic formula.

$$s^2 - 4s = 0$$
$$a = 1, b = -4 \text{, and } c = 0$$

$$s = \frac{-b \pm \sqrt{b^2 - 4ac}}{2a}$$

$$s = \frac{-(\ ) \pm \sqrt{(\ )^2 - 4(\ )(\ )}}{2(\ )}$$

$$s = \frac{-(-4) \pm \sqrt{(-4)^2 - 4(1)(0)}}{2(1)}$$

$$s = \frac{4 \pm \sqrt{16-0}}{2}$$

$$s = \frac{4 \pm 4}{2}$$

$$s = \frac{8}{2} = 4 \text{ or } \frac{0}{2} = 0$$

The solution to the quadratic equation confirms our conjecture that the perimeter of a square and its area are equal if $s = 4$ or $s = 0$. However, a square with side length of zero does not exist. So $s = 0$ is not a reasonable answer to our conjecture; the only time the perimeter and area are numerically equal is when the length of the side of the square is 4.

Sometimes you begin to solve a real-life problem by applying logic. In the following problem, you are not given many facts. You can begin by using logic to help you decide what concepts you would use to solve the problem. As always, it is important to check your answer to see if it is reasonable.

Example: Suppose your dog wanders away from your house. If you guess that he has been missing for 30 minutes and he walks at a rate of one mile an hour in any direction, how large will the search area be at that time?

The fact that we are given a rate and an approximate time suggests that we could calculate the distance our dog travels in the given time by using the formula, "distance equals rate times time" ($d = rt$). Because he can travel in any direction, logic suggests a distance from a given point, which would form a circular search area and the distance is the radius of the circle.

Calculating the distance:

$d = rt$

$d = 1 \cdot \dfrac{1}{2}$

$d = \dfrac{1}{2}$ mile (the radius)

Area of a circle:

$A = \pi r^2$

$A = \pi \left(\dfrac{1}{2} \right)^2$

$A = \pi \cdot \dfrac{1}{4}$

$A \approx \dfrac{22}{7} \cdot \dfrac{1}{4}$

$A \approx \dfrac{11}{14}$ mile^2

This value is almost a square mile. That's a large search area.

Example: You are flying a kite, and a friend says, "The kite looks like it is a mile high in the sky." A quick analysis of the situation, common logic, and some geometry leads me to say that is not correct.

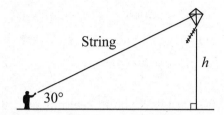

What facts do you know? You know that the roll holds 500 feet of string. You think the angle at which you are holding the string is approximately 30°. Using your knowledge about 30°-60°-90° triangles, you logically can conclude that the height of the kite is approximately 250 feet. Remember in a 30°-60°-90° triangle, the leg opposite the 30-degree angle is half the hypotenuse (the string). Obviously, your friend tends to exaggerate.

Processes, Procedures, and Algorithms

You will be asked questions on the AIMS test that require you to determine whether given procedures are valid. Those procedures might involve simplifying an arithmetic or algebraic expression, or they might involve solving an equation or inequality. Your success in answering these questions depends on applying the skills you have learned in Strands 1 through 4. On the test you will be given a question with several "solutions" from which you will have to choose the correct procedure. Below are a few examples of procedures that have been done correctly and that are similar to those that you might be asked to evaluate.

Examples:

$$-8x+2(3x+4) \geq 10$$
$$-8x+6x+8 \geq 10$$
$$-2x+8 \geq 10$$
$$-2x \geq 2$$
$$x \leq -1$$

- Be careful to use the distributive property accurately.
- Use the laws of signed numbers to combine like terms.
- Add or subtract from both sides of the inequality correctly.
- Remember when dividing by a negative the sign is reversed.

$$7(4^2-5)+2(7+4)$$
$$7(16-5)+2(11)$$
$$7(11)+2(11)$$
$$77+22$$
$$99$$

- Remember to use the rules for order of operations.
- Caution: $4^2 = 4 \cdot 4$
- Use the laws of signed numbers to combine like terms.

$$2\left(a^2 - 2ab + b^2\right) - 4\left(a+b\right)^2$$

$$2\left(a^2 - 2ab + b^2\right) - 4\left(a+b\right)\left(a+b\right)$$

$$2\left(a^2 - 2ab + b^2\right) - 4\left(a^2 + 2ab + b^2\right)$$

$$2a^2 - 4ab + 2b^2 - 4a^2 - 8ab - 4b^2$$

$$-2a^2 - 12ab - 2b^2$$

- Square the binomial.
  $(a + b)^2 = (a + b)(a + b)$
- Use the appropriate procedure to multiply $(a + b)(a + b)$.
- Distribute the coefficient over each polynomial.
- Use the laws of signed numbers to combine like terms.

The AIMS test asks you to determine whether a given procedure is valid. Generally, you will be given a problem that is both correctly and incorrectly solved. You will need to study each procedure to determine which one is correct. Below are examples of typical errors that students often make.

Example: $(2 - 8)3^2 \div 3(2)$

$$\left(2-8\right)3^2 \div 3(2)$$

$$\left(-6\right)9 \div 3(2)$$

$$-54 \div 3(2)$$

$$-18\left(2\right)$$

$$-36$$

This procedure is correct. It contains no mathematical mistakes, and it accurately follows the order of operations.

$$\left(2-8\right)3^2 \div 3(2)$$

$$\left(-6\right)9 \div 3(2)$$

$$-54 \div 3(2)$$

$$-54 \div 6$$

$$-9$$

In this procedure there is an error as you move from line three to line four. Order of operations requires that you complete multiplication and divisions from left to right. On this step, the product of 3 times 2 was incorrectly performed before the division of −54 divided by 3.

$$\left(2-8\right)3^2 \div 3(2)$$

$$\left(-6\right)6 \div 3(2)$$

$$-36 \div 3(2)$$

$$-12\left(2\right)$$

$$-24$$

In this procedure there is an error as you move from line one to line two. The error is that three squared is nine, not six.

Example: $2(x + 4) = 6x - 16$

$$2\left(x+4\right) = 6x - 16$$

$$2x + 4 = 6x - 16$$

$$20 = 4x$$

$$5 = x$$

There is an error as you move from line one to line two. You should take 2 times the x as well as the 4. This is an application of the distributive property.

$$2(x+4)=6x-16$$

$$2x+8=6x-16$$

$$-4x=-8$$

$$x=2$$

In this procedure, there is an error as you move from line two to line three. When subtracting 8 from both sides of the equation, an error was made. Line three should read –4x = –24.

$$2(x+4)=6x-16$$

$$2x+8=6x-16$$

$$8x=-24$$

$$x=-3$$

In this procedure, there is an error as you move from line two to line three. When subtracting 6x from both sides of the equation, an error was made. Line three should read –4x = –24.

$$2(x+4)=6x-16$$

$$2x+8=6x-16$$

$$24=4x$$

$$6=x$$

This procedure is correct. It contains no mathematical mistakes, and it accurately follows the algorithm for solving equations.

Algorithm

An **algorithm** is a set of step-by-step instructions for completing a task.

Algorithm—Steps to follow

Examples:

- The instructions for finding the average of a group of numbers—add the numbers and divide by the number of numbers—is an example of an algorithm.

- If you were to give directions to a friend about how to get to the mall from school, you would describe a path that your friend could follow to get there. Your directions may be different than the directions another person could give to your friend. Many times there is more than one way of accomplishing a task. However, it is still a step-by-step process (turn left, drive 1 mile, turn right, etc.). These directions are the basis of an algorithm.

An algorithm, as mentioned earlier, is a step-by-step process for solving a problem. On the AIMS test, you may be asked questions concerning algorithms. Those questions may be written in different ways. You may be asked to select an appropriate algorithm. Below is an example of this.

Can we show that any three points, *A*, *B*, and *C*, on the same surface determine only one line? Let's use the slope algorithm to show this. Just as a reminder, the algorithm for slope is to find the change in the *y*-coordinates over the change in the *x*-coordinates. First, let's find the slope of the line through *A* and *B*. Second, let's find the slope of the line through *B* and *C*. Our conjecture is that if the slopes are the same, then the points are collinear.

Given three points—*A*(0, 0), *B*(4, 3), and *C*(8, 6).	
Slope $\overline{AB}$ $$m=\frac{3-0}{4-0}$$ $$m=\frac{3}{4}$$	Slope $\overline{BC}$ $$m=\frac{6-3}{8-4}$$ $$m=\frac{3}{4}$$

Our conjecture is true; the points are collinear.

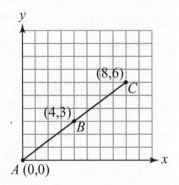

Can we apply the slope algorithm and our conjecture to four points, A, B, C, and D, on the same surface? First, let's find the slope of the line through A and B. Second, let's find the slope of the line through C and D. Our conjecture is that if the slopes are the same, then the points are collinear.

Given four points—$A(1, 2)$, $B(4, 7)$, $C(4, 2)$, and $D(7, 7)$.	
Slope $\overline{AB}$ $m = \dfrac{7-2}{4-1}$ $m = \dfrac{5}{3}$	Slope $\overline{CD}$ $m = \dfrac{7-2}{7-4}$ $m = \dfrac{5}{3}$

Is this sufficient to conclude that A, B, C, and D are collinear? The four points are graphed below. Are they collinear?

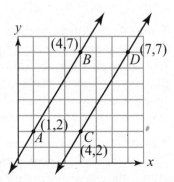

No, they are not. In fact, they are points on parallel lines. The slopes of parallel lines are equal. When using this slope algorithm to prove that points are collinear, the segments must share endpoints.

Example:

To show that A, B, C, and D are collinear using the slope algorithm, you could show that the slopes of $\overline{AB}$, $\overline{AC}$, and $\overline{CD}$ are equal.

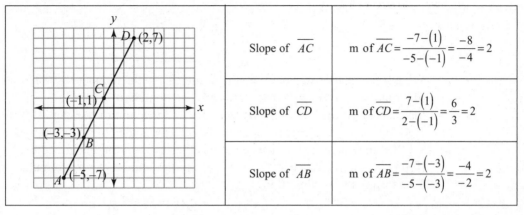

Slope of $\overline{AC}$	m of $\overline{AC} = \dfrac{-7-(1)}{-5-(-1)} = \dfrac{-8}{-4} = 2$
Slope of $\overline{CD}$	m of $\overline{CD} = \dfrac{7-(1)}{2-(-1)} = \dfrac{6}{3} = 2$
Slope of $\overline{AB}$	m of $\overline{AB} = \dfrac{-7-(-3)}{-5-(-3)} = \dfrac{-4}{-2} = 2$

Notice the segments do not all share the same endpoint. It is only necessary that each segment share an endpoint with one other segment.

You may be asked to determine the purpose of a defined algorithm. Follow the steps in the algorithm below and try to discover the purpose of this algorithm.

1. Find the area of the base of a three-dimensional figure by multiplying π times the radius squared.
2. Multiply the altitude of the figure times the product in step 1.
3. Label the units as cubic units.

In step 1, did you notice that the algorithm used a radius and π? That means the base must be a circle. This means you could be working with a cone or a cylinder.

In step 3, the label on the answer is in terms of cubic units. This means you are finding volume. Let's refer to the formulas on the reference sheet that help us find volume.

Volume of a cylinder: $V = \pi r^2 h$	Volume of a cone: $V = \dfrac{1}{3}\pi r^2 h$

Because the algorithm did not include the factor, $\dfrac{1}{3}$, we must be finding the volume of a cylinder.

Here is another algorithm. What is its purpose?

1. Select a vertex of a given polygon. Draw all the diagonals that can be drawn from that vertex.
2. Count the number of triangles formed.
3. Multiply the number of triangles found in step 2 times 180.
4. Label the answer in degrees.

A diagram that includes the diagonals would make it easier to visualize the diagonals and see the triangles that are formed.

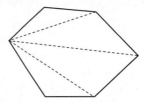

What is the relationship between the triangles and the number, 180? The sum of the angles of a triangle is 180°. When you multiply the number of triangles times 180, you are finding the sum of the angles of the triangles and the sum of the angles of the polygon. This algorithm is used to find the sum of the interior angles of a given polygon.

You may be asked to determine whether two algorithms are equivalent. The formula (found on the reference sheet) for finding the volume of a rectangular prism is $V = Bh$. This formula is an algorithm.

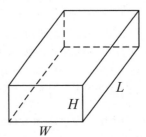

Look at the three algorithms below. Are they an equivalent to the algorithm on the reference sheet?

1. Find the area of the base of the box ($W \cdot L$).
2. Multiply the area of the base by the height H.
3. Label using units cubed.

1. Find the area of the right face of the box ($L \cdot H$).
2. Multiply the area of the right face by the width W.
3. Label using units cubed.

1. Find the area of the front face of the box ($W \cdot H$).
2. Multiply the area of the front face by the length L.
3. Label using units cubed.

All three are equivalent algorithms and can be used to find the volume. A rectangular prism is unique in the family of prisms because any of its faces can be considered to be a base.

The very last performance objective in Strand 5 asks the student to verify characteristics of a given geometric figure. To do this, we use various coordinate geometry formulas to show that segments are parallel, perpendicular, or congruent.

Given quadrilateral $ABCD$ with the following coordinates, verify that the opposite sides of the quadrilateral are parallel.

$A(1, 2)$, $B(4, 7)$, $C(4, 2)$, and $D(7, 7)$.

Slope of $\overline{AB}$	$m = \dfrac{7-2}{4-1} = \dfrac{5}{3}$
Slope of $\overline{CD}$	$m = \dfrac{7-2}{7-4} = \dfrac{5}{3}$
$\overline{AB} \parallel \overline{CD}$	The slopes are equal.
Slope of $\overline{AC}$	$m = \dfrac{2-2}{4-1} = \dfrac{0}{3} = 0$
Slope of $\overline{BD}$	$m = \dfrac{7-7}{7-4} = \dfrac{0}{3} = 0$
$\overline{AC} \parallel \overline{BD}$	The slopes are equal.

Therefore, the opposite sides of quadrilateral $ABCD$ are parallel. The resulting quadrilateral is a parallelogram as shown below.

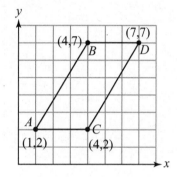

Example:

Given $\triangle ABC$ with vertices $A(0, 0)$, $B(0, 4)$, and $C(4, 3)$. Point D is the midpoint of $\overline{AC}$.

Find the coordinates of D, and verify that it divides $\overline{AC}$ into two congruent segments.

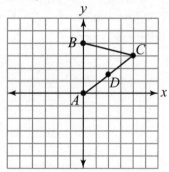

Coordinates of point D (midpoint)	$\left(\dfrac{4+0}{2}, \dfrac{0+3}{2}\right) = \left(2, \dfrac{3}{2}\right)$
Distance formula, AD	$AD = \sqrt{(2-0)^2 + \left(\dfrac{3}{2} - 0\right)^2}$ $AD = \sqrt{(2)^2 + \left(\dfrac{3}{2}\right)^2}$ $AD = \sqrt{4 + \dfrac{9}{4}}$ $AD = \sqrt{\dfrac{25}{4}}$ $AD = \dfrac{5}{2}$
Distance formula, CD	$CD = \sqrt{(4-2)^2 + \left(3 - \dfrac{3}{2}\right)^2}$ $CD = \sqrt{(2)^2 + \left(\dfrac{3}{2}\right)^2}$ $CD = \sqrt{4 + \dfrac{9}{4}}$ $CD = \sqrt{\dfrac{25}{4}}$ $CD = \dfrac{5}{2}$

Since the measures of AD and CD are equal, the segments are congruent. A midpoint divides a segment into two congruent segments.

Example:

Let's look at a new triangle, $\triangle XYZ$, whose vertices are $X(1, 1)$, $Y(7, 5)$, and $Z(2, 6)$. The midpoint of $\overline{XY}$ is given to be $W(4, 3)$. Let's verify that $\overline{ZW}$ is perpendicular to $\overline{XY}$ and, therefore, an altitude of $\triangle XYZ$.

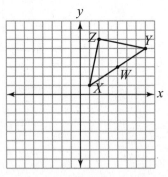

Slope of $\overline{XY}$	$m = \dfrac{5-1}{7-1} = \dfrac{4}{6} = \dfrac{2}{3}$
Slope of $\overline{ZW}$	$m = \dfrac{6-3}{2-4} = \dfrac{3}{-2} = \dfrac{-3}{2}$
$\overline{XY} \perp \overline{ZW}$	The product of the slopes is -1. The slopes are negative reciprocals.

$\overline{XY}$ is perpendicular to $\overline{ZW}$ and, therefore by definition, an altitude of $\triangle XYZ$.

Are you able to do each of the following?

❏ Determine whether a given procedure is valid
  • Simplifying expressions
  • Solving equations
  • Solving inequalities

❏ Work with algorithms

❏ Work with if-then statements

❏ Write, identify, and analyze valid conjectures

❏ Create and critique valid inductive and deductive arguments

❏ Identify or construct counterexamples

❏ State the inverse, converse, or contrapositive of a given statement and determine its truth

❏ Construct a simple formal or informal proof

❏ Verify characteristics of a given geometric figure

Practice Problems—Strand 5

1. This conjecture $(a + b)^2 = a^2 + b^2$ is false. Which counterexample shows that it is indeed false?

   A. Let $a = 1$ and $b = 0$
   B. Let $a = 0$ and $b = 1$
   C. Let $a = 0$ and $b = 0$
   D. Let $a = 1$ and $b = 1$

2. Which procedure is correct for simplifying the expression: $(5 + 3)^2 - 4 \cdot 2 \div 8$?

   A. $(5 + 3)^2 - 4 \cdot 2 \div 8$
   $8^2 - 4 \cdot 2 \div 8$
   $64 - 4 \cdot 2 \div 8$
   $60 \cdot 2 \div 8$
   $120 \div 8$
   15

   B. $(5 + 3)^2 - 4 \cdot 2 \div 8$
   $(25 + 9) - 4 \cdot 2 \div 8$
   $34 - 4 \cdot 2 \div 8$
   $34 - 8 \div 8$
   $34 - 1$
   33

   C. $\left(5 + 3\right)^2 - 4 \cdot 2 \div 8$
   $8^2 \quad - 4 \cdot 2 \div 8$
   $64 \quad - 4 \cdot 2 \div 8$
   $64 \quad - 8 \div 8$
   $64 \quad - 1$
   63

   D. $\left(5 + 3\right)^2 - 4 \cdot 2 \div 8$
   $8^2 \quad - 4 \cdot 2 \div 8$
   $64 \quad - 4 \cdot 2 \div 8$
   $60 \quad - 8 \div 8$
   $52 \quad \div 8$
   $\dfrac{52}{8}$

3. Which procedure is correct for solving the inequality $3(x + 3) - 4 > 20 + x$?

   A. $-3\left(x + 3\right) - 4 > 20 + x$
   $-3x + 9 - 4 > 20 + x$
   $-3x + 5 > 20 + x$
   $-4x + 5 > 20$
   $-4x > 15$
   $x > \dfrac{-15}{4}$

   B. $-3\left(x + 3\right) - 4 > 20 + x$
   $-3x - 9 - 4 > 20 + x$
   $-3x - 13 > 20 + x$
   $-4x > 33$
   $x > \dfrac{-33}{4}$

   C. $-3\left(x + 3\right) - 4 > 20 + x$
   $-3x - 9 - 4 > 20 + x$
   $-3x - 13 > 20 + x$
   $-4x > 33$
   $x < \dfrac{-33}{4}$

   D. $-3\left(x + 3\right) - 4 > 20 + x$
   $-3x - 9 - 4 > 20 + x$
   $-3x - 13 > 20 + x$
   $-2x > 33$
   $x < \dfrac{-33}{2}$

4. Points A (3, 2), B (7, 2), C (6, 9), and D (4, 9) are vertices of a quadrilateral. What is the most specific name for $ABCD$?

   A. Parallelogram
   B. Rectangle
   C. Kite
   D. Trapezoid

5. Consider the following conjecture. If an even number is written $2k$, where k is a whole number, then $2k + 1$ must be an odd number. What can be your next course of action in dealing with this conjecture?

A. Test every possible case to prove the conjecture.
B. Begin looking for a counterexample to disprove the conjecture.
C. Develop an argument to prove the conjecture.
D. Test the first five whole numbers for the variable k.

6. Determine the contrapositive of the following if-then statement:

If three points are noncollinear, then they form a triangle.

A. If three points form a triangle, then they are noncollinear.
B. If three points do not form a triangle, then they are not noncollinear.
C. If three points do not form a triangle, then they are noncollinear.
D. If three points are not noncollinear, then they do not form a triangle.

7. Which statement is the converse of the statement: If two angles have the same measure, then they are congruent?

A. If two angles are not congruent, then they do not have the same measure.
B. If two angles have the same measure, then they are congruent.
C. If two angles do not have the same measure, then they are not congruent.
D. If two angles are congruent, then they have the same measure.

8. How should the following statements be rearranged to form a logical if-then argument?

1. If I go practice, then I will make the team.
2. If I work hard, then I will finish my homework.
3. If I make the team, then I will receive an athletic letter.
4. If I finish my homework, then I will go to practice.
5. If I receive an athletic letter, then I will buy a letter sweater.

A. 3, 5, 2, 4, 1
B. 2, 4, 1, 3, 5
C. 2, 1, 3, 5, 4
D. 4, 1, 3, 5, 2

9. The algorithm below is a step-by-step geometric procedure.

I. Subtract the first y-coordinate value from the second y-coordinate value.
II. Subtract the first x-coordinate value from the second x-coordinate value.
III. Write your answers from step I and step II as a ratio of step I to step II.

Which of the following does the algorithm best represent?

A. Calculating the slope of a line.
B. Calculating the y-intercept of a graph.
C. Calculating the midpoint of a line segment.
D. Calculating the length of a line segment.

10. Which of the following algorithms are equivalent?

    I. Given two lines, draw a transversal. If the alternate interior angles are equal, then you have the answer you are looking for.

    II. Given two lines, compare their slopes. If they are equal, then you have the answer you are looking for.

    III. Given two lines, compare their slopes. If they are negative reciprocals, then you have the answer you are looking for.

    A. I and II
    B. II and III
    C. I and III
    D. I, II, and III

---

11. Given a quadrilateral, consider the following additional facts:

    One pair of opposite sides is parallel. The other pair of sides is also parallel. All of the sides are the same length. An angle of the quadrilateral is not a right angle.

    Which of the following best describes the quadrilateral?

    A. Rectangle
    B. Rhombus
    C. Trapezoid
    D. Square

12. Which of the following could be used to accurately graph $2x + 3y = 6$?

    A. Use the equation in its current form. Plot the point $(0, 6)$. Then plot the next point by moving up 3 units and to the right 2 units. Connect the two points.

    B. Transform the equation to $y = \dfrac{-2}{3}x + 6$. Plot the point $(0, 6)$. Then plot the next point by moving down 2 units and to the right 3 units. Connect the two points.

    C. Transform the equation to $y = \dfrac{-2}{3}x + 2$. Plot the point $(0, 2)$. Then plot the next point by moving up 3 units and to the left 2 units. Connect the two points.

    D. Transform the equation to $y = \dfrac{-2}{3}x + 2$. Plot the point $(0, 2)$. Then plot the next point by moving down 2 units and to the right 3 units. Connect the two points.

---

13. Suppose that you are planning to paint your room. You have determined the surface area to be painted to be 390 square feet and the price per gallon for paint is $12.50. What other information is needed to determine how much the paint needed will cost?

    A. The dimensions of the room
    B. The color of the paint
    C. The number of square feet one gallon of paint will cover
    D. The number of walls

14. The equation, $y = 4 \cdot 7 \cdot x$, could be used to represent which of the following?

   A. The area of a triangle with sides equal to 4 cm, 7 cm, and x cm.
   B. The area of a trapezoid with bases equal to 4 cm and 7 cm and a height of x cm.
   C. The volume of a rectangular prism with a width of 4 cm, a length of 7 cm, and a height of x cm.
   D. The perimeter of a triangle with sides of 4 cm, 7 cm, and x cm.

15. Assuming the following statements are true, which of the following conclusions can you reach?

   If you do not enjoy the outdoors, then you will not enjoy camping.

   Willy does not like the outdoors.

   A. Willy is afraid of bears.
   B. Willy goes camping every chance he gets.
   C. Willy does not enjoy camping.
   D. Willy enjoys camping.

Answer Key

1.	D	4.	D	7.	D	10.	A	13.	C
2.	C	5.	B	8.	B	11.	B	14.	C
3.	C	6.	B	9.	A	12.	D	15.	C

Answers Explained

1. **D** Placing the value of 1 for both a and b into the equation reveals the statement to be false and is a valid counterexample.

$$(a + b)^2 = a^2 + b^2$$
$$(1 + 1)^2 = 1^2 + 1^2$$
$$(2)^2 = 1 + 1$$
$$4 = 2$$

2. **C** Option A is not correct. It does not follow the order of operations in the fourth line when it combines 64 with 4 before the multiplication. Option B is not correct. Line two squares the binomial terms before combining them. Option D is not correct. The fourth line performs a subtraction prior to division.

3. **C** Option A is not correct. The negative three was not correctly distributed to the three of the binomial. Option B is not correct. The inequality sign is not reversed in the last line. When dividing by a negative in an inequality, the inequality sign reverses. Option D is not correct. In line three moving the variable x to the left side requires that you subtract x from both sides.

4. **D** Plotting the points on a grid suggests a trapezoid. To verify this conjecture, find the slopes of the sides using the formula from the reference sheet:

$$m = \frac{y_2 - y_1}{x_2 - x_1}$$

Side AB

$$m = \frac{2-2}{7-3} = \frac{0}{4} = 0$$

Side BC

$$m = \frac{9-2}{6-7} = \frac{7}{-1} = -7$$

Side CD

$$m = \frac{9-9}{4-6} = \frac{0}{-2} = 0$$

Side DA

$$m = \frac{2-9}{3-4} = \frac{-7}{-1} = 7$$

The opposite sides AB and CD have the same slope making them parallel. Sides BC and DA do not have the same slopes. The conjecture is correct.

5. **B** You would begin to search for a counterexample. It would take just one counterexample to make the conjecture false.

6. **B** Recall that in a contrapositive statement, negate both the if and then statements and switch their order.

7. **D** The converse of a conditional is a new sentence obtained by exchanging the hypothesis and the conclusion of the given conditional.

8. **B** To correctly join these if-then statements, look for an if statement that does not appear later as a then statement. This will be the start of the series. In this problem, "if I work hard" does not appear as a conclusion. Look for the then statement that does not appear as an if statement, this will be the last statement. In this problem "I will buy a letter sweater" does not appear as a hypothesis. Use the then statement in a proceeding statement to find the if statement of the next statement.

9. **A** The algorithm describes the process of calculating the slope of a line given two points.

10. **A** Statements I and II both describe the process of showing that two lines are parallel.

11. **B** The description is that of a rhombus. No right angle rules out the square. Both pairs of opposite sides are parallel rules out the trapezoid. All sides the same length ruled out the rectangle.

12. **D** Choice D describes the correct method of transforming the equation into slope-intercept form and drawing the line. Choice A does not change the equation into slope-intercept form. Choice B does not correctly transform the equation. The constant 6 was not divided by the coefficient of y. Choice C does not correctly plot the slope. The slope formula is a vertical change over the horizontal change.

13. **C** Choice A is not important because the surface area has been determined. Choice B will not answer how much paint is needed. Choice D is also not important because the surface area has been determined. Knowing how many cans of paint to buy requires the surface area and the coverage of each gallon of paint.

14. **C** The product of the three given dimensions could be used to represent the volume. Choice A—this product is not the area formula of a triangle. Choice B—this product is not the formula for a trapezoid. Choice D—this product is not the formula for computing perimeter.

15. **C** The statement "Willy does not like the outdoors" satisfies the premise, you will not enjoy camping. Therefore, the conclusion, "Willy does not enjoy camping," is true.

Practice Test 1
Answer Sheet
Fill in the bubble completely. Erase carefully if an answer is changed.

1. A B C D	18. A B C D	35. A B C D
2. A B C D	19. A B C D	36. A B C D
3. A B C D	20. A B C D	37. A B C D
4. A B C D	21. A B C D	38. A B C D
5. A B C D	22. A B C D	39. A B C D
6. A B C D	23. A B C D	40. A B C D
7. A B C D	24. A B C D	41. A B C D
8. A B C D	25. A B C D	42. A B C D
9. A B C D	26. A B C D	43. A B C D
10. A B C D	27. A B C D	44. A B C D
11. A B C D	28. A B C D	45. A B C D
12. A B C D	29. A B C D	46. A B C D
13. A B C D	30. A B C D	47. A B C D
14. A B C D	31. A B C D	48. A B C D
15. A B C D	32. A B C D	49. A B C D
16. A B C D	33. A B C D	50. A B C D
17. A B C D	34. A B C D	

Cut along dotted line.

Practice Test 1
Answer Sheet
Fill in the bubble completely. Erase carefully if an answer is changed.

51. A B C D
52. A B C D
53. A B C D
54. A B C D
55. A B C D
56. A B C D
57. A B C D
58. A B C D
59. A B C D
60. A B C D
61. A B C D
62. A B C D
63. A B C D
64. A B C D
65. A B C D
66. A B C D
67. A B C D

68. A B C D
69. A B C D
70. A B C D
71. A B C D
72. A B C D
73. A B C D
74. A B C D
75. A B C D
76. A B C D
77. A B C D
78. A B C D
79. A B C D
80. A B C D
81. A B C D
82. A B C D
83. A B C D
84. A B C D

85. A B C D
86. A B C D
87. A B C D
88. A B C D
89. A B C D
90. A B C D
91. A B C D
92. A B C D
93. A B C D
94. A B C D
95. A B C D
96. A B C D
97. A B C D
98. A B C D
99. A B C D
100. A B C D

Cut along dotted line.

Practice Test 1

Directions: Choose the best answer for each of the following.

1. Which of the four absolute value expressions does **not** have a value of 7?

   A. $|-13| - |6|$
   B. $|-4| + |-3|$
   C. $|2 - 3| + |3 - 9|$
   D. $|4| - |3|$

2. In the figure below, the circle is inscribed in the square. If a point is chosen at random within the square, what is the probability that it is also within the shaded circular region?

   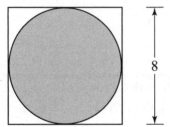

   A. $\dfrac{1}{4}$

   B. $\dfrac{1}{8}$

   C. $\dfrac{\pi}{4}$

   D. $\dfrac{\pi}{8}$

3. The spinner shown is spun once. What is the probability it will land on a multiple of five?

   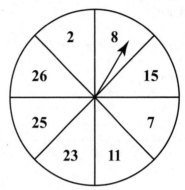

   A. $\dfrac{5}{8}$

   B. $\dfrac{3}{8}$

   C. $\dfrac{1}{4}$

   D. $\dfrac{1}{8}$

GO ON

4. The graph below shows the average number of calls received by the 911 operator in Cave Creek, Arizona. Over the time period from 1993 to 2004, what seems to be the relationship between time and the number of calls?

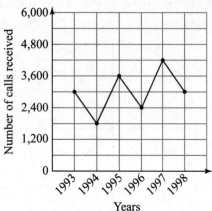

A. Over the time period from 1993 to 1998, there is a constant increase in the number of calls.

B. There were more calls made in 1997 than in 1993.

C. Over the time period from 1993 to 1998, the maximum number of calls in one year was 2,400.

D. During the years 1995 and 1996, the total number of calls was 5,000 calls.

5. Nan went the movies with a group of friends several different times during the summer. She always buys the tickets ahead of time and then her friends pay her back. The table below shows the relationship between the number of tickets she bought and the cost of those tickets.

Number of tickets (n)	2	3	5	8	10	12
Cost of tickets (C) each time	$14	$21	$35	$56	$70	$84

Which of the following equations represents the relationship between the cost (C) and the number of tickets bought (n)?

A. $C = 7 + n$

B. $C = n - 7$

C. $C = 7n$

D. $C = \dfrac{n}{7}$

6. Which of these equations represents a line passing through $(6, 5)$ and $(-3, -1)$?

A. $y = \dfrac{2}{3}x + 1$

B. $y = \dfrac{3}{2}x - 4$

C. $y = \dfrac{4}{3}x - 3$

D. $y = \dfrac{3}{4}x + \dfrac{1}{2}$

GO ON ➡

7. Which of the following best represents the graph of a line with an undefined slope?

A.

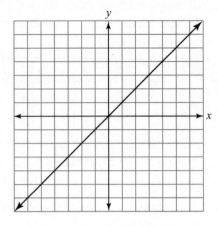

B.

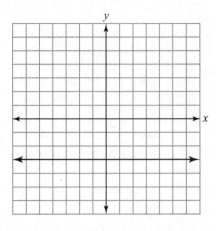

C.

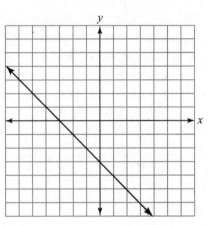

D.

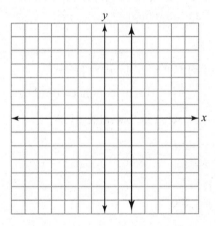

---

8. In circle O, the shaded region is one-sixth of the circle. The radius of the circle is 8. What is the length of chord AB?

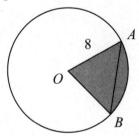

  A. 8

  B. $8\sqrt{2}$

  C. $8\sqrt{3}$

  D. 16

9. The graph of which inequality is shown below?

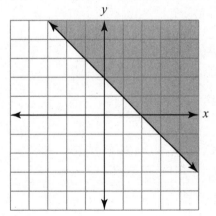

  A. $y < -x + 2$
  B. $y > -x + 2$
  C. $y \le -x + 2$
  D. $y \ge -x + 2$

GO ON ➡

10. What is the converse of the conditional statement?

    If all the sides of a polygon are congruent, then the polygon is equilateral.

    A. If all the sides of a polygon are not congruent, then the polygon is not equilateral.
    B. If all the sides of a polygon are congruent, then the polygon is not equilateral.
    C. If a polygon is not equilateral, then all its sides are not congruent.
    D. If a polygon is equilateral, then all its sides are congruent.

11. The sequence below uses the rule, $A_n = \dfrac{3n+5}{2}$. If the sequence continues and $A_n = 13$, what is the value of n?

    A. $n = 2.6$
    B. $n = 6$
    C. $n = 7$
    D. $n = 8.6$

12. Mel has 2 red socks, 6 blue socks, 2 purple socks, and 8 black socks in a drawer. What is the probability she will choose a black sock if she already took a red sock from the drawer?

    A. $\dfrac{1}{18}$

    B. $\dfrac{1}{8}$

    C. $\dfrac{8}{17}$

    D. $\dfrac{1}{17}$

13. In order to compare the expected probability of flipping a coin, Mary flipped a fair coin 10 times. The results of her experiment are in the table below.

Flip #	1	2	3	4	5	6	7	8	9	10
Results of flip	H	H	T	H	T	T	T	H	H	T

How do the results of her experiment compare with the expected probability of getting heads when flipping a fair coin?

A. Her results match the expected probability exactly.
B. Her results represent a probability less than the expected probability.
C. Her results represent a probability greater than the expected probability.
D. It is impossible to compare her results with the expected probability.

14. The graph below shows below the results when $100 is invested in two different accounts, one with a return rate of 8% and the other with a return rate of 10%.

8% vs. 10% Rate of Return

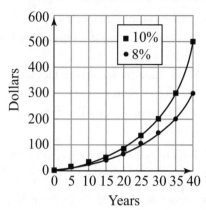

Based on this graph, which statement below is **false**?

A. The money value of both accounts is increasing.

B. After 35 years, the account with a return rate of 8% earned half as much as the account with return rate of 10%.

C. After 40 years, the account with a return rate of 10% earned in excess of $200 more than the account with a return rate of 8%.

D. The 2% difference in the rate of return makes a measurable difference in the growth of the account over a 40-year period.

---

15. Nick purchases a CD on sale for 30% off. The original price was $13.95. How could Nick find by how much the original price would be reduced?

A. Divide $13.95 by 0.30.
B. Multiply $13.95 by 0.30.
C. Divide $13.95 by 0.70.
D. Multiply $13.95 by 0.70.

16. Which is the solution to the following inequality?

$$2x - 3 \geq 11$$

A. $x \leq 4$
B. $x \leq 7$
C. $x \geq 7$
D. $x \geq 4$

---

17. Which equation can be used to find the value of x in the right triangle shown?

A. $\cos 70° = \dfrac{5}{x}$

B. $\cos 70° = \dfrac{x}{5}$

C. $\sin 70° = \dfrac{5}{x}$

D. $\sin 70° = \dfrac{x}{5}$

---

18. The figure below is formed by joining an isosceles right triangle and a 30°-60°-90° triangle. If the hypotenuse of the 30°-60°-90° triangle is 18, what is the measure of the legs of the isosceles right triangle?

A. $9\sqrt{3}$
B. 9
C. $9\sqrt{2}$
D. 36

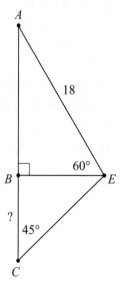

GO ON ➡

19. Which point best represents the solution to the system of linear equations shown in the graph below?

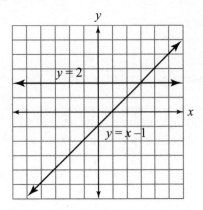

A. (1, 0)
B. (3, 2)
C. (2, 3)
D. (0, -1)

20. Which of the following would be a counterexample for the following statement?

If $m\angle R$ is less than 110°,
then $\angle R$ is an acute angle.

A. $m\angle R = 10°$
B. $m\angle R = 89°$
C. $m\angle R = 100°$
D. $m\angle R = 125°$

21. Which of the following expressions is arranged in descending order?

A. $5x^4 - 6x^3 + 7x - 8 + 9x^2$
B. $5x^4 - 6x^3 + 9x^2 + 7x - 8$
C. $9x^2 - 8 + 7x - 6x^3 + 5x^4$
D. $-8 + 7x + 9x^2 - 6x^3 + 5x^4$

22. Skyway High School is considering adding co-ed volleyball to the sports program for the fall season. In order to get an unbiased sample of interest in co-ed volleyball, the school should survey which group below?

A. All girls in dance class
B. Every third student entering second-period class
C. All students who were elected this year to the student council
D. The varsity football team

23. Which of the following experiments contains a dependent event?

A. From a standard deck of cards, Jose draws a Spade, puts it back in the deck, and draws a Heart.
B. On a spinner with 6 congruent sectors numbered 1 through 6, Greg first spins a 4 and then spins a 2 on the second spin.
C. From a standard deck of cards, Jamie draws a Club, sets it aside, and then draws a Diamond.
D. Monica tosses a fair coin two consecutive times, and it lands on heads both times.

GO ON ➡

24. The following graph shows the number of sunny days by month for a given year in Flagstaff, Arizona. Use the graph to determine which month has the highest number of sunny days.

Sunny Days in Flagstaff, Arizona

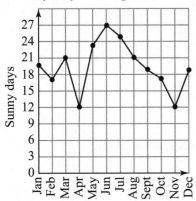

A. March
B. June
C. July
D. December

25. Which of the following is an equation for a line that is parallel to the given line?

$$4x + 3y = 10$$

A. $3x + 4y = 6$

B. $y = -\dfrac{4}{3}x + 6$

C. $-6x - 8y = 7$

D. $y = -\dfrac{3}{4}x + 6$

26. Solve $2(6x - 7) = 6x - 24$.

A. $\dfrac{19}{3}$

B. $-\dfrac{5}{3}$

C. $-\dfrac{3}{5}$

D. $\dfrac{3}{19}$

27. What are the x- and y-intercepts of the equation below?

$$3x + 4y = 12$$

A. The x-intercept is $(4, 0)$ and the y-intercept is $(0, 3)$.
B. The x-intercept is $(3, 0)$ and the y-intercept is $(0, 4)$.
C. The x-intercept is $(0, 4)$ and the y-intercept is $(3, 0)$.
D. The x-intercept is $(0, 3)$ and the y-intercept is $(4, 0)$.

28. The set of real numbers shown below is a subset of which of the following sets?

$$\left\{ -5, -2\tfrac{1}{2}, 0.6694, 3, \sqrt{16} \right\}$$

A. The set of natural numbers
B. The set of integers
C. The set of rational numbers
D. The set of irrational numbers

GO ON ➡

Practice Test 1

29. The graph of $y = 2x + 1$ is shown below.

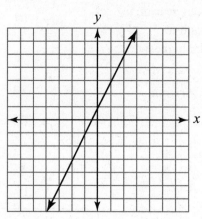

Which of the following would be the graph if the 1 was changed to a 3?

A.

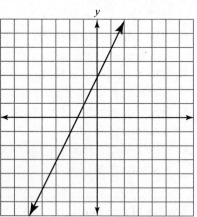

B.

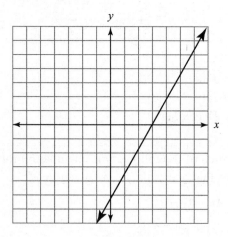

C.

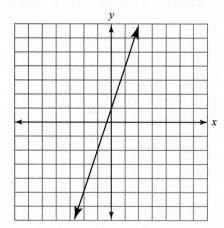

D.

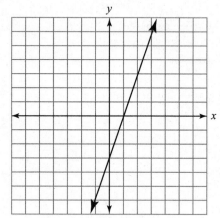

GO ON ➡

30. Assuming the following statements are true, which of the following conclusions can you reach?

> If you like mathematics, then you will like astronomy. Kelly likes mathematics.

A. Kelly should take more mathematics.
B. Kelly will like astronomy.
C. Kelly does not like astronomy.
D. Kelly does like mathematics.

31. What is the value of the following expression?

$$(3.6 \times 10^4)(2.1 \times 10^7)$$

A. 7.56×10^3
B. 7.56×10^{10}
C. 7.56×10^{11}
D. 7.56×10^{28}

32. Which of the following groups could be used to take a census of a school?

A. The junior class
B. All social studies classes
C. The entire student body
D. The Drama Club members

33. Jon is planning a family reunion cruise for his family. He can choose to go on 4 different cruise lines, to 3 different countries, at 7 different departure dates. How many different trip combinations, consisting of one cruise line, one country, and one date, are possible?

A. 14
B. 28
C. 31
D. 84

34. The chart below, which shows the rates for first class mail, represents a function.

Weight not over (ounces)	Rate
1	$0.39
2	0.63
3	0.87
4	1.11
5	1.35

What is the range of this function?

A. {1, 2, 3, 4, 5}
B. {0.39, 0.63, 0.87, 1.11, 1.35}
C. {(1, 0.39), (2, 0.63), (3, 0.87), (4, 1.11), (5, 1.35)}
D. {(0.39, 1), (0.63, 2), (0.87, 3), (1.11, 4), (1.35, 5)}

35. Evaluate the expression $3(x - 2) + 3y$ if $x = 4$ and $y = 2$.

A. 12
B. 15
C. 16
D. 24

36. What is the value of x that will make the following proportion true?

$$\frac{8}{6} = \frac{x+4}{x-1}$$

A. 32
B. 24
C. 16
D. 12

37. Which of these is equivalent to the equation below?

$$A = 4\pi r^2$$

A. $r = \dfrac{2A}{4\pi}$

B. $r = \dfrac{A}{4\pi}$

C. $r = \dfrac{\sqrt{A}}{2\sqrt{\pi}}$

D. $r = \sqrt{\dfrac{4A}{\pi}}$

38. Given the diagram which shows the transformation of figure A to figure A'.

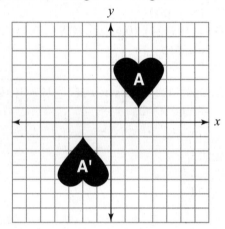

Which of the following is **not** a correct description of the transformation?

A. First reflect the pre-image over the y-axis and then reflect the resulting image over the x-axis.

B. First reflect the pre-image over the x-axis and then reflect the resulting image over the y-axis.

C. Rotate the pre-image 180° clockwise about the origin.

D. Translate the pre-image 4 units to the left and then translate the resulting image 2 units down.

39. In the diagram below, $\angle BAC \cong \angle DAC$ and $\angle BCA \cong \angle DCA$. Which of the methods below would you use to most directly prove the triangles congruent?

A. AAS (Angle-Angle-Side)
B. SSS (Side-Side-Side)
C. SAS (Side-Angle-Side)
D. ASA (Angle-Side-Angle)

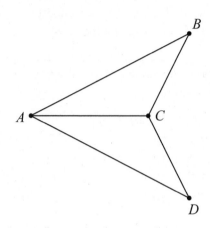

40. Which of the following expressions is equivalent to the one shown below?

$$7x^2y - 2x^2y + 5xy^2 - 3xy^2$$

A. $5x^4y^2 + 2x^2y^4$
B. $5x^2y + 2xy^2$
C. $7x^6y^6$
D. $5x^2y + 5xy^2 - 3xy^2$

41. The function below represents the number of bacteria (p) that 302 cells increases to after t hours.

$$p(t) = 302(2^t)$$

Which of the following is closest to the total number of bacteria cells after 3 hours?

A. 1,200
B. 1,800
C. 2,400
D. 3,000

GO ON ➡

42. Which of the following graphs shows a positive correlation between the two variables?

A.

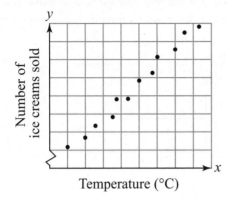

Temperature (°C)

B.

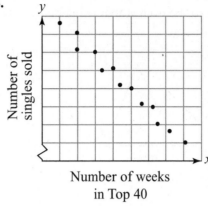

Number of weeks
in Top 40

C.

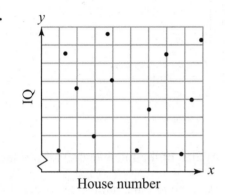

House number

D.

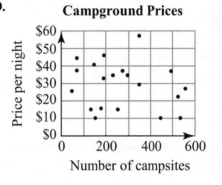

---

43. The judges at the music contest will be awarding trophies for first, second, third, and fourth places out of 15 entries. Which expression gives the number of ways the judges can award these trophies?

A. $\dfrac{4!}{11!}$

B. $\dfrac{15!}{11!}$

C. $\dfrac{11!}{15!}$

D. $\dfrac{11!}{4!}$

GO ON ➡

44. An oven is turned on and set for a certain temperature. The temperature of an oven is a function of the length of time the oven has been on. Once the oven reaches the desired temperature, the thermostat keeps the oven at that temperature. Choose the graph that models the change in oven temperature over time.

A.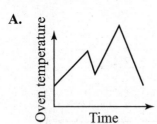

B.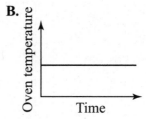

C.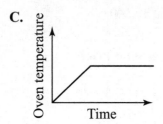

D.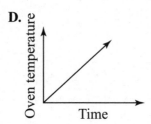

45. Which of the equations below can be written using this table of values?

x	y
1	-2
-2	-5
4	1
0	-3
-5	-8

A. $y = x - 1$
B. $y = x - 3$
C. $y = -2x$
D. $y = \dfrac{x}{4}$

46. Which statement is true about the graphs of the equations below?

$y = 3x + 2$
$y = 3x - 4$

A. The lines are parallel.
B. The lines are perpendicular.
C. The lines intersect but are not perpendicular.
D. The lines coincide (same line).

47. Which of the following sets is a finite set?

A. {the rational numbers between 7 and $\sqrt{144}$ }
B. {the integers between -416 and 124}
C. {irrational numbers less than 12}
D. {the integers greater than 5}

GO ON ➡

48. Which similarity theorem can be used to prove △*BAC* ~ △*DEC*?

    A. SSS (Side-Side-Side)
    B. AA (Angle-Angle)
    C. SAS (Side-Angle-Side)
    D. AAS (Angle-Angle-Side)

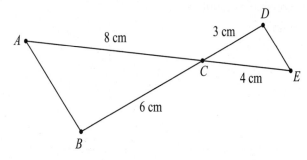

49. What is the distance between the points (−3, −5) and (2, 7)?

    A. $\sqrt{5}$
    B. 13
    C. 17
    D. 169

50. Steps 1 and 2 describe an algorithm.

    Step 1: Isolate the variable term.
    Step 2: Square both sides of the equation—the result is the solution.

    The algorithm above can solve which of these equations?

    A. $x^2 - 2x - 3 = 0$
    B. $x + 5 = 0$
    C. $\sqrt{x} - 9 = 0$
    D. $x^3 + 2x + 6 = 0$

STOP

Now would be a good time for you to take a break. When you actually take the AIMS test at your school, you will be given a break at this time.

Practice Test 1

Directions: Choose the best answer for each of the following.

51. A square has an area of 180 square centimeters. Which of the following measures is the most reasonable estimate of the length of one of its sides?

    A. 90
    B. 60
    C. 45
    D. 13

52. The scatter plot below shows the ages and heights of 20 trees on a tree farm.

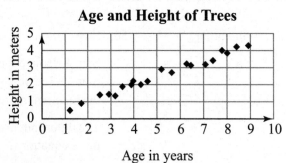

Which of the following graphs represents the best trend line for the data?

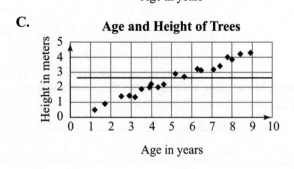

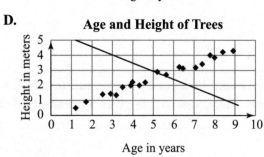

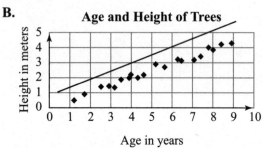

GO ON ➡

53. $\triangle XYZ$ is an isosceles triangle. Which of the statements below is **not** always true for $\triangle XYZ$?

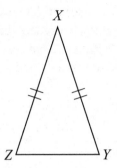

A. The sum of the angles is 180°.
B. The triangle has two equal sides.
C. The triangle has two equal angles.
D. The sum of the squares of the legs of the triangle is equal to the square of the hypotenuse.

---

54. Which of the following procedures correctly simplifies the numerical expression below?

$$4 + 8 \div 2 \times 3 - 1$$

A.
$$4 + 8 \div 2 \times 3 - 1$$
$$12 \div 2 \times 3 - 1$$
$$6 \times 3 - 1$$
$$18 - 1$$
$$17$$

B.
$$4 + 8 \div 2 \times 3 - 1$$
$$12 \div 2 \times 3 - 1$$
$$12 \div 6 - 1$$
$$2 - 1$$
$$1$$

C.
$$4 + 8 \div 2 \times 3 - 1$$
$$4 + 4 \times 3 - 1$$
$$4 + 2 \times 2$$
$$4 + 8$$
$$12$$

D.
$$4 + 8 \div 2 \times 3 - 1$$
$$4 + 4 \times 3 - 1$$
$$4 + 12 - 1$$
$$16 - 1$$
$$15$$

---

55. Which of the following is a set of integers?

A. $\{\frac{1}{2}, \sqrt{4}, 7.3\}$

B. $\{9, \sqrt{3}, 3.6\}$

C. $\{\frac{1}{5}, \sqrt{6}, 5.12\}$

D. $\{-7, \sqrt{9}, 0\}$

---

56. Given: $A = \begin{bmatrix} 1 & -1 \\ 0 & 1 \end{bmatrix}$ $B = \begin{bmatrix} -1 & 1 \\ -1 & 0 \end{bmatrix}$

$C = \begin{bmatrix} -1 & -1 \\ 0 & 1 \end{bmatrix}$, what is the value of

$A + 2B - C$?

A. $\begin{bmatrix} -1 & -1 \\ -1 & 2 \end{bmatrix}$

B. $\begin{bmatrix} -2 & 0 \\ -1 & 2 \end{bmatrix}$

C. $\begin{bmatrix} -1 & -2 \\ 1 & 0 \end{bmatrix}$

D. $\begin{bmatrix} 0 & 2 \\ -2 & 0 \end{bmatrix}$

---

57. Given the following statements about a quadrilateral:

Both pairs of opposite sides are congruent.
Both pairs of opposite sides are parallel.
The diagonals are congruent.

For which one of the following quadrilaterals are all three statements always true?

A. Parallelogram
B. Trapezoid
C. Rhombus
D. Rectangle

58. Jamie is putting up a bookshelf that is 12 inches wide. The shelf will be perpendicular to the wall and the brace supporting the shelf will form a 45° angle with the wall.

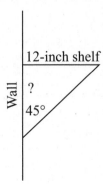

How far below the shelf will Jamie attach the brace?

A. 5 inches
B. 6 inches
C. 12 inches
D. 18 inches

---

59. The right circular cone shown below has a diameter of 6 feet and a height of 9 feet.

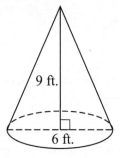

What is the volume of the cone?

A. 81π units^3
B. 54π units^3
C. 27π units^3
D. 18π units^3

60. Connie works for Real Estate For Real, Inc. She wants to reduce a 6-inch by 6-inch square photo to use in a real estate multiple listing on the Internet. She wants the image of the photo to be a square that measures 2 inches by 2 inches. What scale factor does Connie use to make the reduction

$$\frac{\text{Image}}{\text{Pre-image}}?$$

A. $\dfrac{1}{2}$

B. $\dfrac{1}{3}$

C. $\dfrac{1}{4}$

D. $\dfrac{1}{6}$

---

61. Which value is closest to the value of the following expression?

$$\frac{1+\sqrt{15}}{2}$$

A. 2.5
B. 4
C. 7.5
D. 8

62. The table below shows the **median** score on a comprehensive high school math test in a school district.

School #	Median score
1	78
2	73
3	82
4	65

Based on the table, which of the following statements must be true?

A. The highest score came from school #3.
B. At least half of the students at school #4 scored above a 65.
C. Every student from school #1 scored better than 70.
D. Half of the students from school #2 scored at or below 76.

63. In how many ways can you arrange all of the letters of the word MATH?

A. 4
B. 6
C. 12
D. 24

64. Using the recursion formula below,

$$a_1 = 4$$
$$a_n = 2a_{n-1} + 6$$

what are the first 5 terms of the sequence?

A. 4, 12, 30, 66, 138
B. 4, 22, 50, 106, 218
C. 4, 14, 34, 74, 154
D. 4, 14, 32, 68, 131

65. Which of the following is equivalent to this expression?

$$\frac{(6x^{-2}y^{-6})(8x^3y^{-4})}{(4x^{-4}y^{-2})}$$

A. $\dfrac{7x^5}{2y^8}$

B. $\dfrac{12x^5}{y^8}$

C. $12x^5y^8$

D. $\dfrac{7x^5y^8}{2}$

66. Which of the following is the simplified value of the expression below?

$$\sqrt{64x^{16}y^6}$$

A. $32x^4y^3$
B. $8x^8y^3$
C. $32x^8y^3$
D. $8x^4y^3$

67. If $\overline{LO}$ is a diameter of circle P, what is $m\angle ONL$?

A. 90°
B. 60°
C. 45°
D. 30°

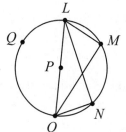

GO ON ➡

68. What is the image of $A(3, -1)$ if it is translated by moving left 2 units and up 3?

   A. $A'(1, 2)$
   B. $A'(-1, -2)$
   C. $A'(-3, 1)$
   D. $A'(-1, 3)$

---

69. Which property of multiplication is modeled below?

$$\frac{1}{4}(x+12)=\frac{x}{4}+3$$

   A. Inverse property
   B. Distributive property
   C. Commutative property
   D. Associative property

---

70. Which of the following inequalities is solved using an **invalid** procedure?

   A. $-4x + 7 \leq 11$
        $-4x \leq 4$
            $x \leq -1$

   B. $-2(x - 6) \leq 16$
      $-2x + 12 \leq 20$
         $-2x \leq 8$
           $x \geq -4$

   C. $5x + 6 \leq x + 2$
         $4x \leq -4$
           $x \leq -1$

   D. $5(x + 1) \leq 3(x + 5)$
      $5x + 5 \leq 3x + 15$
         $2x \leq 10$
           $x \leq 5$

71. Marybeth is planning the fall dance. She is writing a survey to determine the theme of the dance and what refreshments to serve. Which of the following would **not** be a good question to ask?

   A. Do you want the dance to have a Halloween theme?
   B. If we decide to serve a snack at the dance, what would be your choice for that snack?
   C. If we decide to serve a drink in addition to water, what would be your choice for that beverage?
   D. Do you have a curfew time at which you are required to be home?

---

72. The table represents how the air temperature combines with the humidity to form the heat index.

Heat Index Table

Relative Humidity					
60	90	100	114	132	149
55	89	98	110	126	142
50	88	96	107	120	135
45	87	95	104	115	129
40	86	93	101	110	123
35	85	91	98	107	118
30	84	90	96	104	113
	85	90	95	100	105

Air temperature (°F)

Which statement is correct?

   A. As the humidity and temperature decrease, the heat index increases.
   B. As the humidity and temperature increase, the heat index decreases.
   C. As the humidity increases and the temperature decreases, the heat index increases.
   D. As the humidity and temperature increase, the heat index increases.

GO ON ➡

73. The Regal High Debate Club qualified for the state meet. How many possible two-member debate teams can be formed from a pool of six students in the Regal High Debate Club?

   A. 729

   B. 360

   C. 15

   D. 8

---

74. Study the diagram below.

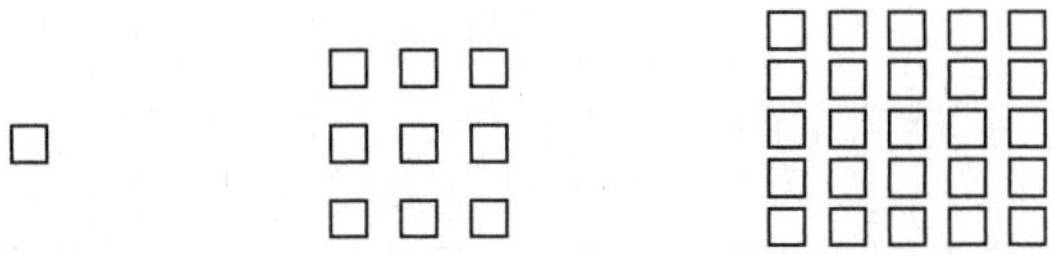

1^2	3^2	5^2
Diagram 1	Diagram 2	Diagram 3

If the above pattern continued, how many squares should the seventh diagram contain?

   A. 169

   B. 121

   C. 49

   D. 36

75. What is the algebraic expression that represents the sum of 10 and 3 times a number x?

   A. $10 + 3x$

   B. $3(10 + x)$

   C. $10 - 3x$

   D. $(10 + 3)x$

---

76. Which of the following expressions is equivalent to $(7ab)^2$?

   A. $14a^2b^2$

   B. $7a^2b^2$

   C. $7ab^2$

   D. $49a^2b^2$

GO ON ➡

77. Which of the following is **not** a possible next step in the process of solving this equation for x?

$$45\pi = \frac{x}{200}(20\pi)$$

A. $45\pi = \frac{x}{20}(2\pi)$

B. $9{,}000\pi = x(20\pi)$

C. $25\pi = \frac{x}{200}$

D. $45\pi = \frac{x\pi}{10}$

78. Which of the following is the graph of $y = \frac{3}{5}x - 1$?

A.

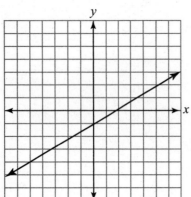

B.

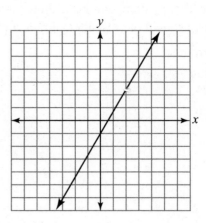

C.

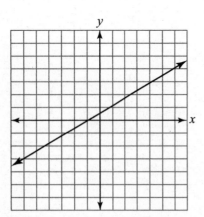

D.

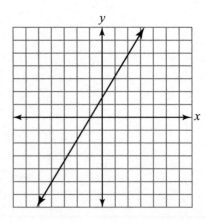

GO ON ➡

79. What is the length of $\overset{\frown}{AB}$ if $m\angle AOB = 60°$ and $OA = 12$?

A. 4π
B. 6π
C. 12π
D. 24π

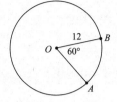

80. What is the total surface area of the cone shown below?

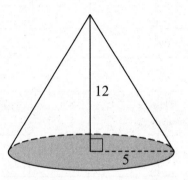

A. 85π sq units
B. 90π sq units
C. 300π sq units
D. 325π sq units

81. The following is a list of the number of people who went to a museum exhibit at different times during the day: 80 people went in at 11:00, 117 people went in at 12:00, 142 people went in at 1:00, 102 people went in at 2:00, 111 people went in at 3:00 and 94 people went in at 4:00. Which table shows the data organized correctly?

A.

Time	Number of people
11:00	102
12:00	142
1:00	111
2:00	94
3:00	117
4:00	80

B.

Time	Number of people
11:00	80
12:00	142
1:00	117
2:00	111
3:00	94
4:00	102

C.

Time	Number of people
11:00	80
12:00	117
1:00	142
2:00	102
3:00	111
4:00	94

D.

Time	Number of people
11:00	142
12:00	102
1:00	117
2:00	111
3:00	80
4:00	94

Practice Test 1

GO ON

82. The table below shows the cost of a new sports utility vehicle. If the trend continues, how much will a new sports utility vehicle cost in 2008?

Year	Cost
1996	$5,000
1998	$10,000
2000	$15,000
2002	$20,000
2004	$25,000

A. $45,000
B. $40,000
C. $35,000
D. $30,000

83. Which expression will define the nth term of the linear pattern defined by the table shown?

1	2	3	4	5	$\cdots$	n
7	10	13	16	19	$\cdots$	?

A. $2n + 10$
B. $n + 9$
C. $4n + 1$
D. $3n + 4$

84. Which of the following table of values represents a relation that is **not** a function?

A.

Number of tickets purchased	Amount of discount
1	15%
2	15%
3	20%
4	20%
5	25%
6	25%

B.

Hours worked	Amount earned
8	$56.00
16	$112.00
24	$168.00
32	$224.00
40	$280.00
48	$336.00

C.

x	y
−2	−1
−1	0
0	1
1	2
2	3

D.

x	y
−2	4
−1	1
0	−2
0	−8
4	−14

GO ON ➡

85. A repair service charges $35 to send a service person on a call and $40 per hour for labor. If *h* stands for the number of hours of labor, which expression below can the company use to compute the charge for the service call?

   A. $\dfrac{35}{40h}$

   B. $35 + 40h$

   C. $35h + 40$

   D. $75h$

86. What is the solution to the following equation?

   $$\sqrt{3x-8} = 2$$

   A. -2

   B. $3\dfrac{1}{3}$

   C. 4

   D. $33\dfrac{1}{3}$

87. Which of the following is **not** a net of a cube?

   A.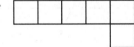

   B.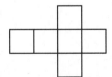

   C.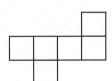

   D.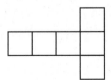

88. Which of the following transformations does not always preserve the dimensions of the figure?

   A. Translation
   B. Dilation
   C. Reflection
   D. Rotation

89. What is the sum of the interior angles of figure *ABCDEF*?

   A. 60°
   B. 120°
   C. 360°
   D. 720°

   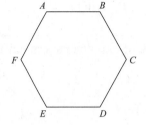

90. What is the midpoint of the line segment joining points (3, 5) and (–6, 1)?

   A. $\left(-1\dfrac{1}{2},\ 3\right)$

   B. $\left(-3,\ 6\right)$

   C. $\left(4\dfrac{1}{2},\ 2\right)$

   D. $\left(4,\ -2\dfrac{1}{2}\right)$

91. Which of the following addition properties justifies the statement below?

   $$3 + (-3) = 0$$

   A. Commutative
   B. Identity
   C. Inverse
   D. Closure

GO ON ➡

92. Kelly asked Mr. Conrad for a letter of recommendation to the scholarship committee. Kelly's scores on her chapter tests were 20, 75, 95, 80, and 90. Which measure of central tendency should Mr. Conrad use to give Kelly the best recommendation (the one with the highest numeric value)?

A. Mean
B. Median
C. Mode
D. Range

93. Study the pattern below.

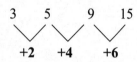

If this pattern continues what will be the seventh term?

A. 21
B. 23
C. 33
D. 45

94. The graph below shows the changes in a bank balance over time. If the bank balance is zero at A, describe what happens during the time period from A to B.

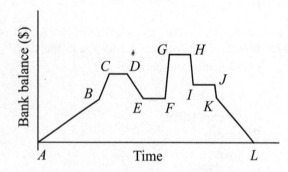

A. Withdrawals were made from the account during that time.
B. Deposits were made to the account during that time.
C. No deposits or withdrawals were made during that time.
D. A service fee was withdrawn from the account.

95. Jamie worked 40 hours per week for 2 weeks. She earned $7.35 an hour. What will be her total pay?

A. $294.00
B. $588.00
C. $2,940.00
D. $5,880.00

96. What are the values of x that make the following equation true?

$$x^2 + 10x - 24 = 0$$

A. $x = -6$ or $x = -4$
B. $x = -12$ or $x = 2$
C. $x = -2$ or $x = 12$
D. $x = 6$ or $x = 4$

97. In the diagram below, line ℓ is parallel to line m. What is the value of x in each of the expressions?

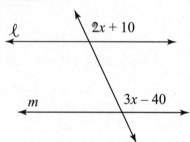

A. 6
B. 10
C. 30
D. 50

GO ON ➡

98. Which of the following is the graph of $y = -x^2 + 3$?

A.

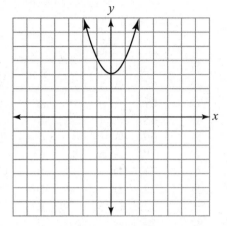

B.

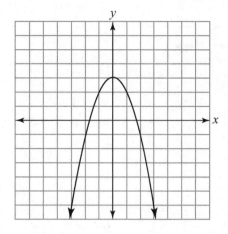

C.

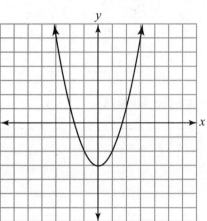

D.

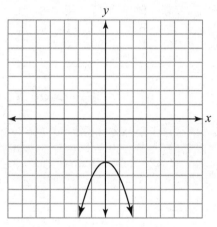

99. Find the area of the shaded region in the figure below.

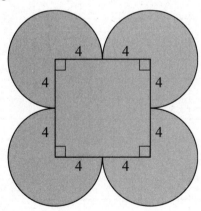

4   4

4   4

4   4

4   4

A. $64 + 64\pi$ sq units

B. $16 + 64\pi$ sq units

C. $64 + 48\pi$ sq units

D. $16 + 48\pi$ sq units

100. Given: Regular square pyramid with slant height GH and altitude $\overline{GF}$ drawn. $GH = 15$ cm and $AB = 18$ cm. What is the length of altitude $\overline{GF}$?

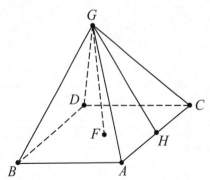

A. $\sqrt{549} = 3\sqrt{61}$

B. $\sqrt{306} = 3\sqrt{34}$

C. $\sqrt{144} = 12$

D. $\sqrt{99} = 3\sqrt{11}$

GO ON ➡

Solutions: Practice Test 1

Answer Key

1. D	14. B	27. A	40. B	53. D	65. B	77. C	89. D
2. C	15. B	28. C	41. C	54. D	66. B	78. A	90. A
3. C	16. C	29. A	42. A	55. D	67. A	79. A	91. C
4. B	17. A	30. B	43. B	56. D	68. A	80. B	92. B
5. C	18. B	31. C	44. C	57. D	69. B	81. C	93. D
6. A	19. B	32. C	45. B	58. C	70. A	82. C	94. B
7. D	20. C	33. D	46. A	59. C	71. D	83. D	95. B
8. A	21. B	34. B	47. B	60. B	72. D	84. D	96. B
9. D	22. B	35. A	48. C	61. A	73. C	85. B	97. D
10. D	23. C	36. C	49. B	62. B	74. A	86. C	98. B
11. C	24. B	37. C	50. C	63. D	75. A	87. A	99. C
12. C	25. B	38. D	51. D	64. C	76. D	88. B	100. C
13. A	26. B	39. D	52. A				

Practice Test 1 Diagnostic Chart

Question #	Correct	Incorrect	Strand to study	Page number(s) to review	Question #	Correct	Incorrect	Strand to study	Page number(s) to review
1			1	48	23			2	87
2			2	89	24			2	73
3			2	83	25			3	121
4			2	73	26			3	137
5			3	134	27			3	120
6			3	136	28			1	39
7			3	118	29			3	120
8			4	209	30			5	252
9			4	219	31			1	50
10			5	254	32			2	75
11			3	108	33			2	91
12			2	88	34			3	112
13			2	79	35			3	127
14			2	73	36			3	139
15			1	45	37			3	142
16			3	138	38			4	231
17			4	147	39			4	197
18			4	178	40			3	128
19			4	221	41			1	55
20			5	258	42			2	77
21			1	373	43			2	92
22			2	64	44			3	124

Diagnostic Chart

Question #	Correct	Incorrect	Strand to study	Page number(s) to review	Question #	Correct	Incorrect	Strand to study	Page number(s) to review
45			3	116	73			2	93
46			3	121	74			3	108
47			1	44	75			3	134
48			4	197	76			3	128
49			4	222	77			5	263
50			5	265	78			4	218
51			1	55	79			4	209
52			2	76	80			4	214
53			4	174	81			2	66
54			5 (1)	263 (49)	82			2	73
55			1	40	83			3	108
56			3	148	84			3	114
57			4	183	85			3	134
58			4	179	86			3	142
59			4	210	87			4	187
60			4	232	88			4	232
61			1	55	89			4	196
62			2	71	90			4	222
63			2	92	91			1	44
64			3	108	92			2	71
65			3	129	93			3	108
66			3	132	94			3	124
67			4	202	95			1	55
68			4	231	96			3	143
69			1	43	97			4	191
70			5	263	98			4	224
71			2	64	99			4	211
72			2	73	100			4	215

# You Got Correct		Your Classification
Grading Scale for AIMS		
Exceeds	100–88	
Meets	87–60	
Approaches	59–48	
Falls Far Below	47–0	

Answers Explained

1. **D** The absolute value of a number is always a positive number:
   Choice A is equal to 7: $|-13| - |6| = 13 - 6 = 7$.
   Choice B is equal to 7: $|-4| + |-3| = 4 + 3 = 7$.
   Choice C is equal to 7: $|2 - 3| + |3 - 9| = |-1| + |-6| = 1 + 6 = 7$.
   Choice D is not equal to 7: $|4| - |3| = 4 - 3 = 1$.

2. **C** To find probabilities related to geometric figures, you find the ratio of the desired area to the possible area. The desired area will be the area of the circle and the possible area will be the total area of the square. The diameter of the circle, 8 units, is the same as the side of the square. Therefore, the radius of the circle is 4 units, and its area is 16π ($A = \pi r^2$). Since the side of the square is 8, its area is 64 ($A = s^2$). The ratio of the area of the circle to the total area of the square is $\frac{16\pi}{64} = \frac{\pi}{4}$. Therefore, choice C is the correct choice.

3. **C** Simple probabilities are determined by finding the ratio of the number of desired outcomes to the total number of possible outcomes. The desired outcomes are the numbers on the spinner that are multiples of five—15 and 25 (two of them). There are eight numbers altogether on the spinner. The desired ratio is $\frac{2}{8} = \frac{1}{4}$, which is choice C. Choices A, B, and D seem to have the correct number of numbers on the spinner but the wrong number of successful or desired numbers.

4. **B** Choice A is incorrect because the increase in calls was not constant from 1993 to 1998. Choice C is incorrect because the maximum number of calls in 1997 was between 3,600 and 4,800. Choice D is incorrect because the total number of calls in 1995 and 1996 was 6,000. Choice B is the correct answer—there were more calls in 1997 than in 1993.

5. **C** The total cost for the tickets is always 7 times the number of tickets purchased. Therefore, choice C is the correct choice.

6. **A** To find the equation of a line, the slope-intercept form of the equation of a line ($y = mx + b$) can be used. In this equation, m represents the slope of the line and b represents the y-intercept of the line. To find the slope of the line, use the two given points and the formula for slope (found in the reference sheet).

$$m = \frac{y_2 - y_1}{x_2 - x_1}$$ Substitute the slope $\frac{2}{3}$ and the x and y

$$m = \frac{5 - (-1)}{6 - (-3)}$$ values of one of the points into the

$$m = \frac{5 + 1}{6 + 3}$$ $y = mx + b$ equation.

$$m = \frac{6}{9}$$ $5 = \frac{2}{3}(6) + b$

$$m = \frac{2}{3}$$ $5 = 4 + b$

  $1 = b$

Therefore, choice A is the correct choice.

7. **D** A line with an undefined slope is vertical. Choice D is the correct choice.

8. **A** The rotation in a circle is 360°. If the shaded region is one-sixth of the circle, then the measure of $\angle AOB = 60°$. Therefore, triangle AOB is an equiangular triangle. If a triangle is equiangular, then it is equilateral. If $OA = 8$, then $AB = 8$. Choice A is the correct choice.

9. **D** Each of the choices has a y-intercept of (0, 2); each of the choices has a slope of -1. Therefore, the correct choice depends on you deciding correctly if the line is dotted or solid and which side of the line to shade. Because the line is solid you should choose an inequality that is $\leq$ or $\geq$. Therefore, choices A and B are incorrect. The next step is to choose an ordered pair that is obviously in the shaded region, say (2, 2). Substitute the values for x and y into choices C and D.

$y \leq -x + 2$ $y \geq -x + 2$
$2 \leq -(2) + 2$ $2 \geq -(2) + 2$
$2 \leq 0$ $2 \geq 0$
False   True

Choice D is the correct choice.

10. **D** The converse of a statement is a statement that exchanges the if and then clauses of the given statement. Choice D is the correct choice. Choice A is the inverse of the statement; choice B is the original statement with the word "not" in the then part of the statement; choice C is the contrapositive of the given statement.

11. **C** In order to find the value of n that makes $A_n = 13$ in the sequence, replace A_n with 13 in the sequence rule.

$$A_n = \frac{3n + 5}{2}$$
$$13 = \frac{3n + 5}{2}$$
$$26 = 3n + 5$$
$$21 = 3n$$
$$7 = n$$

C is the correct choice. Each of the other choices are incorrect because of various algebraic mistakes.

12. **C** Simple probabilities are determined by finding the ratio of the number of desired outcomes to the total number of possible outcomes. There were 18 socks in the drawer originally but now there are only 17 socks in the drawer because one of the red socks was removed and not replaced. You want the probability related to the black socks—there are 8 black socks in the drawer. Therefore, the correct ratio is $\frac{8}{17}$, which is choice C.

13. **A** The expected probability when flipping a fair coin is one-half; you will get heads one-half of the time and tails the other half of the time. Choice A is the correct choice because she got heads five out of ten times—five out of ten is equal to one-half.

14. **B** Choices A, C, and D are each true statements as they relate to the graph. Choice B is a false statement—after 35 years, the account with a return rate of 8% earned $200 and the account with a return rate of 10% earned $300. Two hundred dollars is not half of $300.

15. **B** To find the discount Nick would receive on the CD, you need to find 30% of $13.95. You must change 30% to the decimal, 0.30. The word "of" in the problem

is an indication that you need to multiply. Therefore, choice B is the correct answer—you are multiplying 0.30 times $13.95.

16. **C** The inequality should be solved in the following manner:
$$2x - 3 \geq 11$$
$$2x \geq 14$$
$$x \geq 7$$
Choice C is the correct choice.

17. **A** The definitions of the trigonometric ratios are on the reference sheet. Notice that the sides of the triangle given or identified with x in the diagram. The side marked x is the hypotenuse of the triangle. The side identified as having a length of 5 is adjacent to the angle whose measure is 70°.
The correct ratio for this information is the cosine; the cosine is the ratio of the leg adjacent to the acute angle to the hypotenuse of the triangle. Choice A is the correct choice.

18. **B** In a 30°-60°-90° triangle, the leg opposite the 30° angle is one-half the length of the hypotenuse. Therefore, BE is 9. In $\triangle CBE$, $BE = CE$. Therefore, $BC = 9$. Choice B is the correct choice.

19. **B** The graphical solution to a system of equations can be found by determining the point of intersection of the two lines. The two lines that represent the system intersect at the point (3, 2). Therefore, the correct answer is choice B.

20. **C** A counterexample is an example that contradicts the truth of a statement. The measure of an acute angle, x, must satisfy the inequality $0 \leq x < 90$. Choice C is less than 110°, but it is not an acute angle. It is a counterexample to the given statement.

21. **B** The order of the terms in an algebraic expression, descending or ascending, is determined by the exponents on the variable in the expression. For an expression to be in descending order, the term with the largest exponent must be first, the term with the next smallest exponent second, and so on down to the term containing the smallest exponent. D is incorrect because the terms are in ascending order. C is incorrect because the choices are not arranged in any order. A is incorrect because the last three terms are not in the correct order. B is the correct choice—the exponents are arranged in descending order, 4, 3, 2, 1, 0.

22. **B** Choices A, C, and D are incorrect because they represent very specific portions of the school population. Choice B is correct because it offers the capability of collecting a random sample from the entire school population.

23. **C** A dependent event is an event that is affected by a previous event. Choices B and D are not dependent events because a second spin of the spinner or a second toss of a coin is not affected by the first spin or toss. Choice A is not a dependent event because the first card drawn is put back in the deck of cards and therefore does not affect the second draw of a card. Choice C is a dependent event because the second draw of a card will be affected by the fact that the first card was not put back in the deck.

24. **B** The month that has the most would be the month in which the graph reaches its maximum value, which is June. Choice B is the correct choice.

25. **B** The slopes of parallel lines are equal. When you solve the given equation for y, the slope is $\dfrac{-4}{3}$. Choice B is the only equation that also has a slope of $\dfrac{-4}{3}$.

26.   **B**   The equation should be solved in the following manner:

$$2(6x-7)=6x-24$$
$$12x-14=6x-24$$
$$6x=-10$$
$$x=\frac{-10}{6}$$
$$X=-\frac{5}{3}$$

Choice B is the correct choice.

27.   **A**   You can find the *x*-intercept of a line by letting $y = 0$. If $y = 0$ in the equation, $x = 4$. You can find the *y*-intercept of a line by letting $x = 0$. If $x = 0$ in the equation, $y = 3$. Therefore, the intercepts are (4, 0) and (0, 3); the correct choice is A.

28.   **C**   Choice A is incorrect because the given set contains fractions, decimals, and negative numbers. Those numbers are not members of the set of natural numbers. Choice B is incorrect because the given set contains fractions and decimals. Those numbers are not members of the set of integers. Choice D is incorrect because there are no irrational numbers in the given set. Choice C is the correct answer. All of the numbers in the given set are rational numbers.

29.   **A**   The slope of the given equation is 2 and its *y*-intercept is 1. Notice what the graph of the given equation looks like. The graph crosses the *y*-axis at (0, 1). From that point, the next identified point is up 2 and to the right 1. The only thing that is being changed is the *y*-intercept—it changes to 3. Choice A is the only choice that has a *y*-intercept of 3.

30.   **B**   The premise, "like mathematics," is fulfilled in the statement "Kelly likes mathematics." Therefore, the conclusion in the given statement "you will like astronomy" is true. Choice B is the correct answer—Kelly will like astronomy.

31.   **C**   This expression involves the multiplication of two scientific numbers. That multiplication is carried out by first multiplying the two decimal numbers, 3.6 and 2.1. If you notice the four choices all contain 7.56. That is a clue to you that the product of 3.6 and 2.1 must be 7.56. You don't actually have to do the multiplication. The next thing you must do is determine the exponent on the product of $(10^4)(10^7)$. Exponents count factors—you have 4 factors of ten and 7 factors of ten so you have altogether 11 factors of ten. Choices A and D are incorrect because the exponent on ten is either the difference of 4 and 7 or the product of 4 and 7. Choice B is incorrect because of an arithmetic error in combining 4 and 7. Choice C is the correct choice.

32.   **C**   A census must represent data that is collected from the entire desired population, in this case, the school. Choices A, B, and D represent groups that are just part of the school population. Choice C is the correct choice.

33.   **D**   The number of possible outcomes can be found using an outcome tree, a chart, or the counting principle. It is probably easiest to use the counting principle in this problem. The counting principle says that you can determine the total number of outcomes by multiplying the number of choices for each item in the cruise package. Therefore, we could find the total number of possible outcomes by multiplying 4 • 3 • 7, which is 84. Choice D is the correct choice.

34.   **B**   The range of a function is the set of numbers that the dependent variable can be. In this case, it is the postage rate for the given weights. B, {0.39, 0.63, 0.87, 1.11, 1.35}, is the correct choice. Choice A is incorrect; it is the domain of the function. Choices C and D are incorrect because they are sets of ordered pairs.

35. **A** To evaluate an expression is to substitute the given values into the expression and determine the numeric value of the expression.

$$3(x - 2) + 3y = 3(4 - 2) + 3 \cdot 2 = 3(2) + 3 \cdot 2 = 6 + 6 = 12$$

Choice A is the correct choice.

36. **C** The means and extremes of a proportion are the terms that are diagonally across from each other. To solve a proportion, you find the product of the means and the extremes and solve the resulting equation.

$$\frac{8}{6} = \frac{x+4}{x-1}$$
$$6(x+4) = 8(x-1)$$
$$6x + 24 = 8x - 8$$
$$-2x = -32$$
$$x = 16$$

Therefore, the correct choice is C.

37. **C** Each of the equations in the choices are solved for r. To solve for r means you need to isolate the variable r by performing the opposite of the indicated operations in the correct order.

$$A = 4\pi r^2$$
$$\frac{A}{4\pi} = r^2$$
$$\sqrt{\frac{A}{4\pi}} = r$$
$$\frac{\sqrt{A}}{\sqrt{4} \cdot \sqrt{\pi}} = r$$
$$\frac{\sqrt{A}}{2\sqrt{\pi}} = r$$

Choice C is the correct choice.

38. **D** Choices A, B, and C each reflect or rotate A into A'. The translation described in choice D does not transform A into A'. Using the translation described in D would result in the point of the heart being at the bottom of A', not the top.

39. **D** $\overline{AC} \cong \overline{AC}$ by the reflexive property. Using this segment together with the given congruent angles, the triangles can be proven congruent using ASA. Choice D is the correct choice. The phrase "most directly" is contained in the stem of the problem because you could prove $\angle B \cong \angle D$ and then use AAS.

40. **B** Choice A is incorrect because the exponents are incorrect. Choice C is incorrect because not all of the terms are similar and cannot be combined. Choice D is incorrect because the expression has not been completely simplified. Choice B is the correct choice.

41. **C** The value of t in the problem is 3. $2^3 = 8$. Therefore, you are multiplying 302 times 8.

$$p(t) = 302(2^t)$$
$$p(3) = 302(2^3)$$
$$p(3) = 302(8)$$
$$p(3) = 2,416$$

This answer is closest to choice C.

42. **A** A graph that represents a positive correlation shows one variable increasing as the other one also increases. The points must follow a flow that, when read left to right, follows an upward slope. Choices C and D are incorrect because the data points are all over the graph—there is no upward flow from left to right. Choice B is incorrect because the flow of the data in downward; that is, as one variable increases the other decreases. Choice A is the correct answer—the flow of the data is upward when read left to right.

43. **B** The first question you must ask yourself in this problem is "Is this a combination problem or a permutation problem?" Does order matter or not? Order does matter because the trophies would be awarded differently if someone was in first place rather than second place. Therefore, this problem is a permutation problem. The formula for finding permutations is on the reference sheet,

$$_nP_r = \frac{n!}{(n-r)!}$$

The variable n represents the total number of items, which is the 15 entries. The variable r represents the number of items in the group which is four. Therefore, the number of ways the judges can award the trophies is

$$\frac{15!}{(15-4)!} = \frac{15!}{11!}$$

which is choice B.

44. **C** Choice C is the correct choice. It shows the oven heating up to the set temperature and the thermostat maintaining that temperature.

45. **B** Choose any two ordered pairs from the table, say (1, –2) and (4, 1). Find the slope of the line containing them.

$$\frac{1-(-2)}{4-1} = \frac{1+2}{4-1} = \frac{3}{3} = 1$$

Choices A and B are the only equations that have a slope of 1. One of the ordered pairs in the table, (0, –3), is the y-intercept of the line. The y-intercept in choice B is –3. Choice B is the correct choice.

46. **A** If the slopes of two lines are equal and their equations are unique equations, the lines are parallel. If the slopes of two lines are opposite reciprocals (their product is –1), the lines are perpendicular. If the slopes of two lines are different but not opposite reciprocals, the lines intersect but are not perpendicular. In this problem, the slope of both lines is 3, and they are unique equations (their y-intercepts are different); the two lines are parallel. Choice A is the correct choice.

47. **B** A set is considered to be finite if you can count the number of elements in the set. Choices C and D are incorrect because they go on forever in one direction. You cannot count the number of elements in the sets. Choice A is incorrect because there are infinitely many rational numbers, fractions, and decimals, between 7 and $\sqrt{144}$. Think of all of the fractions you could write between 7 and

12: $7\frac{1}{2}, 7\frac{1}{4}, 7\frac{1}{9}, 7\frac{2}{11}, 7\frac{5}{19}, 8\frac{3}{4}, 8\frac{3}{5}, 9\frac{3}{13}$, etc. Choice B is correct because you can count the number of integers between −416 and 124.

48. **C** The ratio of the given corresponding sides is $\frac{4}{8} = \frac{3}{6}$ which equals $\frac{1}{2}$. Therefore the corresponding sides are proportional. $\angle ACB \cong \angle DCE$ because they are vertical angles. Therefore, the triangles are similar because of the SAS similarity theorem. Choice C is the correct choice.

49. **B** The formula for the distance between two points can be found in the reference sheet.

$$AB = \sqrt{(x_2 - x_1)^2 + (y_2 - y_1)^2}$$
$$AB = \sqrt{(2 - (-3))^2 + (7 - (-5))^2}$$
$$AB = \sqrt{(5)^2 + (12)^2}$$
$$AB = \sqrt{25 + 144}$$
$$AB = \sqrt{169}$$
$$AB = 13$$

The correct answer is choice B.

50. **C** The algorithm described in Steps 1 and 2 is the algorithm used to solve radical equations. Choice C is the correct answer.

$$\sqrt{x} - 9 = 0$$
$$\sqrt{x} = 9$$
$$x = 81$$

Choices A and D are quadratic equations and would be solved by factoring or the quadratic formula. Choice B is a linear equation requiring you to subtract 5 from both sides of the equation.

51. **D** The area of a square is found using the formula, $A = s^2$. Therefore, $180 = s^2$ and $\sqrt{180} = s$. Your task is to determine a reasonable estimate of $\sqrt{180}$. Choice D is the best choice. $\sqrt{100} = 10$, $\sqrt{121} = 11$, $\sqrt{144} = 12$, $\sqrt{169} = 13$, and $\sqrt{196} = 14$.

$\sqrt{180}$ is closer to 13 than any of the other choices. Choice D is the correct choice.

52. **A** A trend line, or line of best fit, should follow the flow of the data, with approximately half of the data points above the line and half below the line. Choice B is incorrect because the trend line is above all of the data points. Choices C and D are incorrect because they do not follow the flow of the data. Choice A is the correct answer.

53. **D** Choice D is correct. The only triangle in which the sum of the squares of the legs of the triangle is equal to the square of the hypotenuse is a right triangle. Even though an isosceles triangle can sometimes be a right triangle it is not always a right triangle. The choice that is not always true is choice D.

54. **D** Choice D is the correct answer because all of the rules for order of operations were followed. Choices A and B are incorrect because in the second line 4 + 8 was

added before the multiplication and division were completed. Choice C is incorrect because in the third line, 3 – 1, was subtracted before the multiplication and division was completed.

55. **D** The set of integers does not include fractions or decimals. It also does not include nonperfect square roots. Choices A, B, and C each have a decimal number, a fraction, and/or a nonperfect square root. The numbers in the set for letter D are all integers.

56. **D** The expression $A + 2B - C$ means you are to add matrix A to two times matrix B and then subtract matrix C. Order of operations must be followed with matrices as with real numbers.

$$\begin{matrix} A & + & 2B & - & C \end{matrix}$$

$$\begin{bmatrix} 1 & -1 \\ 0 & 1 \end{bmatrix} + 2\begin{bmatrix} -1 & 1 \\ -1 & 0 \end{bmatrix} - \begin{bmatrix} -1 & -1 \\ 0 & 1 \end{bmatrix}$$

$$\begin{bmatrix} 1 & -1 \\ 0 & 1 \end{bmatrix} + \begin{bmatrix} -2 & 2 \\ -2 & 0 \end{bmatrix} - \begin{bmatrix} -1 & -1 \\ 0 & 1 \end{bmatrix}$$

$$\begin{bmatrix} -1 & 1 \\ -2 & 1 \end{bmatrix} - \begin{bmatrix} -1 & -1 \\ 0 & 1 \end{bmatrix} = \begin{bmatrix} 0 & 2 \\ -2 & 0 \end{bmatrix}$$

Therefore, choice D is the correct choice.

57. **D** The hierarchy of quadrilaterals and the characteristics of those quadrilaterals is an important part of geometry. Choice D is correct because a rectangle always has opposite sides congruent and parallel, and it always has congruent diagonals. Choice A is incorrect because a parallelogram does not always have congruent diagonals. Choice B is incorrect because a trapezoid only has congruent diagonals when it is an isosceles trapezoid. Choice C is incorrect because a rhombus only has congruent diagonals when it is a square.

58. **C** The triangle formed by the shelf, the wall, and the brace is a 45°-45°-90° triangle. The length of the shelf and the distance between the shelf and the place where the brace is attached to the wall are the legs of the isosceles right triangle; those legs are congruent. If the shelf is 12 in., then the brace will be attached 12 in. below the shelf. Choice C is the correct answer.

59. **C** The formula for finding the volume of a cone can be found on the reference sheet. The first step is finding the radius of the circular base; it is 3 feet.

$$V = \frac{1}{3}\pi r^2 h$$

$$V = \frac{1}{3}\pi (3)^2 (9)$$

$$V = \frac{1}{3}\pi \cdot 3 \cdot 3 \cdot 9$$

$$V = \frac{1}{3}\pi \cdot 3 \cdot 3 \cdot 9$$

$$V = 27\pi \text{ units}^3$$

The volume of the cone is 27π units^3; the correct choice is C.

60. **B** The scale factor will be the ratio of the size of the image (2 in.) to the size of the pre-image (6 in.).

$$\frac{2}{6} = \frac{1}{3}$$

The correct answer is choice B.

61. **A** To find a reasonable value for the expression, $\dfrac{1+\sqrt{15}}{2}$, you must determine

a reasonable estimate of $\sqrt{15}$. $\sqrt{15}$ is very close to $\sqrt{16}$, which is equal to 4.

Therefore, $\dfrac{1+\sqrt{15}}{2}$ is very close to $\dfrac{1+4}{2}$, which is equal to $\dfrac{5}{2}$ or 2.5. Choice A

is the correct answer.

62. **B** Because all of the scores in the table are median scores, they only indicate the middle score at each school. Choice A is incorrect because you do not know what the highest score from school #1 is—you only know the middle score. The highest score may or may not have come from that school. Choice C is incorrect because even though the median score is 78 there could be several students who scored below 70. Choice D is incorrect—the median score is 73 which means half of the students scored above 73 and half scored below 73, not 76. Choice B is correct—the median score at school #4 is 65, which means half of the students scored at or above 65.

63. **D** Because order is important this problem requires the use of permutations to solve. The formula for permutations from the reference sheet is

$$_nP_r = \frac{n!}{(n-r)!}$$

The variable n represents the total number of items, which is the four letters. The

variable r represents the number of items in the group which is also four.

Therefore, the number of ways the judges can award the trophies is

$\dfrac{4!}{(4-4)!} = \dfrac{4!}{0!} = 4!$. $4! = 4 \cdot 3 \cdot 2 \cdot 1 = 24$, which is choice D. Remember, that 0! is

defined as 1.

64. **C** Each of the choices correctly has four as the first term. Each subsequent term is found by multiplying the preceding term times two and adding six. Therefore, the second term should be $2 \cdot 4 + 6$, which is 14. Choices A and B are incorrect because the second term in each of them is not 14. The third term is found by multiplying 14 times two and adding six. Therefore, the third term should be $2 \cdot 14 + 6$, which is 34. The third term of choice D is not 34. The third term of choice C is 34. Choice C is the correct choice.

65. **B**

$$\frac{(6x^{-2}y^{-6})(8x^3y^{-4})}{(4x^{-4}y^{-2})} = \frac{(6)(8x^3)(x^4y^2)}{(4)(x^2y^6)(y^4)}$$

Begin to simplify the expression by simplifying the variable factors with negative exponents. This requires changing the sign of the exponent and moving the factor.

$$\frac{(6)(8x^3)(x^4y^2)}{(4)(x^2y^6)(y^4)} = \frac{(6)(8)(x^7y^2)}{(4)(x^2y^{10})}$$

Then multiply the variable expressions by adding the exponents.

$$\frac{(6)(8)(x^7y^2)}{(4)(x^2y^{10})} = \frac{48x^7y^2}{4x^2y^{10}} = \frac{12x^5}{y^8}$$

Then multiply and divide the numerical coefficients and the like variable factors. Choice B is the correct choice.

Choice B is the correct choice.

66. **B** $\sqrt{64x^{16}y^6} = \sqrt{8\cdot8\cdot x^8\cdot x^8\cdot y^3\cdot y^3} = 8x^8y^3$. Therefore, choice B is the correct choice.

67. **A** $\angle ONL$ is an inscribed angle and inscribed angles are one-half of their intercepted angles. $\angle ONL$ is one-half of $\overset{\frown}{LQO}$ which is 180°. $m\angle ONL = 90°$, which is choice A.

68. **A** If A is translated 2 units left, the x-coordinate of A will be 2 less than 3. If A is translated 3 units up, the y-coordinate will be 3 units more than -1. Therefore, the image of A is $A'(1, 2)$, which is choice A.

69. **B** The property being modeled is the distributive property. Don't let the fraction scare you. The distributive property requires you to multiply $\frac{1}{4}$ times x and 12.

70. **A** Choice A is the correct choice. The last line of the solution is incorrect—the sense of the inequality should have been reversed. All of the other choices contain correct solutions.

71. **D** The questions that Marybeth put on her survey should be related to the theme of the dance and the refreshments that will be served at the dance. The only choice that does not contain a question about the theme or the refreshments is in choice D. Choice D is the correct answer.

72. **D** As you start to solve this problem, it is important to look closely at the table. Looking at the table along the left side (the vertical scale), as the relative humidity increases from 30 to 60, the first column of temperatures increases from 84 to 90. Looking at the table along the bottom (the horizontal scale), as the temperatures increase from 85 to 105, the first row of temperatures increases from 84 to 113. Throughout the whole table as the humidity increases the temperature increases, and as the temperature increases the humidity increases. Choice D is the correct choice.

73. **C** This question is a combination question because the two member team is the same no matter in which order the team members are chosen. The formula for finding combinations, which can be found on the reference sheet, is

$$_nC_r = \frac{n!}{(n-r)! \cdot r!}$$

The variable n represents the total number of students in the pool, which is six. The variable r represents the number of students who will be on the committee, which is two. Therefore, the number of ways the judges can award the trophies is

$$_nC_r = \frac{6!}{(6-2)!\cdot 2!} = \frac{6!}{4!\cdot 2!} = \frac{\overset{3}{\cancel{6}}\cdot 5\cdot \cancel{4\cdot 3\cdot 2\cdot 1}}{\cancel{4\cdot 3\cdot 2\cdot 1}\cdot \cancel{2}\cdot 1} = 15$$, which is choice C.

74. **A** The pattern of squares can represented with the numeric pattern, $1^2, 3^2, 5^2, 7^2, 9^2, 11^2, 13^2, 15^2, \ldots$. The seventh number in the pattern is 13^2, which is 169. Choice A is the correct choice.

75. **A** The word sum means addition. Three times the number x is represented by $3x$. Choice A is the correct answer.

76. **D** $(7ab)^2 = (7ab)(7ab) = 7 \cdot 7 \cdot a \cdot a \cdot b \cdot b = 49a^2b^2$. Therefore, choice D is the correct choice.

77. **C** Choice A is acceptable—the numerator/denominator division by ten is okay.

$$45\pi = \frac{x}{\underset{20}{\cancel{200}}}(\overset{2}{\cancel{20}}\,\pi)$$

Choice B is acceptable—both sides of the equation were multiplied by 20.
Choice D is acceptable—the numerator/denominator division by 20 is okay.

$$45\pi = \frac{x}{\underset{10}{\cancel{200}}}(\overset{1}{\cancel{20}}\,\pi)$$

Choice C is the correct answer—it appears that 20π is being subtracted from both sides of the equation. Multiplication is the indicated operation—subtraction is not the inverse of multiplication.

78. **A** From the equation, $y = \frac{3}{5}x - 1$, you can get the slope and the y-intercept. The slope is $\frac{3}{5}$ and the y-intercept is -1. To graph the equation, start by placing a point at -1 on the y-axis. From that point, do what the slope says, up 3 and to the right 5. Choices C and D are incorrect because the y-intercept value was placed on the x-axis. Choice B is incorrect because the slope was interpreted backwards. Choice A is the correct choice.

79. **A** The length of $\overset{\frown}{AB}$ will be one-sixth the circumference of circle O. The formula from the reference sheet is

$$\text{length of } \overset{\frown}{AB} = \frac{m\overset{\frown}{AB}}{360°}\left(2\pi r\right)$$

One-sixth of the circle is $60°$. Therefore,

$$\text{length of } \overset{\frown}{AB} = \frac{60°}{360°}\left(2\pi \cdot 12\right)$$

$$\text{length of } \overset{\frown}{AB} = \frac{1}{6}\left(24\pi\right) = 4\pi \text{ units}$$

The correct answer is choice A.

80.   B   The total surface area is the lateral surface plus the circular base. The formula taken from the reference sheet is $T = \frac{1}{2}(2\pi r)\ell + \pi r^2$. The radius, r, is 5 but we must find ℓ, the slant height. Using the Pythagorean theorem and the given right triangle,

$$5^2 + 12^2 = \ell^2$$
$$25 + 144 = \ell^2$$
$$169 = \ell^2$$
$$13 = \ell$$

$$T = \frac{1}{2}(2\pi \cdot 5)13 + \pi\left(5\right)^2$$
$$T = (5\pi)13 + 25\pi$$
$$T = 65\pi + 25\pi$$
$$T = 90\pi$$

The correct choice is B.

81.   C   The information in the question contains the times 11:00, 12:00, 1:00, 2:00, 3:00, and 4:00. Check each of the tables to be sure the times are correct. Then check to see that each of the times is matched with the correct number of people. Choice A is incorrect because 11:00 is matched with 102—it should be matched with 80. Choice B is incorrect because 12:00 is matched with 142—it should be matched with 117. Choice D is incorrect because 11:00 is matched with 142—it should be matched with 80. Choice C is the correct choice—all of the times match with the correct number of people.

82.   C   The cost of the vehicle increases by $5,000 every two years. From 2004 to 2008, there would be two increases, one in 2006 and the second in 2008. Therefore, the car would increase in cost by $10,000 for a total cost of $35,000. Choice C is the correct choice.

83.   D   The values of n, given in the table, are 1, 2, 3, 4, and 5. One way of solving this problem is to substitute each of the values of n into each of the expressions to see if you get the given value of the expression. Using $n = 1$, the value of the expressions in the choices would be

$$2n + 10 = 2(1) + 10 = 12$$
$$n + 9 = (1) + 9 = 10$$
$$4n + 1 = 4(1) + 1 = 5$$
$$3n + 4 = 3(1) + 4 = 7$$

The only value that matches the value in the table is 7 which is from the expression in choice D.

84.   D   A function maps the values of the independent variable into one and only one value of the dependent variable. Choice D is not a function because zero is mapped into two different values, –2 and –8. Look for multiple occurrences of independent values.

85.   B   If the cost of labor is $40 per hour, the cost of h hours of labor can be represented by $40h$. The cost of the service call, $35, is added to $40h$. Choice B is the correct answer.

86.   **C**   The first step in solving a radical equation or an equation containing a radical is to isolate the radical expression. The equation in this problem is already in that form. The next step is to remove the radical symbol by squaring both sides of the equation.

$$\sqrt{3x-8} = 2$$
$$\left(\sqrt{3x-8}\right)^2 = 2^2$$
$$3x-8 = 4$$
$$3x = 12$$
$$x = 4$$

Choice C is the correct choice.

87.   **A**   Choice A is not the net of a cube. If the faces shown below are folded, they will form a five-faced figure.

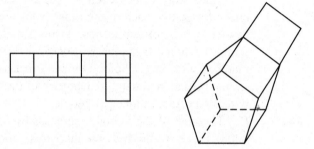

The sixth face would be attached to one of the five faces.

88.   **B**   The dilation of a figure causes a figure to get larger or smaller. Therefore, the dimensions will not be preserved. Choice B is the correct choice.

89.   **D**   The sum of the interior angles is found by using the formula $S = (n - 2)180$, where n is the number of sides. The polygon has six sides.

$$S = (n - 2)180$$
$$S = (6 - 2)180$$
$$S = 4 \cdot 180$$
$$S = 720$$

Therefore choice D is the correct choice.

90.   **A**   The x-coordinate of the ordered pair for the midpoint of a segment can be found by dividing the sum of the x-coordinates by 2; the y-coordinate can be found by dividing the sum of the y-coordinates by 2. You can also use the formula for midpoint from the AIMS reference sheet.

$$\left(\frac{3+(-6)}{2}, \frac{5+1}{2}\right)$$
$$\left(\frac{-3}{2}, \frac{6}{2}\right)$$
$$\left(-1\frac{1}{2}, 3\right)$$

The correct answer is choice A.

91.   **C**   The given statement combines two numbers that are opposites, or inverses. Choice C is the correct answer—it is the property of inverses.

92. **B** If the scores are arranged in order, largest to smallest—95, 90, 80, 75, and 20—the middle score, or the median, is 80. Each of the scores appears only one time so there is no mode. The range is the difference between the largest score and the smallest score, 95 – 20 = 75. The mean (or average) is found by adding the scores (95 + 90 + 80 + 75 + 20 = 360) and dividing that sum by five (360 ÷ 5 = 72). The median is 80, the range is 75, the mean is 72, and there is no mode. The largest of these is the median, 80. Choice B is the correct choice.

93. **D** Notice the numbers in the pattern increase by an increasing difference. The difference between the first and second is two, the difference between the second and third is four, the difference between the third and fourth is six. If the pattern continues, the difference between the subsequent terms will be eight, ten, and twelve. Add each difference to the listed term to get the next term.

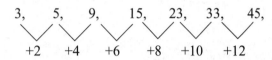

The correct choice is choice D.

94. **B** The portion of the graph from A to B shows an increase in the bank balance. A deposit will increase the balance in a bank account. Choice B is correct.

95. **B** If Jamie worked 40 hours for 2 weeks, she worked 80 hours altogether. If she earned $7.35 for each of those 80 hours, you would multiply 80 times $7.35 to get her total pay. Even though you cannot use a calculator on this test, this is a reasonably easy computation. The correct choice is choice B. Choices C and D are not even reasonable answers. Choice A is only Jamie's pay for one week.

96. **B** To solve this equation, use the quadratic formula from the reference sheet or factoring.

Quadratic Formula Method
Identify the value of a, b, and c.

$a = 1$, $b = 10$, and $c = -24$

$$x = \frac{-b \pm \sqrt{b^2 - 4ac}}{2a}$$

$$x = \frac{-(\ ) \pm \sqrt{(\ )^2 - 4(\ )(\ )}}{2(\ )}$$

$$x = \frac{-(10) \pm \sqrt{(10)^2 - 4(1)(-24)}}{2(1)}$$

$$x = \frac{-10 \pm \sqrt{100 + 96}}{2}$$

$$x = \frac{-10 \pm \sqrt{196}}{2}$$

$$x = \frac{-10 \pm 14}{2}$$

$$x = \frac{-10 + 14}{2} \text{ or } \frac{-10 - 14}{2}$$

$$x = \frac{4}{2} \text{ or } \frac{-24}{2}$$

$$x = 2 \text{ or } -12$$

Factoring Method

$$x^2 + 10x - 24 = 0$$

$$(x + 12)(x - 2) = 0$$

$$x = -12 \text{ or } x = 2$$

97.   **D**   The algebraic expressions, $2x + 10$ and $3x - 40$, represent corresponding angles. If two parallel lines are cut by a transversal, corresponding angles are congruent. Therefore, $2x + 10 = 3x - 40$, and $x = 50$. Choice D is the correct choice.

98.   **B**   The vertex of the graph of the quadratic is the y-intercept of the graph. The y-intercept of the graph is $(0, 3)$. Therefore, choices A and B are incorrect. Because the equation has $-x^2$ as its first term, the graph opens downward. Choice B opens downward and has a y-intercept of $(0, 3)$. Choice B is the correct answer.

99.   **C**   The side of the square is 8 units; the area of the square is 64 units squared. At the vertices of each of the squares is three-fourths of a circle with radius 4. There are four of these three-fourth circles.

$$A = \pi r^2$$
$$A = \pi (4)^2$$
$$A = 16\pi$$
$$\frac{3}{4}(16\pi) = 12\pi \text{ sq units}$$

$4(12\pi) = 48\pi$ sq units
The square plus the partial circles is $64 + 48\pi$ sq units.
The correct answer is choice C.

100.   **C**   Draw line segment $\overline{FH}$. $\overline{GF}$ is a leg of right triangle GFH. Applying the Pythagorean theorem,
$(GF)^2 + 9^2 = (15)^2$
$(GF)^2 + 81 = 225$
$(GF)^2 = 144$
$GF = 12$

Practice Test 2
Answer Sheet
Fill in the bubble completely. Erase carefully if an answer is changed.

1. A B C D
2. A B C D
3. A B C D
4. A B C D
5. A B C D
6. A B C D
7. A B C D
8. A B C D
9. A B C D
10. A B C D
11. A B C D
12. A B C D
13. A B C D
14. A B C D
15. A B C D
16. A B C D
17. A B C D

18. A B C D
19. A B C D
20. A B C D
21. A B C D
22. A B C D
23. A B C D
24. A B C D
25. A B C D
26. A B C D
27. A B C D
28. A B C D
29. A B C D
30. A B C D
31. A B C D
32. A B C D
33. A B C D
34. A B C D

35. A B C D
36. A B C D
37. A B C D
38. A B C D
39. A B C D
40. A B C D
41. A B C D
42. A B C D
43. A B C D
44. A B C D
45. A B C D
46. A B C D
47. A B C D
48. A B C D
49. A B C D
50. A B C D

Cut along dotted line.

Practice Test 2
Answer Sheet
Fill in the bubble completely. Erase carefully if an answer is changed.

51. Ⓐ Ⓑ Ⓒ Ⓓ
A B C D

52. Ⓐ Ⓑ Ⓒ Ⓓ
A B C D

53. Ⓐ Ⓑ Ⓒ Ⓓ
A B C D

54. Ⓐ Ⓑ Ⓒ Ⓓ
A B C D

55. Ⓐ Ⓑ Ⓒ Ⓓ
A B C D

56. Ⓐ Ⓑ Ⓒ Ⓓ
A B C D

57. Ⓐ Ⓑ Ⓒ Ⓓ
A B C D

58. Ⓐ Ⓑ Ⓒ Ⓓ
A B C D

59. Ⓐ Ⓑ Ⓒ Ⓓ
A B C D

60. Ⓐ Ⓑ Ⓒ Ⓓ
A B C D

61. Ⓐ Ⓑ Ⓒ Ⓓ
A B C D

62. Ⓐ Ⓑ Ⓒ Ⓓ
A B C D

63. Ⓐ Ⓑ Ⓒ Ⓓ
A B C D

64. Ⓐ Ⓑ Ⓒ Ⓓ
A B C D

65. Ⓐ Ⓑ Ⓒ Ⓓ
A B C D

66. Ⓐ Ⓑ Ⓒ Ⓓ
A B C D

67. Ⓐ Ⓑ Ⓒ Ⓓ
A B C D

68. Ⓐ Ⓑ Ⓒ Ⓓ
A B C D

69. Ⓐ Ⓑ Ⓒ Ⓓ
A B C D

70. Ⓐ Ⓑ Ⓒ Ⓓ
A B C D

71. Ⓐ Ⓑ Ⓒ Ⓓ
A B C D

72. Ⓐ Ⓑ Ⓒ Ⓓ
A B C D

73. Ⓐ Ⓑ Ⓒ Ⓓ
A B C D

74. Ⓐ Ⓑ Ⓒ Ⓓ
A B C D

75. Ⓐ Ⓑ Ⓒ Ⓓ
A B C D

76. Ⓐ Ⓑ Ⓒ Ⓓ
A B C D

77. Ⓐ Ⓑ Ⓒ Ⓓ
A B C D

78. Ⓐ Ⓑ Ⓒ Ⓓ
A B C D

79. Ⓐ Ⓑ Ⓒ Ⓓ
A B C D

80. Ⓐ Ⓑ Ⓒ Ⓓ
A B C D

81. Ⓐ Ⓑ Ⓒ Ⓓ
A B C D

82. Ⓐ Ⓑ Ⓒ Ⓓ
A B C D

83. Ⓐ Ⓑ Ⓒ Ⓓ
A B C D

84. Ⓐ Ⓑ Ⓒ Ⓓ
A B C D

85. Ⓐ Ⓑ Ⓒ Ⓓ
A B C D

86. Ⓐ Ⓑ Ⓒ Ⓓ
A B C D

87. Ⓐ Ⓑ Ⓒ Ⓓ
A B C D

88. Ⓐ Ⓑ Ⓒ Ⓓ
A B C D

89. Ⓐ Ⓑ Ⓒ Ⓓ
A B C D

90. Ⓐ Ⓑ Ⓒ Ⓓ
A B C D

91. Ⓐ Ⓑ Ⓒ Ⓓ
A B C D

92. Ⓐ Ⓑ Ⓒ Ⓓ
A B C D

93. Ⓐ Ⓑ Ⓒ Ⓓ
A B C D

94. Ⓐ Ⓑ Ⓒ Ⓓ
A B C D

95. Ⓐ Ⓑ Ⓒ Ⓓ
A B C D

96. Ⓐ Ⓑ Ⓒ Ⓓ
A B C D

97. Ⓐ Ⓑ Ⓒ Ⓓ
A B C D

98. Ⓐ Ⓑ Ⓒ Ⓓ
A B C D

99. Ⓐ Ⓑ Ⓒ Ⓓ
A B C D

100. Ⓐ Ⓑ Ⓒ Ⓓ
A B C D

Cut along dotted line.

Practice Test 2

Directions: Choose the best answer for each of the following.

1. Which number(s) in the set below are natural numbers?

$$\left\{ \sqrt{7}, -2, 0, \frac{4}{9}, -2\frac{1}{4}, 2.5, 4 \right\}$$

   A. 4
   B. 0, 4
   C. 0
   D. −2, 0

2. Subscripts are used to identify the placement of numbers in matrices. In the matrix below, the subscripts are written with the row number first, then the column number.

$$\begin{bmatrix} a_{11} & a_{12} \\ a_{21} & a_{22} \end{bmatrix}$$

   In the matrix below, which number is in the place identified by a_{23}?

$$\begin{bmatrix} 2 & -1 & 3 \\ 7 & 5 & -4 \\ 6 & -8 & 9 \end{bmatrix}$$

   A. −8
   B. −4
   C. 6
   D. 9

3. Which of the values below is closest to the value of the expression below?

$$3\sqrt{15}$$

   A. 9
   B. 12
   C. 15
   D. 21

GO ON ➡

4. You flip a fair coin that shows heads or tails and spin the spinner shown. Which of the following tables represents the number of possible outcomes in the sample space?

A.

H1	T2
H3	T4
H5	T6
H7	T8

B.

H1	T1
H2	T2
H3	T3
H4	T4
H5	T5
H6	T6
H7	T7
H8	T8

C.

H1
H2
H3
H4
T5
T6
T7
T8

D.

H1	T9
H2	T10
H3	T11
H4	T12
H5	T13
H6	T14
H7	T15
H8	T16

5. Evaluate the expression

$$\frac{4a^2 - y}{x+2} \text{ if } x = -2, y = 3, \text{and } a = -4$$

A. Undefined

B. 0

C. $-\dfrac{67}{4}$

D. $\dfrac{61}{4}$

6. Which number is in the solution set to the following inequality?

$$-(5 - x) > 2x - 16$$

A. 9

B. 11

C. 15

D. 26

GO ON ➡

7. A flower garden is in the shape of a triangle with a height of 39 feet and a base of 21 feet. Which is closest to the area of the garden?

   A. 800 sq. ft.
   B. 600 sq. ft.
   C. 400 sq. ft.
   D. 200 sq. ft.

8. Which of these is equivalent to the equation below?

$$A = \frac{1}{2}h(b_1 + b_2)$$

   A. $b_1 = \dfrac{2A}{h} - b_2$
   B. $b_1 = 2A - h - b_2$
   C. $b_1 = 2Ah - b_2$
   D. $b_1 = \dfrac{2A}{b_2} - h$

9. Which statement describes the relationship between the pre-image $ABCD$ and its image $A'B'C'D'$?

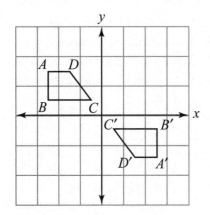

   A. Translation right 3 units, down 3 units
   B. Reflection over the line $y = x$
   C. Rotation of 180° about the origin
   D. Rotation of 180° about the point (−1, 1)

10. What is the total surface area of a rectangular prism with dimensions of 2 in., 3 in., and 4 in.?

   A. 8 sq. in.
   B. 24 sq. in.
   C. 26 sq. in.
   D. 52 sq. in.

11. The junior class used an arch made of balloons tied to a frame to decorate the entrance to the ballroom for the junior-senior prom. The shape of the frame can be modeled using the graph below. What is the maximum height (in meters) of the arch of balloons?

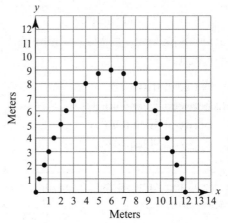

   A. 3 meters
   B. 6 meters
   C. 9 meters
   D. 12 meters

GO ON ➡

Practice Test 2

12. Marci got a new fish tank. She added gravel to the tank, and then she added a decorative rock. Next she added water, let the water rest for a while, and then added the fish.

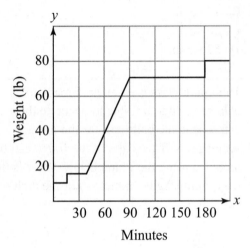

Minutes

In which interval was Marci filling the tank with water?

A. between 0 and 15
B. between 15 and 40
C. between 40 and 90
D. between 90 and 180

13. What is the value of the following expression?

$$3\left[1^2\left(5-3+1\right)^3 \div 9\right]$$

A. $\dfrac{1}{3}$
B. 6
C. 9
D. 18

14. In which of the following problems would you need to use the formula for finding the number of combinations to complete the solution to the problem?

A. Twenty students from your class of 500 will be randomly selected to receive a trip to Disneyland. How many groups of twenty students are possible?
B. Ten students are auditioning for 3 different roles in a play. In how many ways can the 3 roles be filled?
C. In how many ways can 5 people sit in a car that holds 5 passengers?
D. On an exam, you are asked to list 5 historical events in the order in which they occurred. If you don't know the answer and have to guess, how many different orders would you have to choose from?

15. Given lines a, b, and c, which of the following statements is true?

Line a: $3x - 4y = -4$
Line b: $3x - 4y = 8$
Line c: $3x + 4y = 8$

A. Lines a and b are parallel.
B. Lines a and c are parallel.
C. Lines a and c are perpendicular.
D. Lines a and b are perpendicular.

GO ON ➡

16. Which of these equations represents the line graphed below?

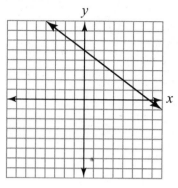

A. $y = \frac{3}{4}x - 5$

B. $y = -\frac{3}{4}x + 5$

C. $y = 5x + \frac{3}{4}$

D. $y = 5x - \frac{3}{4}$

17. Terri bought a bicycle helmet. The original price of the helmet was $39.95. The helmet was on sale for 20% off. Which of the following operations would Terri perform to find the sale price?

A. Divide $39.95 by 0.20
B. Multiply $39.95 by 0.20
C. Divide $39.95 by 0.80
D. Multiply $39.95 by 0.80

18. What is the measure of angle B in the figure below?

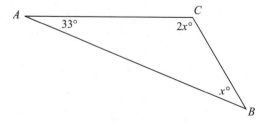

A. 33°
B. 49°
C. 57°
D. 60°

19. Your bank account has $54 in it when you write checks for $56, $57, and $18. You then deposit $50 and $16. How much is in the account? Are you overdrawn?

A. $115, no
B. –$115, yes
C. $5, no
D. –$11, yes

20. The graph of which inequality is shown below?

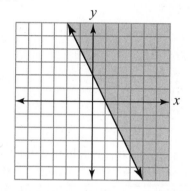

A. $y < -2x + 2$
B. $y > -2x + 2$
C. $y \leq -2x + 2$
D. $y \geq -2x + 2$

21. Which of the following is a rule for the function represented by the graph below?

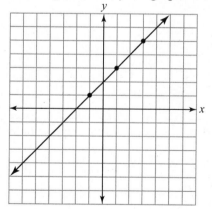

A. Multiply the x value times 3 to get the y value.
B. Multiply the x value times 3 and add 2 to get the y value.
C. Add 3 to the x value to get the y value.
D. Add 2 to the x value to get the y value.

GO ON ➡

Practice Test 2

22. Which property of multiplication is modeled below?

$$\frac{5}{4} \cdot \frac{4}{5} = 1$$

A. Inverse property
B. Distributive property
C. Commutative property
D. Associative property

23. Which monomial should come next in the pattern below?

$$3y, 6y^2, 12y^3, 24y^4, \ldots$$

A. $36y^5$
B. $48y^5$
C. $30y^6$
D. $36y^6$

24. The judges at the marching band contest award prizes for first, second, and third place out of 11 entries. Which expression gives the number of ways the judges can award first, second, and third place?

A. $\dfrac{11!}{8!}$

B. $\dfrac{8!}{11!}$

C. $\dfrac{11!}{3!}$

D. $\dfrac{3!}{11!}$

25. What is the value of the expression below?

$$-3 + |-3| + 7$$

A. 13
B. 7
C. 1
D. 0

26. What is the value of x that will make the following proportion true?

$$\frac{6}{x-5} = \frac{3}{x+1}$$

A. -2

B. $-\dfrac{16}{5}$

C. -7

D. $\dfrac{7}{3}$

27. Given that $\overline{AB} \| \overline{DC}$ and $\overline{AD} \| \overline{BC}$, which of the following is the most specific name that can be given to the quadrilateral $ABCD$?

A. Trapezoid
B. Rectangle
C. Parallelogram
D. Kite

28. What is the range of possible lengths x for the third side of a triangle in which the other two sides are 64 in. and 43 in.?

A. $21 \le x \le 107$
B. $21 < x < 107$
C. $0 < x < 107$
D. $43 \le x \le 64$

29. What is the area of the sector bounded by $\overparen{AB}$ and radii $\overline{OB}$ and $\overline{OA}$ if $m\angle AOB = 60°$ and $OA = 12$.

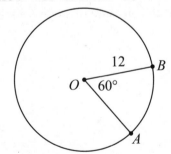

A. 4π units squared
B. 6π units squared
C. 12π units squared
D. 24π units squared

GO ON ➡

30. In order to make some decisions about opening a new movie theater, a theater owner is collecting data and information. Which of the following would not be an appropriate question to include in that information-gathering document?

A. How many times a month do you go to a theater to see a movie?

B. Do you have children who go to see a movie in a theater at least once a month?

C. Would you be more likely to attend a movie in a theater if it was a discount theater ($2–$4 a movie)?

D. Have you ever seen a movie about jet skiing?

31. Which of the following represents an equation that contains the points on the graph below?

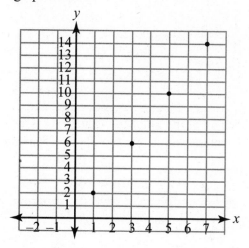

A. $y = 2x + 1$

B. $y = x + 1$

C. $y = \dfrac{x}{2}$

D. $y = 2x$

32. Which of the input/output diagrams does not represent a function?

A.

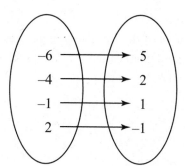

B.

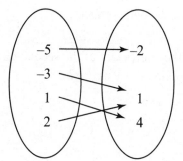

C.

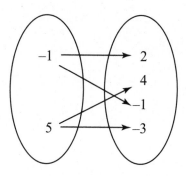

D.

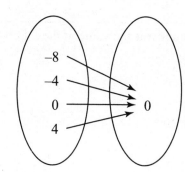

GO ON ➡

Practice Test 2

33. What is the *n*th term in the pattern shown below?

$$1, 4, 7, 10, 13, \ldots$$

A. $n + 3$
B. $4n$
C. $3n + 1$
D. $3n - 2$

---

34. Which of the following values of *z* is a counterexample for the following statement?

> If $x > y$, then $xz > yz$.

A. 3
B. $\dfrac{1}{3}$
C. 3.3
D. -3

---

35. Which of the following sets is a finite set?

A. {all the fractions whose numerator is 1}
B. {the whole numbers less than 5}
C. {irrational numbers less than 12}
D. {all of the odd numbers}

---

36. What is the coefficient of the second term in the expression $2x^3 + 6x + 8$?

A. 2
B. 3
C. 6
D. 8

---

37. Which number is the value of the following expression?

$$(2.5 \times 10^3)(3.2 \times 10^{-5})$$

A. 8.0×10^{-15}
B. 8.0×10^{-8}
C. 8.0×10^{-2}
D. 8.0×10^{2}

---

38. Kathy's average for the first 4 tests in American History was 83. Her fifth test score was 97. Which of these is a reasonable conclusion about Kathy's average in American History after the fifth test?

A. Kathy's average after 5 tests was closer to 83 than 97.
B. Kathy's average after 5 tests was closer to 97 than 83.
C. Kathy always makes high grades in American History.
D. Kathy's average after 5 tests was 90.

---

39. Given the following four statements and quadrilateral *ABCD*.

1. If *ABCD* is a parallelogram, then $\overline{AB} \| \overline{DC}$.

2. If *ABCD* is a square, then *ABCD* is a rectangle.

3. If *ABCD* is a rectangle, then *ABCD* is a parallelogram.

4. If $\overline{AB} \| \overline{DC}$, then $\angle A$ is supplementary to $\angle D$.

Which of the following is the correct logical arrangement of these related if-then statements?

A. 4, 2, 3, 1
B. 2, 3, 1, 4
C. 2, 4, 1, 3
D. 4, 1, 2, 3

GO ON ➡

40. What is the volume of a sphere whose radius is 3 inches?

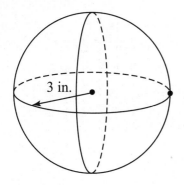

A. 36π in.^{3}
B. 72π in.^{3}
C. 144π in.^{3}
D. 216π in.^{3}

41. What is the algebraic expression that represents twice the sum of 5 times a number x and 6?

A. $(2 + 5)(x + 6)$
B. $5(x + 4)$
C. $2(5x + 6)$
D. $2[5(x + 6)]$

42. A quality tester tests a random sample of 20 items. The graph shows the results. Each item should be 16 inches long, but any length in the interval 16 ± 0.1 inch is acceptable.

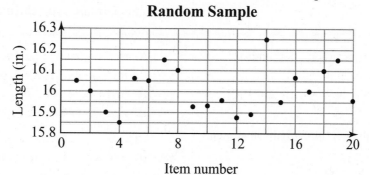

Using the information in the graph, how many items fall within the acceptable interval?

A. 12
B. 14
C. 16
D. 18

GO ON ➡

43. Which of the following is an approximation of a normal distribution?

A.

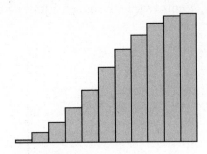

B.

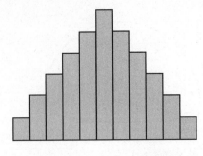

C.

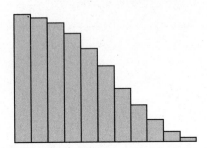

D.

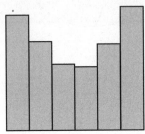

---

44. Which of the following sets is an infinite set?

A. {0, 1, 2, 3, 4, . . ., 9, 10}
B. {even numbers between –200 and –2}
C. {fractions between 0 and 1}
D. {whole numbers between 10 and 100}

---

45. What is the eighth term in the pattern shown below?

1, 8, 15, 22, 29, . . .

A. 36
B. 43
C. 50
D. 57

---

46. Which of the following is the simplified value of the expression below?

$$\sqrt[3]{27x^3y^{12}}$$

A. $9xy^6$
B. $3xy^6$
C. $9xy^4$
D. $3xy^4$

---

47. If m∠A is 24° greater than m∠B. If m∠A and m∠B are supplementary, find m∠A and m∠B.

A. m∠A = 102°, m∠B = 78°
B. m∠A = 24°, m∠B = 66°
C. m∠A = 24°, m∠B = 156°
D. m∠A = 66°, m∠B = 34°

---

48. If $\overline{AO}$ is a radius of circle O and $\overleftrightarrow{BC}$ is tangent to the circle at A, what is the relationship between radius $\overline{AO}$ and tangent $\overleftrightarrow{BC}$?

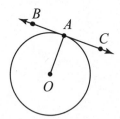

A. $AO = \dfrac{1}{2}BC$

B. $\overline{AO} \parallel \overleftrightarrow{BC}$

C. $AO = 2BC$

D. $\overline{AO} \perp \overleftrightarrow{BC}$

GO ON ➡

49. A computational algorithm is shown below using the ordered pairs (3, 4) and (–2, 6).

$$\frac{4-6}{3-(-2)} = \frac{-2}{5} = -\frac{2}{5}$$

Which of the following does that algorithm accomplish?

A. It finds the slope of a line.
B. It finds the length of a segment
C. It finds the midpoint of a segment.
D. It finds the mean of some data points.

50. What is the area of the figure if all angles are right angles?

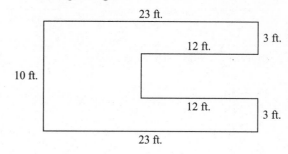

A. 230 sq. ft.
B. 194 sq. ft.
C. 182 sq. ft.
D. 146 sq. ft.

Now would be a good time for you to take a break. When you actually take the AIMS test at your school, you will be given a break at this time.

STOP

Directions: Choose the best answer for each of the following.

51. Solve $\dfrac{2}{3}x + 6 = -18$

    A. −36
    B. −18
    C. 18
    D. 36

52. The hourly parking fees for the local airport from 1996 to 2006 are shown on the line graph. Using this information, what will the hourly parking fee be for 2007?

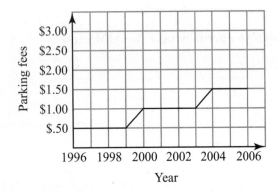

    A. $1.50
    B. $2.00
    C. $2.50
    D. $3.00

53. The following data gives the number of applicants that applied for a job at a given company each month of 1999.

    62, 63, 65, 68, 68, 72,
    76, 78, 79, 82, 83, 85

    What is the median of the data?

    A. 68
    B. 72
    C. 74
    D. 76

54. A point is chosen from the region bounded by quadrilateral *ABCD*. Which of the following represents the probability that it is **not** in △*XYZ*?

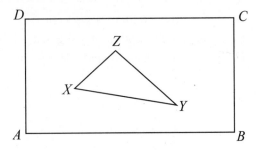

    A. $\dfrac{\text{Area of } ABCD - \text{Area of } \triangle XYZ}{\text{Area of } ABCD}$

    B. $\dfrac{\text{Area of } \triangle XYZ}{\text{Area of } ABCD - \text{Area of } \triangle XYZ}$

    C. $\dfrac{\text{Area of } \triangle XYZ}{\text{Area of } ABCD}$

    D. $\dfrac{\text{Area of } ABCD}{\text{Area of } \triangle XYZ}$

55. Which of the following represents the first four terms in the pattern defined by the recursive rule below?

    $a_1 = 1$
    $a_n = a_{n-1} + 3$

    A. 1, 3, 6, 9
    B. 1, 4, 7, 10
    C. 1, 3, 9, 27
    D. 1, 4, 5, 9

GO ON ➡

56. What is the solution to the following equation?

$$6\sqrt{x-2} + 5 = 35$$

A. 7
B. 23
C. 27
D. 152

57. A wagon wheel has 12 spokes. What is the measure of the angle between any two spokes?

A. 15°
B. 18°
C. 30°
D. 60°

58. Given the conditional: If $m\angle A = 98$, then $\angle A$ is an obtuse angle. Which of the following identifies the relationship between the given statement and the statement below?

  If $\angle A$ is an obtuse angle, then $m\angle A = 98$.

A. Converse
B. Contrapositive
C. Inverse
D. Negation

59. Which of the following inequalities is solved incorrectly?

A. $2x > 4$
  $x > 2$

B. $3x > -6$
  $x > -2$

C. $-10 > 2x$
  $-5 > x$

D. $-3x \leq -12$
  $x \leq 4$

60. Which similarity theorem can be used to prove $\triangle BAC \sim \triangle EDF$?

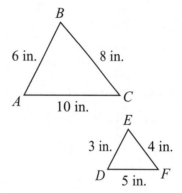

A. SSS
B. AA
C. SAS
D. AAS

61. A theater ticket for adults is A dollars, and the price of a child's ticket is C dollars. If 80 adults and 100 children attend the theater one night, how much money did the theater take in on the night in question?

A. $80C + C \cdot A$ dollars
B. $80A + 100C$ dollars
C. $100A + 80C$ dollars
D. $8{,}000AC$ dollars

62. Using the information in the table and the accompanying graph, what would be the closest approximation of how many calories are in a hamburger that contains 32 grams of fat?

Fat	Calories
25	470
39	680
19	420
34	590
43	660
39	640
35	570

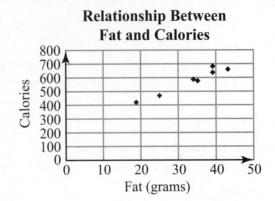

Relationship Between Fat and Calories

A. 720 calories
B. 650 calories
C. 545 calories
D. 400 calories

---

63. Which of the following graphs represents a negative correlation between the two variables?

A.

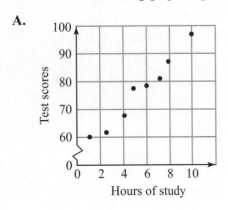

B.

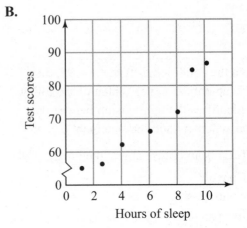

C.

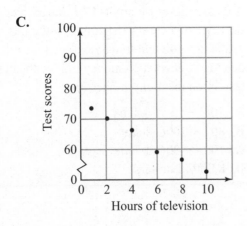

D.

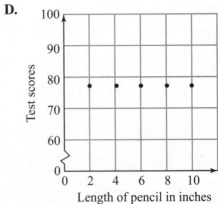

GO ON ➡

64. You write the letters of the word WISDOM on pieces of paper and place them in a bag. You randomly choose an M, do not replace it, and then choose a second letter. What is the probability that you choose the letter S on the second draw?

A. $\dfrac{1}{30}$

B. $\dfrac{1}{6}$

C. $\dfrac{1}{5}$

D. $\dfrac{1}{2}$

---

65. The following graphs show the position of two runners over time as they participate in a Special Olympics fun run. Which one of the graphs shows the two runners moving at the same speed?

A.
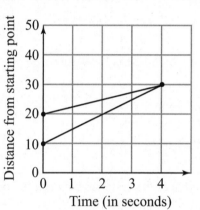

B.
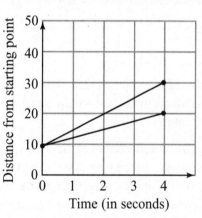

C.
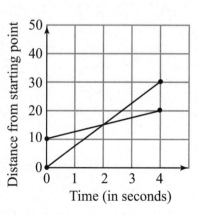

D.
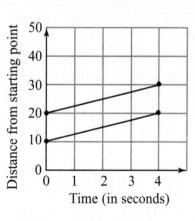

GO ON ➡

66. What are the values of x that make the following equation true?

$$x^2 + 12x - 13 = 0$$

A. $x = -13$ or $x = -1$
B. $x = -13$ or $x = 1$
C. $x = 13$ or $x = -1$
D. $x = 13$ or $x = 1$

---

67. The pre-image, $\triangle ABC$, is transformed into the image, $\triangle A''B''C''$, using two transformations, one following the other.

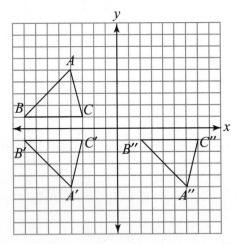

Which of the following describes two possible transformations that would transform $\triangle ABC$ into $\triangle A''B''C''$?

A. $\triangle ABC$ is reflected over the y-axis, and its image is then translated 5 units right.
B. $\triangle ABC$ is reflected over the x-axis, and its image is then translated 5 units right.
C. $\triangle ABC$ is reflected over the y-axis, and its image is then translated 10 units right.
D. $\triangle ABC$ is reflected over the x-axis, and its image is then translated 10 units right.

68. A 15-foot long water slide is attached to a 9-foot high platform. How far is the bottom of the water slide from the base of the platform?

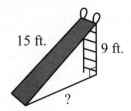

A. 4 feet
B. 8 feet
C. 12 feet
D. 24 feet

---

69. The solution of the equation below contains an error.

$$-5(x + 2) + 7x = 8 + 4x$$

$-5x - 10 + 7x = 8 + 4x$	Line 1
$2x - 10 = 8 + 4x$	Line 2
$2x = 18$	Line 3
$x = 9$	Line 4

In which line is the error?

A. Line 1
B. Line 2
C. Line 3
D. Line 4

---

70. What is the distance between the points $(6, -7)$ and $(-2, 8)$?

A. $\sqrt{17} \approx 4.1$

B. $\sqrt{65} \approx 8.1$

C. $\sqrt{241} \approx 15.5$

D. $\sqrt{289} = 17$

GO ON ➡

71. Which of the equations below can be written using this table of values?

x	y
−1	4
0	2
1	0
2	−2
3	−4

A. $y = 2x - 2$
B. $y = 2x + 2$
C. $y = -2x - 2$
D. $y = -2x + 2$

72. An ounce of low fat milk contains 32.940 mg of calcium and 36.295 mg of potassium. An ounce of reduced fat milk contains 33.855 mg of calcium and 42.700 mg of potassium. An ounce of whole milk contains 30.805 mg of calcium and 40.565 mg of potassium. Which matrix below represents the relationship between the calcium and potassium in low fat milk, reduced fat milk, and whole milk?

A. $\begin{bmatrix} 32.940 & 40.565 \\ 33.855 & 42.700 \\ 30.805 & 36.295 \end{bmatrix}$

B. $\begin{bmatrix} 32.940 & 42.700 \\ 33.855 & 36.295 \\ 30.805 & 40.565 \end{bmatrix}$

C. $\begin{bmatrix} 32.940 & 36.295 \\ 33.855 & 40.565 \\ 30.805 & 42.700 \end{bmatrix}$

D. $\begin{bmatrix} 32.940 & 36.295 \\ 33.855 & 42.700 \\ 30.805 & 40.565 \end{bmatrix}$

GO ON ➡

73. Which of the following graphs shows the placement of the line of best fit that will generate an equation to predict future sightings of bald eagle's nests in Thompson County?

A.

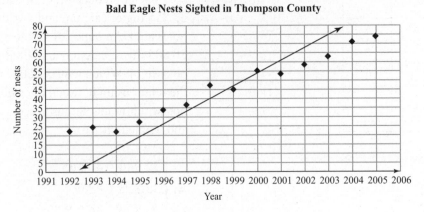

B.

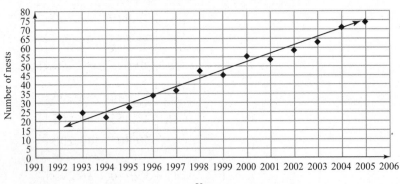

C.

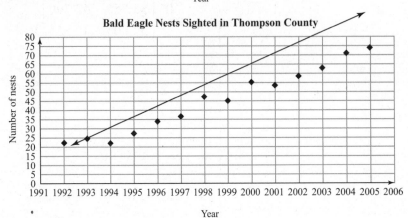

D.

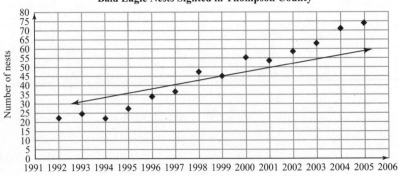

GO ON ➡

74. Which of the following represents an independent event?

   A. A box of flowers holds 4 red roses and 2 yellow roses. You randomly choose a rose and give it to one of the queen candidates. Then you randomly choose another rose and give it to a second queen candidate.

   B. A bag contains 4 blue marbles and 8 yellow marbles. You choose one marble at random, do not replace it, and then choose a second marble at random.

   C. In order to create random two-digit numbers, you spin a spinner containing the digits 0 through 9 twice.

   D. You write the letters of the word ALGEBRA on pieces of paper and place them in a bag. You randomly draw one letter, do not replace it, then randomly draw a second letter.

---

75. What is the range of $f(x) = x - 2$ if the domain is $\{2, 3, 4, 5, 6\}$?

   A. $\{4, 5, 6, 7, 8\}$
   B. $\{0, 1, 2, 3, 4\}$
   C. $\{-4, -6, -8, -10, -12\}$
   D. $\{-1, 1, 2, 3, 4\}$

---

76. What is the sine of $\angle A$?

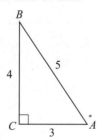

   A. $\sin A = \dfrac{3}{5}$

   B. $\sin A = \dfrac{3}{4}$

   C. $\sin A = \dfrac{4}{5}$

   D. $\sin A = \dfrac{4}{3}$

---

77. Which of the following equations is represented by the graph below?

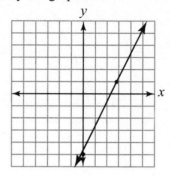

   A. $2y - x = 10$
   B. $2x + y = -5$
   C. $2x - y = 5$
   D. $2y + x = -5$

---

78. In the diagram below, the congruent parts are marked. Using only the congruent parts as marked, which of the methods below would you use to prove the triangles congruent?

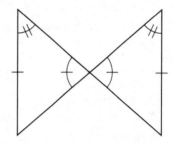

   A. AAS
   B. SSS
   C. SAS
   D. ASA

---

79. Using the rules for order of operations, the first two steps in evaluating the expression below are which of the following?

   $60 - 4x^3$ for $x = 3$

   A. Substitute 3 for x and then subtract 4 from 60.

   B. Substitute 3 for x and then multiply 4 times 3.

   C. Substitute 3 for x and then find the cube of 3.

   D. Substitute 3 for x and then subtract 60 from 4.

GO ON ➡

80. The graph of $y = 2x + 1$ is shown below.

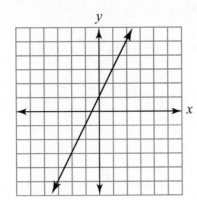

What would the equation have to change to for the graph of the equation to be the graph below?

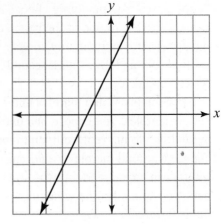

A. $y = 2x + 3$
B. $y = -2x + 1$
C. $y = 2x - 1$
D. $y = -2x + 1$

---

81. An advertising company charges an additional $150,000 each time a 30-second commercial is aired. The cost C (in thousands of dollars) to produce the commercial and air it x times is shown in the table below.

Number of air times (x)	1	2	3	4	5
Cost of the commercial (C)	$450,000	$600,000	$750,000	$900,000	$105,000

Which of the following equations represents the relationship between the cost (C) and the number of air times (x)?

A. $C = 450,000x$
B. $C = 150,000x + 300,000$
C. $C = 150,000x + 150,000$
D. $C = 450,000x + 150,000$

GO ON ➡

82. Given:

$$A = \begin{bmatrix} 1 & -1 \\ 0 & 1 \end{bmatrix} \quad B = \begin{bmatrix} -1 & 1 \\ -1 & 0 \end{bmatrix} \quad C = \begin{bmatrix} -1 & -1 \\ 0 & 1 \end{bmatrix},$$

what is the value of $A + 2B$?

A. $\begin{bmatrix} 0 & 0 \\ -1 & 1 \end{bmatrix}$

B. $\begin{bmatrix} -1 & 1 \\ -2 & 1 \end{bmatrix}$

C. $\begin{bmatrix} -0 & -1 \\ 1 & -1 \end{bmatrix}$

D. $\begin{bmatrix} 1 & 3 \\ -2 & -1 \end{bmatrix}$

83. Which of the following survey methods is **least likely** to create a biased survey?

A. A family wants to gather information from other residents on their street about forming a neighborhood watch. They survey every third house on both sides of the street.

B. In a survey about Americans' interest in soccer, the first 25 people admitted to a high school soccer game were asked, "How interested are you in the world's most popular sport, soccer?"

C. Owners of a computer store survey customers to see whether they should expand their game selection. They conduct the survey in the game aisle.

D. In an effort to gather information on passenger satisfaction during a flight, first-class passengers are given a written survey that is collected at the end of the flight.

84. A spinner is divided into four equal sections that are colored with the primary colors red, yellow, blue, and green. The spinner is spun 20 times. The table below shows the results. For which color is the experimental probability the same as the theoretical probability?

Spinner results

Red	Yellow	Blue	Green
4	8	5	3

A. Red
B. Yellow
C. Blue
D. Green

85. Which of the following expressions is equivalent to the one shown below?

$$9p^2 + 6p^3 - 5p^2 - 3p^3$$

A. $14p^2 + 3p^3$
B. $4p^2 + 3p^3$
C. $7p^2p^3$
D. $7p^2$

86. What is the slope of the line graphed?

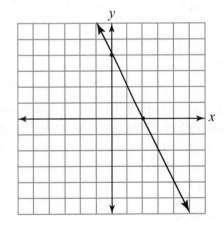

A. 2
B. −2
C. $\dfrac{1}{2}$
D. $-\dfrac{1}{2}$

GO ON →

87. In the graph below, C is going to be translated to C'(3, –2). Using the same translation rule, what will be the image of A?

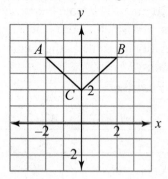

A. A' = (0, 0)
B. A' = (1, 2)
C. A' = (–2, 2)
D. A' = (1, 0)

88. What is the value of x?

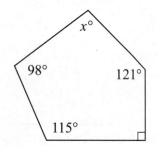

A. 105°
B. 110°
C. 115°
D. 116°

89. The school's principal noticed that a 54-foot tree casts a 63-foot shadow. Using similar triangles, how long would the shadow be for a 6-foot girl standing next to the tree at the same time?

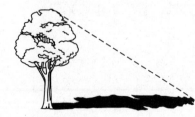

A. 6.6 feet
B. 6.7 feet
C. 6.8 feet
D. 7 feet

90. What are the coordinates of the midpoint of the segment whose endpoints are (8, –5) and (–6, 11)?

A. (1, 3)
B. (3, 1)
C. (–1, 3)
D. (1, –3)

91. Which of the following is the y value in the solution to the linear system below?

$$5x + 3y = 22$$
$$4x - 3y = -4$$

A. –6
B. –1
C. 2
D. 4

92. Which of the following expressions is equivalent to $\sqrt{\dfrac{16x^6}{25y^4}}$?

A. $\dfrac{4x^3}{5y^2}$

B. $\dfrac{4x^2}{5y^2}$

C. $\dfrac{4xy^2}{5}$

D. $\dfrac{4}{5xy^2}$

93. If a number is randomly selected from {0, 1, 2, 3, 4, 5, 6, 7, 8, 9}, what is the probability that the number is greater than 5?

A. $\dfrac{1}{5}$

B. $\dfrac{3}{10}$

C. $\dfrac{1}{3}$

D. $\dfrac{2}{5}$

GO ON →

94. Jamie is planning a plane trip from Phoenix to New York with a stopover in Chicago. There are 5 different flights to Chicago and 3 flights from Chicago to New York. How many possible flight arrangements can be made to fly from Phoenix to New York?

  A. 3
  B. 5
  C. 8
  D. 15

95. Which of the following is equivalent to this expression?

$$\frac{2x}{5} \cdot \frac{10y}{6x}$$

  A. $\dfrac{2y}{3}$

  B. $\dfrac{2x^2 y}{3}$

  C. $\dfrac{10y}{3}$

  D. $\dfrac{6x^2}{25y}$

96. What is the x-intercept of the equation below?

$$2x - 3y = -6$$

  A. (–2, 0)
  B. (–3, 0)
  C. (0, 2)
  D. (0, 3)

97. Which function is represented by the graph shown?

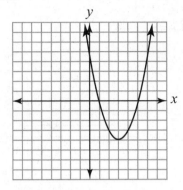

  A. $y = x^2 - 3x - 4$
  B. $y = x^2 - 6x + 5$
  C. $y = -x^2 - 3x - 4$
  D. $y = -x^2 - 6x + 5$

98. What is the measure of the hypotenuse of a right triangle if the measures of the legs are 4 and $4\sqrt{3}$?

  A. $8\sqrt{3}$
  B. 8
  C. $4\sqrt{2}$
  D. 4

99. Given a regular square pyramid with slant height $\overline{GH}$ and height $\overline{GF}$ drawn. $GH = 5$ cm, and the area of base $ABCD$ is 64. What is the length of altitude $\overline{GF}$?

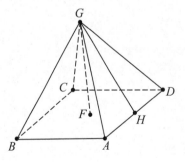

  A. $\sqrt{89} \approx 8.7$

  B. $\sqrt{39} \approx 6.2$

  C. $\sqrt{64} = 8$

  D. $\sqrt{9} = 3$

GO ON ➡

100. Which figure shows the graphic solution to the linear system below?

$$\begin{cases} y = x+1 \\ y = -2x+4 \end{cases}$$

A.

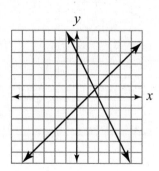

B.

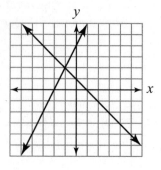

C.

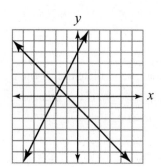

D.

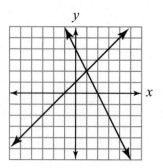

STOP

Solutions: Practice Test 2

Answer Key

1.	A	14.	A	27.	C	40.	A	53.	C	65.	D	77.	C	89.	D
2.	B	15.	A	28.	B	41.	C	54.	A	66.	B	78.	A	90.	A
3.	B	16.	B	29.	D	42.	B	55.	B	67.	D	79.	C	91.	D
4.	B	17.	D	30.	D	43.	B	56.	C	68.	C	80.	A	92.	A
5.	A	18.	B	31.	D	44.	C	57.	C	69.	C	81.	B	93.	D
6.	A	19.	D	32.	C	45.	C	58.	A	70.	D	82.	B	94.	D
7.	C	20.	D	33.	D	46.	D	59.	D	71.	D	83.	A	95.	A
8.	A	21.	D	34.	D	47.	A	60.	A	72.	D	84.	C	96.	B
9.	C	22.	A	35.	B	48.	D	61.	B	73.	B	85.	B	97.	B
10.	D	23.	B	36.	C	49.	A	62.	C	74.	C	86.	B	98.	B
11.	C	24.	A	37.	C	50.	C	63.	C	75.	B	87.	D	99.	D
12.	C	25.	B	38.	A	51.	A	64.	C	76.	C	88.	D	100.	D
13.	C	26.	C	39.	B	52.	A								

Practice Test 2 Diagnostic Chart

Question #	Correct	Incorrect	Strand to study	Page number(s) to review	Question #	Correct	Incorrect	Strand to study	Page number(s) to review
1			1	39	23			3	108
2			1	54	24			2	92
3			1	55	25			1	48
4			2	90	26			3	139
5			3	127	27			4	184
6			3	137	28			4	176
7			4	211	29			4	209
8			3	142	30			2	64
9			4	231	31			3	116
10			4	214	32			3	114
11			3	125	33			3	108
12			3	124	34			5	258
13			1	49	35			1	44
14			2	93	36			3	127
15			3	121	37			1	50
16			4	218	38			2	71
17			1	45	39			5	256
18			4	175	40			4	213
19			1	45	41			3	134
20			4	219	42			2	74
21			3	116	43			2	78
22			1	44	44			1	44

Diagnostic Chart

Question #	Correct	Incorrect	Strand to study	Page number(s) to review	Question #	Correct	Incorrect	Strand to study	Page number(s) to review
45			3	108	73			2	76
46			3	134	74			2	87
47			4	189	75			3	112
48			4	203	76			3	147
49			5 (3)	265 (117)	77			3	116
50			4	211	78			4	197
51			3	137	79			5 (3)	265 (127)
52			2	73	80			4	219
53			2	71	81			3	136
54			2	89	82			3	148
55			3	108	83			2	64
56			3	143	84			2	79
57			4	206	85			3	128
58			5	254	86			3	117
59			5 (3)	263 (138)	87			4	231
60			4	198	88			4	196
61			3	134	89			4	200
62			2	73	90			4	222
63			2	77	91			3	140
64			2	92	92			3	130
65			2	67	93			2	79
66			3	143	94			2	91
67			4	231	95			3	127
68			4	177	96			3	120
69			5	263	97			4	224
70			4	222	98			4	177
71			3	116	99			4	215
72			3	148	100			4	221

# You Got Correct		Your Classification
Grading Scale for AIMS		
Exceeds	100–88	
Meets	87–60	
Approaches	59–48	
Falls Far Below	47–0	

Answers Explained

1. **A** Natural numbers, which are also called counting numbers, are the numbers contained in the set $\{1, 2, 3, 4, 5, \ldots\}$. The only choice that contains only natural numbers is choice A. Choices B, C, and D are incorrect because 0 is not a natural number. Choice D is also incorrect because -2 is not a natural number.

2. **B** The variable expression a_{23} is referring to the element of the matrix in the second row, third column. The element in that position is -4. Choice B is the correct choice.

3. **B** To determine an approximate value of $3\sqrt{15}$ you need to approximate $\sqrt{15}$.

   $\sqrt{15}$ is very close to $\sqrt{16}$, which is equal to 4. Therefore, $\sqrt{15}$ is very close

   to 4 and $3\sqrt{15}$ is very close to $3 \cdot 4$ or 12. Choice B is the correct choice.

4. **B** It is possible to get heads with each of the numbers on the spinner and tails with each of the numbers on the spinner. Choice B is the correct choice.

5. **A** To evaluate this expression, you must substitute the given values for the variables and simplify the expression using the rules for order of operations.

   $\dfrac{4a^2 - y}{x+2}$ if $x = -2$, $y = 3$, and $a = -4$

   $\dfrac{4(-4)^2 - (3)}{(-2)+2}$

   $\dfrac{4 \cdot 16 - 3}{(-2)+2}$

   $\dfrac{64 - 3}{(-2)+2}$

   $\dfrac{61}{0}$

   The denominator of the fraction is zero. Division by zero is undefined. Therefore, the correct choice is A.

6. **A** Work through the steps shown.

$-(5-x) > 2x - 16$	Distribute the negative sign to both terms.
$-5 + x > 2x - 16$	Move terms to each side of the inequality using opposites.
$-x > -11$	Clear negative coefficient on x and switch the
$x < 11$	inequality sign.

   Choice A is correct. It is the only number that makes the inequality true.

7.   **C**   An estimate for the area of this triangular-shaped garden can be found by rounding the height and base respectively, to 40 feet and 20 feet. Then it is easy to find the area by using the formula, $A = \frac{1}{2}bh$, from the reference sheet.

$$A = \frac{1}{2}bh$$
$$A \approx \frac{1}{2} \cdot 20 \cdot 40$$
$$A \approx 400\,\text{ft}^2$$

Notice the use of the approximate sign—the base and height were rounded before we worked the problem.

8.   **A**   Notice that each of the choices for answers are solved for the variable b_1. Work the steps as shown to isolate the equation for the desired variable.

$$A = \frac{1}{2}h(b_1 + b_2)$$ Multiplying both sides by 2 will clear the fraction.
$$2A = h(b_1 + b_2)$$

$$\frac{2A}{h} = b_1 + b_2$$ Dividing both sides by h isolates the binomial.

$$\frac{2A}{h} - b_2 = b_1$$ Moving the b_2 by subtracting.

9.   **C**   The pre-image (*ABCD*) is a trapezoid with the longest parallel side as its base. The translation described in choice A would move the trapezoid to the fourth quadrant, but the longest parallel side would remain on the bottom. Choice A is incorrect. The reflection described in choice B would also place the trapezoid into the fourth quadrant, but the longest parallel side would now be vertical. Choice B is incorrect. For a rotation, the point of rotation is at the center of the two figures. The point (–1, 1) is not at the center, but the point (0, 0) is at the center. The correct choice is C.

10.   **D**   Follow the steps using the total surface area formula from the AIMS reference sheet.
$T = 2B + Ph$
$T = 2(2 \cdot 3) + (2 + 3 + 2 + 3)4$
$T = 12 + 40$
$T = 52$ in.^{2}

11.   **C**   The arch of the balloons is represented as points on the graph of some quadratic equation. The maximum value of the dependent variable is the maximum height of the balloons. The maximum value of the dependent variable is 9 meters. Choice C is the correct answer.

12.   **C**   The weight of the tank and its components will gradually increase as it is being filled with water. That gradual increase in weight is represented in the interval from 40 to 90, which is choice C.

13.   **C**   You must follow the rules for order of operations in the question. Remember PEMDAS.

$3\left[1^2(5-3+1)^3 \div 9\right]$ $3\left[1^2(3)^3 \div 9\right]$	Perform the operations inside the parentheses first. Remember to do the additions and subtractions from left to right.
$3\left[1 \cdot 27 \div 9\right]$	Simplify the exponential numbers.
$3\left[27 \div 9\right]$ $3\left[3\right]$ 9	Simplify the expression inside the bracket, remembering to do the multiplication and division from left to right. Then complete the last multiplication.

14.   **A**   Choices B, C, and D each involve groups in a certain order—in choice B student x in the lead role would be different than student x in a supporting role; in choice C, student x sitting in the driver's seat would be different than student x in the middle back seat; in choice D, listing the Revolutionary War before the War of 1812 would be different than listing them in reverse order. Choice A is the correct choice—order doesn't matter because the students are going as a group.

15.   **A**   To decide which of the choices is true, we must know the slopes of the lines. We can find their slopes by solving them for y (slope-intercept form— $y = mx + b$).

$$3x-4y=-4 \qquad\qquad 3x-4y=8 \qquad\qquad 3x+4y=8$$
$$-4y=-3x-4 \qquad\quad -4y=-3x+8 \qquad\quad 4y=-3x+8$$
$$y=\frac{-3}{-4}x-\frac{4}{-4} \qquad\quad y=\frac{-3}{-4}x+\frac{8}{-4} \qquad\quad y=\frac{-3}{4}x+\frac{8}{4}$$
$$y=\frac{3}{4}x+1 \qquad\qquad y=\frac{3}{4}x-2 \qquad\qquad y=\frac{-3}{4}x+2$$

The slope is $\frac{3}{4}$. $\qquad$ The slope is $\frac{3}{4}$. $\qquad$ The slope is $-\frac{3}{4}$.

The slope lines of a and b are equal. The lines are parallel—choice A is the correct answer. Remember for lines to be perpendicular the slopes must be opposites and be reciprocals.

16.   **B**   The graph has a y-intercept at the point (0, 5). The b value in $y = mx + b$ is 5.

Counting grid corners, we see that the slope has a vertical change of down 3 and to the right 4. The slope is $-\frac{3}{4}$. Choice B is the correct equation.

17.   **D**   If the helmet is on sale for 20% off, then Terri will pay 80% of the original price; the sale price is 80% of $39.95. Choice D is the correct choice. An estimate of the answer could be found by rounding $39.95 to $40.00 and taking 80% of $40.00.

18.   **B**   The sum of the measures of the angles of a triangle is 180°.

$$x + 2x + 33 = 180$$
$$3x + 33 = 180$$
$$3x = 147$$
$$x = 49°$$

19.   **D**   It's important to remember that deposits add to the balance of your account and checks subtract from your balance. Therefore, the calculations required to find your balance are $54 - 56 - 57 - 18 + 50 + 16$. The rules for order of operations require you to perform these operations from left to right. The balance of your account is –$11, and you are overdrawn. The correct choice is D.

20.   **D**   For each choice, the inequalities have the same slope and y-intercept. The correct solution is based on the inequality sign and the side of the line that is shaded. The graph has a solid line which means that the inequality sign includes the "equal to" bar. Choices A and B do not have the equal to sign and are incorrect. Next, pick a point that is not on the line. If your point is in the shaded region of the graph, that point will make a true statement when substituted into the inequality. If you choose a point not in the shaded region and place it into the inequality, the resulting statement will be false. Work through the steps picking (3, 3) from the shaded region and looking for a true statement.

Choice C	Choice D
$y \le -2x + 2$	$y \ge -2x + 2$
$3 \le -2(3) + 2$	$3 \ge -2(3) + 2$
$3 \le -6 + 2$	$3 \ge -6 + 2$
$3 \le -4$ False	$3 \ge -4$ True

21.   **D**   The points on the graph seem to be (1, 3), (2, 4), (3, 5), (4, 6), and (5, 7). In each of these ordered pairs, you can add two to the x value to get the y value. Choice D is the correct choice.

22.   **A**   The numbers $\dfrac{5}{4}$ and $\dfrac{4}{5}$ are reciprocals, or inverses, of each other. Choice A is the correct choice.

23.   **B**   The coefficients of the terms and the variable factors of the terms each represent a different pattern. The coefficients are 3, 6, 12, 24, . . . and the variables are $y, y^2, y^3, y^4, \ldots$. The terms in the coefficients sequence are each two times the preceding term, and the exponents on y in the variable sequence increase by one each term. Twenty-four times two is 48 and one more factor of y would make the variable factor y^5. The next term in the sequence is $48y^5$. Therefore, choice B is the correct answer.

24.   **A**   Because prizes are awarded specifically to first, second, and third place, order matters in this question. The solution to the question requires the use of the permutation formula (on the reference sheet). You are looking for the permutation of 11 things taken three at a time.

$$_{11}P_3 = \frac{11!}{(11-3)!} = \frac{11!}{8!}$$

Choice A is the correct answer.

25.   **B**   This problem requires you to follow the rules for order of operations and you have to understand absolute value. $|-3| = 3$. Therefore, you are simplifying the sum, $-3 + 3 + 7$ which is 7. Choice B is the correct choice.

26.   **C**   Work through the steps shown

$$\frac{6}{x-5} = \frac{3}{x+1}$$

Use the means-extremes product rule to clear fractions.

$$6(x+1) = 3(x-5)$$

Distribute the coefficients for each binomial.

$$6x+6 = 3x-15$$

Move terms to each side of the equality by opposites

$$3x = -21$$

Divide by three to solve for the variable x.

$$x = -7$$

Choice C is correct.

27.   **C**   Given that the opposite pairs of sides of a quadrilateral are parallel, the most specific name that can be used is parallelogram. Choice B, the rectangle, requires right angles. Choice D, the kite, requires consecutive pairs of sides to be congruent. Choice A, the trapezoid, only has one pair of opposite sides parallel.

28.   **B**   The triangle inequality property for sides states that the sum of any two sides is greater than the third side. Set up three inequalities using the sides.

$x < 64 + 43$ $43 < x + 64$ $64 < x + 43$

$x < 107$ $-21 < x$ $21 < x$

The middle inequality is included in the inequality on the right. Values of x greater than 21 will also be greater than -21. Placing the inequalities together yields choice B.

29.   **D**   Follow the steps using the area of a sector formula from the AIMS reference sheet.

$$A = \pi r^2 \left(\frac{\text{Degrees in corresponding arc}}{360°} \right)$$

$$A = \pi (12)^2 \left(\frac{60}{360°} \right)$$

$$A = 144\pi \left(\frac{\cancel{60}^{1}}{\cancel{360°}_{6}} \right)$$

$$A = \frac{144\pi}{6}$$

$$T = 24\pi \text{ units}^2$$

30.   **D**   The choices all include questions about movies or movie theaters, but only choice D is not appropriate to help the proposed owner make decisions about actually opening or running a movie theater.

31.   **D**   It appears that the line passing through these points will cross the axes at the origin. The y-intercept of that line is zero. Choices A and B have y-intercepts of 1—they cannot be the correct answer. One of the points is the point $(1, 2)$. If you substitute this point into choices C and D, it satisfies only the equation in choice D. Choice D is the correct answer.

32.   **C**   This type of input-output diagram is called a mapping. The independent values (on the left) map into the dependent values (on the right). A function maps one independent value into one and only one dependent value. Choice A maps each of the independent values into one and only one dependent value. Choice B might be confusing because the arrows cross each other, but each independent value is still mapped into one and only one dependent value. Choice D might also be confusing in that each of the independent values maps into the same dependent value. However, each independent value maps into only one dependent value (–8 only maps to 0, –4 only maps to 0, 0 only maps into 0, and 4 only maps into 0). Choice C is the mapping that does not represent a function. Negative one maps into both two and negative one, and five maps into both four and negative three.

33. **D** The value of n for the first term is 1. You can test that value in each of the choices.

If $n = 1$, then $n + 3 = 4$.
If $n = 1$, then $4n = 4$.
If $n = 1$, then $3n + 1 = 4$.
If $n = 1$, then $3n – 2 = 1$.

The only choice that generates a first term of 1 is choice D.

34. **D** When you multiply both sides of an inequality by a negative number the inequality must reverse the inequality symbol. Choices A, B, and C all are positive and do not change the sense of the inequality.

35. **B** Choice B is the correct answer; there are five whole numbers less than five (0, 1, 2, 3, and 4). The other choices are sets containing infinitely many elements.

36. **C** A coefficient is a numerical factor in front of a variable factor in a term. The algebraic expression given has three terms; $2x^3$, $6x$, and 8. The middle term is $6x$, and the coefficient in that term is 6. The correct choice is C.

37. **C** The algorithm for multiplying numbers in scientific notation is to first multiply the factors between zero and ten and then multiply the powers of ten by adding the exponents.

$$(2.5 \times 3.2)(10^{-5} \times 10^3) = 8.0 \times 10^{-2}$$

Choice C is the correct choice.

38. **A** The algorithm for finding the average of a group of numbers is to add the numbers and divide by the number of numbers. Therefore, the sum of the first four tests must be the average times four, or $(83)(4) = 332$. If Kathy's score on the fifth test is 97, then the sum of her first five tests would be $332 + 97 = 429$. Her average after five tests is $429 \div 5 = 85.8$. Therefore, choice D is incorrect. Choice C is incorrect—we do not have enough information to comment on all of Kathy's grades in American History. The fifth test score will bring up Kathy's average of 83, but it will not bring it up enough so that is closer to 97 then 83. Choice A is the correct choice.

39. **B** The second statement "$ABCD$ is a square" is only used as an if conditional and not as a conclusion. This statement must come first in the logical argument. The conclusion of this statement, "$ABCD$ is a rectangle" becomes the second statement. This is statement number three. The conclusion for "$ABCD$ is a rectangle" becomes the third statement. This is statement number one. The conclusion of statement number one, "$AB \| DC$," is the beginning of the last statement. A good check is to notice that the conclusion for this last statement is not used as an if statement in this argument.

40. **A** Follow the steps using the volume for a sphere formula from the AIMS reference sheet.

$$V = \frac{4}{3}\pi r^3$$

$$V = \frac{4}{3}\pi (3)^3$$

$$V = \frac{4}{\cancel{3}}\pi \cancel{27}^{9}$$

$$V = 36\pi \text{ in.}^3$$

41. **C** The expression, twice the sum of 5 times a number x and 6, can be thought of as two times the sum of $5x$ and 6. Choice C is the correct answer.

42. **B** The interval in which you want to count data points is between 16.1 and 15.9, including 16.1 and 15.9. There are 14 data points in that interval. Choice B is the correct choice.

43. **B** Normal distribution is represented by a graph in which the mean, median, and mode are all equal. Also, it is bell-shaped and symmetrical about the mean value. Choice B is the correct choice. Choice A and C are skewed to the left and right, and choice D does not represent a normal distribution.

44. **C** A set is infinite if it is not possible to count the number of elements in the set; some sets are very large, but it is possible to count the number of elements in the set. Choice A is a set that contains 11 elements; Choice B contains a large number of elements, but you can count how many there are; and choice D is also countable. Choice C is not countable; even though the largest and smallest numbers in the set are close together on the number line, there are infinitely many fractions between 0 and 1.

45. **C** Each term is 7 more than the preceding term. If you carry the sequence out to the eighth term you will get choice C.

$$1, 8, 15, 22, 29, 36, 43, 50$$

46. **D** Simplify the radical by removing perfect cubes from underneath the radical symbol (common factors in groups of three). In this problem, each factor is a perfect cube: $\sqrt[3]{27x^3y^{12}} = \sqrt[3]{3^3 x^3 y^3 y^3 y^3 y^3} = 3xy^4$.

47. **A** Translating the given information yields the following two equations. This system of equations can be solved by substitution as shown.

$A + B = 180$	Replace the angle A with the second equation
$A = B + 24$	(solved for A).
$B + 24 + B = 180$	Combine like terms and solve.
$2B + 24 = 180$	
$2B = 156$	
$B = 78°$	Replace the value for angle B to find angle A.
$A = B + 24$	
$A = 78 + 24 = 102°$	

48. **D** A tangent line is perpendicular to the radius at the point of tangency by definition. Choice D uses mathematical notation to state this fact.

49. **A** Subtracting the y-coordinates and placing this difference over the difference of the x-coordinates is the algorithm for finding the slope.

50.   **C**   Find the area of the large rectangle ignoring the cut out section. The area is (23 times 10) 230 square feet.

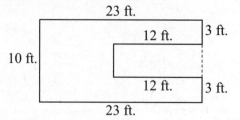

Find the area of the cut out rectangle. To determine the height of the small rectangle, subtract the two lengths of 3 feet from the overall height of ten feet. The height of the small rectangle is 4 feet. This area is (12 times 4) 48 square feet. Subtract the small area from the large area (230 − 48) for the correct area of 182 square feet. You can also divide the original diagram into smaller rectangles to find the total area.

51.   **A**   Work through the steps shown

$$\frac{2}{3}x+6=-18$$ Subtract 6 from both sides.

$$\frac{2}{3}x=-24$$ Multiply by reciprocal.

$$\frac{\cancel{3}}{\cancel{2}}\,\frac{\cancel{2}}{\cancel{3}}\,x=\frac{3}{\cancel{2}}\left(\frac{-12}{\cancel{-24}}\right)$$

$$x=-36$$ Simplify.

Choice A is correct.

52.   **A**   The graph seems to illustrate a type of pattern in which the cost for parking remains the same for three years and then increases by $0.50 during the next year. Therefore, the cost for parking will remain at $1.50 until 2007 and then increase by $0.50 in 2008. The question is asking for the cost in 2007. The correct choice is A.

53.   **C**   The median is the middle score of the 12 scores. There is no single middle score—the scores 72 and 76 are the middle scores. You have to find the average of those two scores. The average is 74—choice C is the correct choice. Choice A is the mode, and choices B and D are the two middle scores between which you must find the average.

54.   **A**   Probability is the ratio of the number of successful choices to the number of possible choices. In this case, the successful points would be in the rectangle but not in the triangle—the area of the rectangle minus the area of the triangle. The area of the rectangle would represent the possible positions for the points. The probability is the area of the rectangle minus the area of the triangle over the area of the rectangle. Choice A is the correct choice.

55.   **B**   Each of the choices has a first term of 1, which is correct. The rule states that each term is 3 more than the preceding term. Choices A and C are incorrect because the second term is only 2 more than the first term. Choice D has a second term of 4, which is correct, but the third term is not 3 more than the second term. Choice B is correct—each term is 3 more than the preceding term.

56.   **C**   Work through the steps as shown

$$6\sqrt{x-2}+5=35$$
$$6\sqrt{x-2}=30$$
$$\sqrt{x-2}=5$$
$$x-2=25$$
$$x=27$$

Isolate the radical by first subtracting the 5 and then dividing the coefficient 6.

With the radical isolated, square both sides and solve for x.

57.   **C**   A circle has a measure of 360°. To find the measure of the angles between the equally spaced spokes, divide 360 by 12. There is 30 degrees between the spokes. Choice C is correct.

58.   **A**   The correct choice is A. The converse of a conditional is a new sentence obtained by exchanging the hypothesis and the conclusion of the given conditional. Choice B, the contrapositive, is a new sentence obtained by exchanging the negation of the conclusion with the negation of the hypothesis of the given conditional. Choice C, the inverse, is a new sentence obtained by negating both the hypothesis and the conclusion of the given conditional. Choice D, the negation of a statement is the opposite of the original statement.

59.   **D**   The fourth inequality has an error. When dividing both sides of an inequality by a negative, the inequality sign switches.

60.   **A**   The ratio of the corresponding sides are equal. SSS means the ratio of the corresponding sides are equal. Similarity can be shown by SSS (Side–Side–Side) similarity.

$$\frac{3}{6}=\frac{5}{10}=\frac{4}{8}$$

61.   **B**   Eighty adult tickets can be represented as $80A$ and 100 children's tickets can be represented as $100C$. The total of these is the amount of money the theater took in on the night in question. Choice B is the correct answer.

62.   **C**   Imagine a line of best fit placed on the graph to represent the given data. The number of fat grams is the independent variable; it is graphed on the horizontal axis. Also, look at the table for a number in the fat column that is close to 32. Using both the table and the graph, decide which one of the choices would be the most reasonable value for the calories in 32 grams of fat. Choice C is the most reasonable answer and therefore is the correct choice.

63.   **C**   Choices A and B are graphs that each represent positive correlation, and choice D is a graph that has neither positive nor negative correlation. Choice C is the correct choice—it represents negative correlation.

64.   **C**   After the M is chosen and not replaced, there are five letters remaining in the bag (W, I, S, D, O). You are asked to find the probability of choosing the letter S from those five letters. Choice C is the correct choice.

65.   **D**   Although the runners do not all start at the same place, the runners in choice D (the correct answer) are running at the same speed, 10 meters in 4 seconds. One runner in each of choices A and B runs 10 meters in 4 seconds and the other runs 20 meters in 4 seconds. In choice C, one runner runs 10 meters in 4 seconds and the other runs 30 meters in 4 seconds.

66. **B** Work through the steps as shown.

$$x^2 + 12x - 13 = 0$$

$$a = 1, b = 12, \text{ and } c = -13$$

$$x = \frac{-b \pm \sqrt{b^2 - 4ac}}{2a}$$

$$x = \frac{-(12) \pm \sqrt{(12)^2 - 4(1)(-13)}}{2(1)}$$

$$x = \frac{-12 \pm \sqrt{144 + 52}}{2}$$

$$x = \frac{-12 \pm \sqrt{196}}{2}$$

$$x = \frac{-12 \pm 14}{2}$$

$$x = \frac{-12 + 14}{2}, \frac{-12 - 14}{2}$$

$$x = \frac{2}{2} \text{ or } \frac{-26}{2}$$

$$x = 1 \text{ or } -13$$

Use the quadratic formula (found on the AIMS reference sheet). Identify the coefficients a, b, and c. Place these values into the formula and simplify.

If you have learned how to solve quadratic equations using the factoring method, your solution would look like the following.

$$x^2 + 12x - 13 = 0$$

$$(x + 13)(x - 1) = 0$$

$$x = -13 \text{ or } x = 1$$

67. **D** $\triangle ABC \rightarrow \triangle A'B'C' \rightarrow \triangle A''B''C''$. $\triangle ABC$ is transformed into $\triangle A'B'C'$ using a reflection over the x-axis. $\triangle A'B'C'$ is transformed into $\triangle A''B''C''$ using a translation to the right. To determine the number of units in the translation, count the number of units C' is to the right of C'. These moves correspond with choice D.

68. **C** To find the length of the base of this slide, recognize that the slide forms a right triangle. Using the Pythagorean theorem, the base can be found.

$$a^2 + b^2 = c^2$$

$$(9)^2 + b^2 = (15)^2$$

$$81 + b^2 = 225$$

$$b^2 = 144$$

$$b = \pm 12$$

Replace the given values into the formula.

Square the terms and solve by isolating the variable b.

Notice the solution yields two solutions. The positive solution represents a length.

69. **C** The third line contains the error. Moving the $4x$ from the right side of the equation to the left requires a subtraction of $4x$ from both sides. The $2x$ on the left side would become $-2x$.

70. **D** Follow the steps using the distance formula from the AIMS reference sheet.

$$d = \sqrt{(x_2 - x_1)^2 + (y_2 - y_1)^2}$$

$$d = \sqrt{((-2) - 6)^2 + (8 - (-7))^2}$$

$$d = \sqrt{(-8)^2 + (15)^2}$$

$$d = \sqrt{64 + 225} = \sqrt{289} = 17$$

71.   **D**   Pick any two values to determine the slope of the line, for example, (0, 2) and (–1, 4).

$$m = \frac{4-2}{-1-0} = \frac{2}{-1} = -2$$

Notice that one of the given points (0, 2) is the *y*-intercept. This means that the *b* value in $y = mx + b$ is given. Choice D has a slope of –2 and an *y*-intercept of 2.

72.   **D**   The important thing to remember when setting up a matrix to represent given data is to place the elements in corresponding positions, rows or columns. The second row of choice A has the reduced fat milk values in the same row but the low fat milk and the whole milk are in mixed rows. The third row of choice B has the whole milk values in the same row but the low fat milk and the reduced fat milk are in mixed rows. The first row of choice C has the low fat milk values in the same row but the reduced fat milk and the whole milk are in mixed rows. Choice D is the correct answer; it has each of the elements related to the types of milk in corresponding positions.

73.   **B**   Choices A and D do not follow the flow of the data points—A is too steep and D is not steep enough. Choice C follows the flow, but it is above most of the data points. Choice B is the most reasonable choice for a line of best fit.

74.   **C**   Choices A, B, and D all contain information indicating that the first event will affect the second—the candidate keeping the rose, the marble not being replaced, and the letter drawn not being replaced. Choice C is correct because the results of the first and second spin of the spinner are independent of each other.

75.   **B**   The range of a function is all of the values that the dependent variable can be (in this case, *y*) when the domain (in this case, *x*) is all of the given values.

If $x = 2$, $x - 2 = 0$.
If $x = 3$, $x - 2 = 1$.
If $x = 4$, $x - 2 = 2$.
If $x = 5$, $x - 2 = 3$.
If $x = 6$, $x - 2 = 4$.

Therefore, choice B is the correct choice.

76.   **C**   The trigonometric ratio in the question is sine of the angle *A*. The ratio for sine is the opposite leg divided by the hypotenuse. The side opposite the angle *A* has a length of four and the length of the hypotenuse is 5. The $\sin(A) = \dfrac{4}{5}$

77.   **C**   Notice that the answer equations are not given in slope-intercept form. A method for finding this solution would be to change each equation into slope-intercept form. Another method would be to pick a couple points from the graph and place into each equation. The points will always yield a true statement in the correct equation. Work through the equations with (0, –5) and (3, 1). Notice the importance of using two points. Choice C is correct.

Choice A $2y - x = 10$
Using (0, –5)   Using (3, 1)
    $2y - x = 10$ $2y - x = 10$
$2(-5) - 0 = 10$ $2(1) - 3 = 10$
       $-10 = 10$ False $-1 = 10$ False

Choice B $\qquad 2x + y = -5$

Using (0, –5)

$2x + y = -5$

$2(0) + (-5) = -5$

$-5 = -5$ True

Using (3, 1)

$2x + y = -5$

$2(3) + 1 = -5$

$7 = -5$ False

Choice C $\qquad 2x - y = 5$

Using (0, –5)

$2x - y = 5$

$2(0) - (-5) = 5$

$5 = 5$ True

Using (3, 1)

$2x - y = 5$

$2(3) - (1) = 5$

$5 = 5$ True

Choice D $\qquad 2y + x = -5$

Using (0, –5)

$2y + x = -5$

$2(-5) + 0 = -5$

$-10 = -5$ False

Using (3, 1)

$2y + x = -5$

$2(1) + 3 = -5$

$5 = -5$ False

78. **A**   Using the congruency markings only, the two triangles can be shown congruent by AAS (Angle–Angle–Side).

79. **C**   The first step in evaluating an expression is to substitute in the given values. All four choices start with this step. The problem becomes an order of operations problem (PEMDAS). Cubing the 3 comes next. The correct choice is C.

80. **A**   The slope of the line stayed the same. The y-intercept changed from (0, 1) to (0, 3). This will change the b value from the equation of the line $y = mx + b$. Choice A is correct.

81. **B**   There is a constant increase in the cost. The equation will be linear. Calculate the constant increase (slope)

$$m = \frac{600,000 - 450,000}{2 - 1} = \frac{150,000}{1} = 150,000$$

Choices A and D have the wrong slope value and are incorrect. Choose a value for the air time and place into either choice B or C. Only choice B correctly yields the table values.

82. **B**   Work through the steps shown:

$A + 2B =$

$$\begin{bmatrix} 1 & -1 \\ 0 & 1 \end{bmatrix} + 2 \begin{bmatrix} -1 & 1 \\ -1 & 0 \end{bmatrix} =$$

$$\begin{bmatrix} 1 & -1 \\ 0 & 1 \end{bmatrix} + \begin{bmatrix} -2 & 2 \\ -2 & 0 \end{bmatrix} = \begin{bmatrix} -1 & 1 \\ -2 & 1 \end{bmatrix}$$

Multiply the matrix B by the scalar 2 and then combine the matrices by corresponding rows and columns.

83. **A**   Choices B, C, and D will most likely create biased surveys because they propose gathering the data from a very specific population. Choice A is correct (least likely to create bias) because a random, overall population is questioned.

84. **C** The theoretical (expected) probability is $\frac{1}{4}$ because there are four equal sections on the spinner. The results of the experimental probabilities are shown below. Choices A, B, and D are incorrect.

$$P(Red) = \frac{4}{20} = \frac{1}{5}$$

$$P(Yellow) = \frac{8}{20} = \frac{2}{5}$$

$$P(Green) = \frac{3}{20} = \frac{1}{5}$$

Choice C is the correct answer because $P(Blue) = \frac{5}{20} = \frac{1}{4}$ is the expected probability.

85. **B** To simplify this expression you must combine like terms. The terms that are like are the terms that have the same exponent on p. Therefore, $9p^2$ and $-5p^2$ are like terms, and $6p^3$ and $-3p^3$ are like terms. Combining $9p^2$ and $-5p^2$ yields $4p^2$, and combining $6p^3$ and $-3p^3$ yields $3p^3$. The correct answer is choice B.

86. **B** Count the grid corners where the line crosses. Using the x- and y-intercepts and starting from the y-intercept, the vertical change is down 4 and the horizontal change is right 2. The slope is -2.

87. **D** Moving point C (0, 2) of the triangle to the point (3, –2) requires the x value to increase by 3 units and the y value to decrease by 4 units. This change needs to happen to each point from the original triangle. The initial coordinates for point A are (–2, 4). Making the changes causes the point A to become (1, 0). This is choice D.

88. **D** Using the AIMS reference sheet the formula for the sum of the interior angles of a pentagon are shown:
$S = (n - 2)(180°)$
$S = (5 - 2)(180°)$
$S = 3(180°) = 540°$
The sum of the given angles is 424°. The difference, represented by the variable x, is 116°. The correct choice is D.

89. **D** Set up similar triangles representing the tree and its shadow with the girl and her shadow. Use the means-extremes product rule to solve for the shadow length. Notice that after setting up the proportions, you can reduce the fraction to make your arithmetic easier.

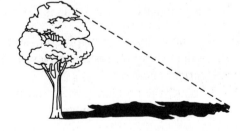

$$\frac{54}{63} = \frac{6}{x}$$

$$\frac{\overset{6}{\cancel{54}}}{\underset{7}{\cancel{63}}} = \frac{6}{x}$$

$$6x = 42$$

$$x = 7$$

90. **A** Follow the steps using the midpoint formula from the AIMS reference sheet.

$$\left(\frac{x_1+x_2}{2}, \frac{y_1+y_2}{2}\right)$$

$$\left(\frac{8+(-6)}{2}, \frac{(-5)+11}{2}\right)$$

$$\left(\frac{2}{2}, \frac{6}{2}\right) = (1,3)$$

91. **D** Use the elimination method to solve this system of linear equations. Combining the equations, we can see that the y variable drops off, and the remaining equation can be solved as shown.

$$5x + 3y = 22 \qquad \text{Combine equations.}$$
$$4x - 3y = -4$$
$$9x = 18 \qquad \text{Solve for } x.$$
$$x = 2$$

Take caution as the question asks for the y value. Place the value of x into either equation for the y solution.

$$5(2) + 3y = 22 \qquad \text{Combine equations.}$$
$$10 + 3y = 22$$
$$3y = 12 \qquad \text{Solve for } y.$$
$$y = 4$$

Choice D is correct. To check your solution, you can place both x and y into the other equation.

$$4(2) - 3(4) = -4$$
$$8 - 12 = -4$$
$$-4 = -4$$

92. **A** Simplify the radical by removing perfect squares from underneath the radical symbol (common factors in groups of two). In this problem, each factor is a perfect square.

$$\sqrt{\frac{16x^6}{25y^4}} = \sqrt{\frac{4^2 x^2 x^2 x^2}{5^2 y^2 y^2}} = \frac{4x^3}{5y^2}$$

93. **D** Probability is the ratio of the number of successful choices to the number of possible choices. There are four numbers greater than five in the set (6, 7, 8, 9). There are ten possible numbers from which you can choose. The probability is $\frac{4}{10}$, which equals $\frac{2}{5}$. Choice D is the correct choice.

94. **D** Use the counting principle to answer this question. There are $5 \cdot 3$ ways to arrange the trip from Phoenix to New York with a stopover in Chicago. Choice D is the correct choice.

95. **A** To make the problem easier, you might want to divide out like factors from the numerators and denominators.

$$\frac{2\!\!\!/x}{\not{5}} \cdot \frac{\overset{2y}{\cancel{10}y}}{\underset{3}{\not{6}x}} = \frac{2y}{3}$$

The correct answer is choice A.

96.   **B**   The *x*-intercept is the point $(x, 0)$ and can be found by placing a zero in for the variable *y* in the given equation.

$$2x - 3(0) = -6$$
$$2x = -6$$
$$x = -3$$

The correct choice is B.

97.   **B**   There are features of the quadratic that will help to match the equations with the graphs. Look at the direction of opening. This graph opens up. This is only true if the leading coefficient (coefficient of the *x* squared term) is positive. Both choices C and D have negative coefficients. The *y*-intercept is another feature to identify. The graph crosses the *y*-axis at the point $(0, 5)$. The five is represented by the constant in the equation. Choice A has a constant of -4 and so it is incorrect. The correct choice is B.

98.   **B**   The measure of the hypotenuse of a right triangle can be determined by Pythagorean theorem. Work through the steps as shown.

$$a^2 + b^2 = c^2$$

The given legs are 4 and $4\sqrt{3}$.

$$\left(4\right)^2 + \left(4\sqrt{3}\right)^2 = c^2$$

Place these values into the formula (found on the AIMS reference sheet).

$$16 + 16(3) = c^2$$

$$16 + 48 = c^2$$

Square the terms and solve.

$$64 = c^2$$

Notice the answer yields two solutions. The positive solution represents a length.

$$\pm 8 = c$$

99.   **D**   With a square base area of 64, each side has a length of 8. $\overline{GF}$ is an altitude and perpendicular to the base. This altitude forms a right ΔGFH with a hypotenuse of 5 cm and the base of the triangle of 4 cm. Using the Pythagorean theorem, the height of $\overline{GF}$ is 3 cm.

$$a^2 + b^2 = c^2$$
$$a^2 + (4)^2 = (5)^2$$
$$a^2 + 16 = 25$$
$$a^2 = 9$$
$$a = \pm 3$$

The solution to the equation yielded two solutions, but the problem involves a length and the measure is positive. The value of the height is 3 cm. The correct choice is D.

100.   **D**   There are several methods for solving systems of linear equations, one of which is the graphical method. The point at which the two lines, representing the equations, intersect is the graphical solution. The solution is the ordered pair $(1, 2)$ and that corresponds to the graph D.

Sample AIMS Reference Sheet

<table>
<tr><td colspan="2">Formulas for Area</td></tr>
<tr><td>Triangle</td><td>$A = \frac{1}{2}bh$</td></tr>
<tr><td>Rectangle</td><td>$A = lw$</td></tr>
<tr><td>Trapezoid</td><td>$A = \frac{1}{2}h(b_1 + b_2)$</td></tr>
<tr><td>Parallelogram</td><td>$A = bh$</td></tr>
<tr><td>Circle</td><td>$A = \pi r^2$</td></tr>
</table>

<table>
<tr><td colspan="2">Key</td></tr>
<tr><td>b = base</td><td>d = diameter</td></tr>
<tr><td>h = height</td><td>r = radius</td></tr>
<tr><td>l = length</td><td>ℓ = slant height</td></tr>
<tr><td>w = width</td><td>B = area of base</td></tr>
<tr><td></td><td>P = perimeter of base</td></tr>
<tr><td colspan="2">Use 3.14 or $\frac{22}{7}$ for π.</td></tr>
</table>

Formulas for Volume and Area of Solids		
Solid	**Volume**	**Total Surface Area**
Right circular cone	$V = \frac{1}{3}\pi r^2 h$	$T = \frac{1}{2}(2\pi r)\ell + \pi r^2 = \pi r\ell + \pi r^2$
Pyramid	$V = \frac{1}{3}Bh$	$T = B + \frac{1}{2}P\ell$
Sphere	$V = \frac{4}{3}\pi r^3$	$T = 4\pi r^2$
Right circular cylinder	$V = \pi r^2 h$	$T = 2\pi rh + 2\pi r^2$
Right prism	$V = Bh$	$T = 2B + Ph$

Linear Equation Forms	Coordinate Geometry
Point-slope form: $y - y_1 = m(x - x_1)$ **Standard or General Form:** $Ax + By = C$ **Slope-intercept form:** $y = mx + b$	Given: Points $A\,(x_1, y_1)$, $B(x_2, y_2)$ **Distance between two points:** $AB = \sqrt{(x_2 - x_1)^2 + (y_2 - y_1)^2}$ **Midpoint between two points:** Midpoint of $\overline{AB} = \left(\dfrac{x_2 + x_1}{2},\ \dfrac{y_2 + y_1}{2} \right)$ **Slope of a line through two points:** (x_1, y_1) and (x_2, y_2) $m = \dfrac{y_2 - y_1}{x_2 - x_1}$
Pythagorean Theorem	**Quadratic Formula**
$a^2 + b^2 = c^2$	$x = \dfrac{-b \pm \sqrt{b^2 - 4ac}}{2a}$
Sum of the measures of the interior angles of a convex polygon with n sides: $S = (n - 2)(180°)$	Distance, rate, time formula, where d = distance, r = rate, t = time: $d = rt$
Permutations of n objects taken r at a time: $_nP_r = \dfrac{n!}{(n-r)!}$	Combinations of n objects taken r at a time: $_nC_r = \dfrac{n!}{(n-r)!\,r!}$

Right Triangle Relationships		
Trigonometric Ratios	**30°-60°-90° Triangle Relationships**	**45°-45°-90° Triangle Relationships**
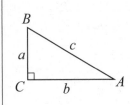 $\sin A = \dfrac{a}{c}$ $\cos A = \dfrac{b}{c}$ $\tan A = \dfrac{a}{b}$	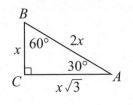	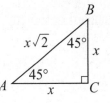

Additional Formulas	
Circumference $= \pi d = 2\pi r$	Use 3.14 or $\dfrac{22}{7}$ for π.
Area of a sector: $A = \pi r^2 \left(\dfrac{\text{Degrees in corresponding arc}}{360°} \right)$	Length of circular arc: Length of $\widehat{AB} = \dfrac{\text{m}\widehat{AB}}{360°}(2\pi r)$

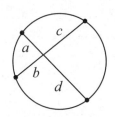

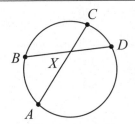

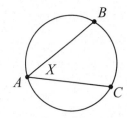

$\dfrac{a}{b} = \dfrac{c}{d}$ or $a \cdot d = b \cdot c$ $\text{m}\angle X = \dfrac{1}{2}\left(\text{m}\widehat{AB} + \text{m}\widehat{CD}\right)$ $\text{m}\angle X = \dfrac{1}{2}\left(\text{m}\widehat{BC}\right)$

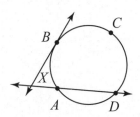

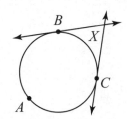

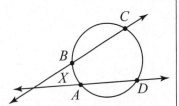

$\text{m}\angle X = \dfrac{1}{2}\left(\text{m}\widehat{BCD} - \text{m}\widehat{AB}\right)$ $\text{m}\angle X = \dfrac{1}{2}\left(\text{m}\widehat{BAC} - \text{m}\widehat{BC}\right)$ $\text{m}\angle X = \dfrac{1}{2}\left(\text{m}\widehat{CD} - \text{m}\widehat{AB}\right)$

Glossary

absolute value   a number's distance from zero on a number line; the absolute value of −4 is 4; the absolute value of 4 is 4 symbolically, |−4| = 4 and |4| = 4

actual measure   the exact measurement of an object

acute angle   an angle whose measure is between 0° and 90°

addends   numbers used in the mathematical operation of addition

addition   a mathematical operation based on "putting things together"

additive inverses   two numbers whose sum is zero; the numbers are opposites

adjacent angles   two coplanar angles that share a common side and a common vertex but do not share common interior points

algebraic expression   a group of numbers, symbols, and variables that express a single or series of operations; mathematical phrase with one or more terms, one or more variables

algebraic sentence   an equation or inequality that represents a relationship between two expressions

algorithm   a set of step-by-step instructions for completing a task

alternate exterior angles   angles formed by a transversal intersecting two lines; angles on opposite sides of the transversal, having two different vertices, and outside the lines

alternate interior angles   angles formed by a transversal intersecting two lines; angles on opposite sides of the transversal, having two different vertices, and between the lines

analog clock   a device, with an hour, minute, and second hand, which shows a continuous sweep of time passing rather than in "jumps"(digital)

angle   a geometric figure consisting of two rays with a common endpoint (vertex)

angle bisector   a line or ray that divides an angle into two congruent angles

appropriate math terminology   vocabulary that accurately defines mathematical concepts, operations, and content at a given grade level

appropriate measure of accuracy   the degree of accuracy required for a given mathematical task (i.e., approximating the number of cubic inches needed in determining the volume of space for packing would have a need for less accuracy than, say, the measurement of a piece of molding to fit precisely on a door frame)

approximation   a value that is sufficiently exact for a specified purpose

arc   a part of a circle that consists of two points, called endpoints, and all points of the circle between them

area   The two-dimensional space enclosed by the perimeter is called the area

arithmetic fact   any of the basic addition and multiplication numerical statements and the corresponding subtraction and division relationships

arithmetic sequence   a set of ordered terms in which the difference between consecutive terms is constant

array   a rectangular arrangement of objects in rows and columns

ascending order   a listing in which numbers or terms are organized in increasing value

associative property   the property that states for real numbers a, b, and c, $(a + b) + c = a + (b + c)$ and $(ab)c = a(bc)$; essentially this property is a grouping of three terms where the sum and product of the first two with the third is the same as the sum or product of the last two and the first

attribute   a common feature of a set of objects or numbers

average   See *mean*

axiom   a self-evident truth; a truth without proof and from which further statements, or theorems, can be derived

axis   either of two perpendicular number lines used to form a coordinate plane

bar graph   a graph in which horizontal or vertical bars represent data

base a term used as a factor for repeated multiplication (i.e., in 4^7, 4 is the base)

base of a polygon the side(s) that is (are) perpendicular to the height

base of a polyhedron either of the two congruent parallel faces of a prism; the face of a pyramid that does not have to be a triangle

biased sample a sample that is not representative of a population

biconditional a logical statement containing the phrase "if and only if" (iff); both the statement and its converse are true

binomial an expression consisting of two terms connected by a plus or minus sign (i.e., $4a + 6$)

bisect to divide into two congruent parts

box and whisker plot a graph that uses a rectangle to represent the middle 50% of a set of data and line segments (or whiskers) where each represents 25% of the data; a line segment representing the median value divides the rectangle so that each section represents 25% of the data

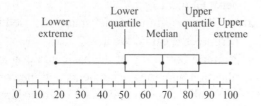

calculation action, process, or result of a mathematical computation

capacity a measure of how much (volume) a container can hold

causation an action that produces an effect

Celsius metric measurement of temperature (i.e., 32 degrees Celsius, 32°C)

census data collected from every member of the identified population

centimeter a metric unit of length equivalent to $\frac{1}{100}$ of a meter

chord of a circle a segment joining any two points on the circle

circle a set of points in a plane equidistant from a given point called the center

circle graph a graph in which a circle is divided into sectors in order to compare different parts of a data set to the entire set (i.e., pie graph)

circumference the perimeter of a circle

closure property a set is closed under an operation if the application of the operation of any members in the set always results in a member of that set

coefficient the numerical factor in an algebraic term (i.e., in $7x$, 7 is the coefficient)

collinear a set of points is said to be collinear if they lie on a single straight line

combinations a group of unordered items or events taken from a larger group (i.e., the number of three-person committees that can be chosen from a group of 21)

common denominator any nonzero number that is a multiple of the denominators of two or more fractions

common factor any number that is a factor of two or more numbers (i.e., 4 is a common factor of 8 and 12)

common multiple a term that contains two or more terms as factors

commutative property the property in addition and multiplication that states the order in which two terms are added or multiplied does not change the results. For real numbers a and b, $a + b = b + a$ and $ab = ba$

complementary angles two angles, the sum of whose measure is 90°

complex fraction a fraction that contains one or more fractions in the numerator or denominator

complex number a number that can be written in the form $a + bi$, where a and b are real numbers and i is the imaginary number; $i = \sqrt{-1}$

composite number a number that has more than two numerical factors

concave polygon a polygon with one or more diagonals that have points outside the polygon

conclusion the *then* clause of a conditional statement

conditional statement a statement in "if-then" form, where the "if" clause is called the hypothesis and the "then" clause is called the conclusion

cone a three-dimensional figure generated by rotating a right triangle about one of its legs

concrete objects physical objects used to represent mathematical situations

congruent coinciding exactly when superimposed

conjecture an unproven statement based on observations

consecutive in order, with nothing missing

consecutive exterior angles angles formed by a transversal intersecting two lines; angles on the same side of the transversal, having two different vertices and outside the two lines; if the two lines are parallel, the same side exterior angles are supplementary

consecutive interior angles angles formed by a transversal intersecting two lines; angles on the same side of the transversal, having two different vertices, and inside the two lines; if the two lines are parallel, the same side interior angles are supplementary

constant a quantity that always stays the same

construct a conclusion or result built or put together systematically

contextual situation relating a mathematical problem to a real modeled or illustrated circumstance

continuous data data in which there are no gaps, jumps, or holes; data that can be measured and broken down into smaller parts and still have meaning; temperature and time data are continuous

contrapositive of a statement a new statement obtained by exchanging the negation of the conclusion with the negation of the hypothesis of a conditional statement

converse of a statement a new statement obtained by exchanging the hypothesis and the conclusion of a conditional statement

convex polygon a polygon with each interior angle measuring less than 180°; all diagonals of a convex polygon lie inside the polygon

coordinate system (Cartesian) a two-dimensional system in which the coordinates of a point are its distances from the origin, the intersection of the x and y axes

coordinates of a point an ordered pair of real numbers that locates a point in a plane

coplanar in the same plane

correlation an association between two variables

corresponding angles angles formed by a transversal intersecting two lines; angles on the same side of the transversal, having two different vertices and in the same relative position; if the two lines are parallel, the corresponding angles are congruent

cosine in a right triangle, the ratio of the length of the leg adjacent to an acute angle to the length of the hypotenuse

counterexample an example that shows that a conjecture is not always true

counting numbers the set of numbers consisting of 1, 2, 3, 4, 5, 6, ... (natural numbers)

cube the third power of a number; a regular three-dimensional figure having six congruent square faces

customary system of measurement the measuring system used most often in the United States (i.e., inches, pounds, gallons)

cylinder a three-dimensional figure composed of two congruent and parallel circular regions joined by a curved surface

data information gathered by observation, questioning, or measurement, usually expressed with numbers

data sets a defined group of information, especially numerical

decimal number system a place-value number system based on groupings by powers of ten

decimal point the point used to write values less than one in the base ten number system

deductive reasoning a series of logical steps in which a conclusion is drawn directly from a set of statements (premises) that are assumed to be true

degree a unit of measure for angles based on dividing a circle into 360 equal parts; or a unit of measure for temperature

denominator the number of equal parts into which a whole is divided (i.e., in the fraction ¾, 4 is the denominator)

density property between any pair of rational numbers there is another number

dependent events two events in which the outcome of the second event is affected by the outcome of the first event

dependent variable in a function, the variable that is determined by the value of the related independent variable

descending an order in which numbers or terms are organized in decreasing value

diagonal a line segment joining two nonadjacent vertices of a polygon

diameter a chord that contains the center of the circle

difference the result of a subtraction

digit in the base ten numeration system, one of the symbols 0, 1, 2, 3, 4, 5, 6, 7, 8, 9

digital clock a device for telling time, makes jumps from number to number (digital clocks usually use numbers with a colon separating the hour from the minutes, 6:30)

dilation a transformation that either enlarges or reduces a geometric figure proportionately

dimension a measure in one direction (i.e., length or width)

discrete data involves a count of data items that can't be broken down into smaller units, such as number of defects, people, or items

discrete mathematics the study of mathematics dealing with objects that can assume only certain "discrete" values; discrete objects can be characterized by integers, whereas continuous objects require real numbers

dissection to separate into parts, usually equal

distance the length of the shortest line segment joining two points

distance formula a formula used to find the distance between two points identified by their ordered pairs: $d = \sqrt{(x_2 - x_1)^2 + (y_2 - y_1)^2}$

distortions of sets of data the use of incorrect proportion, design variation in comparing to sets of data, lack of context, or insignificant data used in direct comparison with meaningful data

distributive property the distributive property of multiplication over addition or subtraction is a multiplication of a group of terms such that the multiplier is multiplied by each and every term in the group a, b, and c, $a(b + c) = ab + ac$ and $a(b - c) = ab - ac$

dividend in a division problem, the quantity to be divided

divisibility one whole number is divisible by another whole number if the result of the division is a whole number without a remainder

division a mathematical operation based on separating into equal parts

divisor in a division problem, the quantity by which another quantity is divided

domain the set of values for the independent variable of a function (i.e., usually, the x-values of a function)

edge of a polyhedron a line segment where two faces of a polyhedron meet

edge (vertex-edge graph) the path that joins two vertices

elapsed time time between two events

ellipsis the mark ". . ." to indicate the continuance of a pattern

empty set a set that contains no elements

endpoint the point at either end of a line segment; also, the initial point of a ray

equation a mathematical sentence in which equivalent values are separated by an equal sign

equivalent equal in value, but in a different form

equilateral triangle a triangle with three congruent sides

estimate a close rather than exact answer

evaluate to find the numerical value of a mathematical expression

even number an integer that is divisible by two without a remainder

event one of the many occurrences that can take place during a probability activity

expanded notation a way to write numbers that shows the place value of each digit. (i.e., 343 = 300 + 40 + 3)

experimental (empirical) probability relating to the outcomes of an actual performance of a probability activity

exponent a number placed to the right of and above a nonzero base that indicates how many times the base is used as a factor; a base with a zero exponent is equal to 1 (i.e., $5^0 = 1$, $5^3 = 5 \cdot 5 \cdot 5$ and $5^{-3} = \dfrac{1}{5^3} = \dfrac{1}{5 \cdot 5 \cdot 5}$)

exponential function a function commonly used to study growth and decay; it has a form $y = a^x$

expression a mathematical phrase containing one or more terms linked by operation symbols

extremes given the proportion $\dfrac{a}{b} = \dfrac{c}{d}$, a and d are the extremes

face of a polyhedron a flat surface on a three-dimensional object

fact family a collection of related addition and subtraction facts, or multiplication and division facts, made from the same numbers

factor (noun) a number or expression that evenly divides another quantity (i.e., 4 is a factor of 12; and $(x + 1)$ is a factor of $x^2 + 3x + 2$)

factor (verb) to represent a number as a product of factors

factorial the product of the given number and every natural number less than itself

Fahrenheit the customary scale system for temperature measurement (32°F)

finite set a set that contains a countable number of elements

formula a general mathematical rule using variables

fractal an algebraically generated complex geometric shape having the property of being endlessly self-similar under magnification

fraction   a number in the form $\frac{a}{b}$, where b is not zero

fractional part   part of a whole or part of a group that is less than a whole

frequency table   a collection of data that specifies the number of occurrences in each of several categories

function   (input–output) a dependent relationship between two sets of numbers in which a value in the first set determines one and only one element in the second set

geometric model   a model of mathematical concepts using geometric representations

geometric sequence   a set of ordered terms in which the ratio between consecutive terms is constant

geometric solid   a three-dimensional shape bounded by surfaces (i.e., rectangular prism, pyramid, cylinder, cone, and sphere)

graph   a pictorial device that shows a relationship between variables or sets of data

greatest common factor   largest factor that two or more numbers have in common (GFC) (i.e., the GFC of 8 and 12 is 4)

grouping symbols   symbols of inclusion; parentheses, brackets, braces, or bars (i.e., ( ) , [ ] , {} , $\overline{}$)

height   the perpendicular distance to a base from a vertex or between bases

hexagon   a polygon with six sides

histogram   a vertical bar graph with each bar representing a certain interval of data

horizontal   parallel to or in the plane of the horizon; in a coordinate grid, the x-axis is a horizontal line

hypotenuse   the side opposite the right angle in a right triangle

hypothesis   the *if* clause of a conditional statement

identity element   a number when used in an operation with a given number leaves the given number unchanged; the identity element for addition is 0; the identity element for multiplication is 1

image   a figure created as the result of a transformation

imaginary numbers   the square root of a negative number expressed using i; $i = \sqrt{-1}$

improper fraction   a fraction in which the numerator is greater than the denominator

independent events   two events in which the outcome of the second event does not relate to the outcome of the first event

indirect proof   a deductive proof using contradiction or elimination to rule out all possible conclusions except the desired one

inductive reasoning   making a generalization based on observation of specific cases or patterns (i.e., formulating a rule after considering several parts of a pattern)

inequality   a statement indicating that two quantities are not equal

inference   a conclusion drawn from data

infinite set   the set in which the number of elements cannot be counted or determined (never ending)

inscribed angles   an angle with its vertex on the circle and with sides that are chords of the circle

integers   the set of numbers consisting of the whole numbers and their opposites . . . $-2, -1, 0, 1, 2 \ldots$

interval   the set of numbers between two numbers a and b; the interval may include a or b

inverse operation   a related but opposite process (i.e., multiplication is the inverse of division)

inverse of a statement   a new statement obtained by negating both the hypothesis and the conclusion of a conditional statement

irrational numbers   a set of numbers that cannot be expressed as a ratio of two integers (i.e., $\pi, \sqrt{2}$)

isosceles triangle   a triangle that has at least two congruent sides

iterative pattern   a pattern generated by using an initial value and repeatedly applying an operation (i.e., 4, 7, 10, 13 is adding 3 each time)

kite   a quadrilateral with two distinct pairs of adjacent, congruent sides

lateral surface   in a prism or a pyramid, it is the face that is not a base

least common multiple   the smallest number for which two or more numbers are factors (i.e., the LCM of 3, 4, and 6 is 12)

line   an undefined geometric term; a straight path that extends infinitely in opposite directions; a line that has no thickness

line graph   a graph in which points are connected by line segments to represent data

line of best fit a line drawn on a scatter plot to estimate the relationship between two sets of variables in a set of data

line of symmetry a line that divides a figure into two congruent parts that are mirror images of each other

line plots a sketch of data in which check marks, x's, or other marks above a number line shows the frequency of each value

line segment a part of a line that consists of two points, called endpoints, and all the points between them

linear equation a polynomial equation containing one or more terms in which the variable is raised to the power of one but no higher

linear function a function that has a constant rate of change and can be modeled by a straight line

liter a metric unit of capacity, equal to the volume of a cube that measures 10 centimeters on a side

logic a system of reasoning used to validate arguments

lowest common denominator the least common multiple of the denominators of every fraction in a given collection of fractions

magnitude size or quantity

manipulatives a wide variety of physical materials, objects, and supplies that students use to foster the learning of abstract ideas in mathematics

matrix a rectangular array of numbers or letters arranged in rows and columns (matrices)

mass matter within an object

maximum the greatest value

mean a measure of central tendency where the sum of a set of numbers is divided by the number of elements in the set; often referred to as the average

means given the proportion $\frac{a}{b} = \frac{c}{d}$, b and c are the means

means-extremes product rule given the proportion $\frac{a}{b} = \frac{c}{d}$, $ad = bc$

measures of central tendency numbers that communicate the "center" or "middle" of a set of data. The mean, median, and mode are statistical measures of central tendency

median a measure of central tendency that identifies a value such that half the data is above the value and half the data is below the value when the data is listed in order

metric system of measurement a measurement system based on the base-10 numeration system (i.e., meter, liter, gram)

midpoint a point on a geometric figure halfway between two points

minimum the least value

mixed number a number that is equal to the sum of a whole number and a fraction

mode a measure of central tendency that is the value or values that occur most frequently in a given set of numbers

model (noun) a representation of concrete materials, objects, or drawings

model (verb) use of concrete materials and the use of the symbolic

monomial an expression consisting of a single term (i.e., $5y$)

multiple of a number a number into which the given number may be divided with no remainder

multiplication the operation of repeated addition (i.e., 4×3 is the same as $4 + 4 + 4$)

natural numbers the set of counting numbers consisting of 1, 2, 3, 4, 5, 6 . . .

negative number a number less than zero

net of a polyhedron a two-dimensional representation of the surface of a three-dimensional figure that has been unfolded

normal curve in statistics, the distribution of data along a bell-shaped curve that reaches its maximum height at the mean

normal distribution a bell-shaped probability distribution; there are as many values that are less than the mean as there are values that are greater than the mean

number line a diagram that represents numbers as points on a line with a uniform scale

number sentence an equation or inequality with numbers

numerator the number or expression written above the line in a fraction; it tells how many equal parts of a total number of parts are described by a fraction

obtuse angle an angle whose measure is greater than 90° and less than 180°

octagon a polygon with eight sides

odd number an integer that is not divisible by 2

open sentence a statement that contains at least one unknown (i.e., $6 + x = 14$)

operation an action performed on some set of quantities (i.e., addition, raising to a power)

order of operations the sequence in which operations are performed when evaluating an expression

ordered pair a pair of numbers used to locate points in the coordinate plane

ordinal number a whole number that names the position of an object in a sequence

origin the intersection of the x- and y-axes in a coordinate plane; the origin is described by the ordered pair $(0, 0)$

outcome one of the possible events in a probability situation

outcome set set of all outcomes of a given situation

outliers/extreme values piece of numerical data that is significantly larger or smaller than the rest of the data in a set

parallel lines lines in the same plane that never intersect and are always the same distance apart

parallelogram a quadrilateral with opposite sides parallel and congruent

pattern a set or sequence of shapes or numbers that are repeated in a predictable manner

pentagon a polygon with five sides

percent a ratio that compares a number to 100 (%)

perfect square a whole number whose square root is a whole number

perimeter the distance around a shape or figure

permutation an ordered arrangement of a set of events or items (if you put the items or events into a different order, you have a different permutation)

perpendicular lines two lines that intersect to form right angles

pi the ratio of the circumference of a circle to its diameter. Pi (π) is an irrational number and approximately equal to 3.14 or $\dfrac{22}{7}$

pictograph a graph that uses pictures or symbols to represent data

place value the value of the position of a digit in a numeral

plane an undefined geometric term; a flat surface that extends infinitely in all directions

point an undefined geometric term; denotes an exact location in space; a point has no size

polygon a closed two-dimensional figure made up of segments, called sides, which intersect only at their endpoints, called vertices

polyhedron a closed three-dimensional figure in which all the surfaces are polygons

polynomial an expression consisting of two or more terms

population in statistics, an entire set of objects, observations, or scores that have something in common

postulate a mathematical statement that is accepted as true without proof

predictions use of base information to produce an approximation of change or result

pre-image a picture or object before it undergoes a transformation

premise a statement that is given to be true

prime number a positive integer that has exactly two different positive factors, itself and 1; 1 is not a prime number

prime factorization a composite number expressed as the product of factors that are prime numbers

prism three-dimensional figures that have two congruent and parallel faces that are polygons; the remaining faces are parallelograms

probability the measure of the likelihood of an event occurring

product the result of multiplication

proof a logical argument that shows why a statement must be true

proper fraction a fraction whose numerator is an integer smaller than its integral denominator

properties of operations mathematical principals that are always true (i.e., commutative, associative, distributive, and inverse properties)

proportion the statement of equality between two ratios

proportionality the concept of having equivalent ratios

pyramid a three-dimensional figure whose base is a polygon and whose other faces are triangles that share a common vertex

Pythagorean theorem in a right triangle, the sum of the squares of the lengths of the legs is equal to the square of the length of the hypotenuse ($a^2 + b^2 = c^2$)

quadrant one of the four sections into which the coordinate plane is divided by the x- and y-axes

quadratic equation a polynomial equation containing one or more terms in which the variable is raised to the second power but no higher

quadratic formula the formula used to find the roots of quadratic equations;

If $ax^2 + bx + c = 0$, $a \neq 0$, then $x = \dfrac{-b \pm \sqrt{b^2 - 4ac}}{2a}$.

quadratic function a function that has an equation of the form: $f(x) = ax^2 + bx + c$, $a \neq 0$; a function with a degree of two

quadrilateral a polygon with four sides

quartiles the quartiles divide an ordered set of data into four groups of the same size

quotient the result of division of one quantity by another (dividend ÷ divisor = quotient)

radius of a circle a segment whose endpoints are the center of the circle and a point on the circle (radii)

random sample each item or element of the population has an equal chance of being chosen as part of a sample of the population

range the set of output values for a function

range (of data set) the difference between the greatest and least number in a set of numbers

rate a ratio comparing two different units (i.e., miles per hour or cents per pound)

ratio a comparison of two values by division; a ratio can be expressed as a to b, $\dfrac{a}{b}$, or $a{:}b$

rational number a number that can be expressed as a ratio of two integers

ray a geometric figure that extends infinitely along a straight path from a point, called its endpoint

real numbers the set of numbers combining rational and irrational numbers

reasonable estimations approximations based on mathematical reasoning that are within the desired degree of accuracy (e.g., $35 + 43 = x$ a reasonable estimation for x would be 75 or 80, not 100 or 700)

reciprocals two numbers whose product is equal to one; the numbers are multiplicative inverses

rectangle a quadrilateral with two pairs of congruent, parallel sides and four right angles (square, parallelogram, quadrilateral, polygon)

recursive pattern a pattern that uses the solution from previous steps to generate the solution to the next step. (i.e., 2, 2, 4, 6, 10, 16 . . .)

reflection a transformation creating a mirror image of a figure on the opposite side of a line

reflex angle an angle that is greater than 180° and less than 360°

reflexive property the property that states that a quantity is equal to itself; the property that states that an object is congruent to itself

regular polygon a convex polygon in which the angles are equiangular and sides are equilateral

repeating decimal a decimal in which one or more digits repeat without termination

rhombus a parallelogram with four congruent sides (plural: rhombi)

right angle an angle whose measure is 90°

right triangle a triangle that contains a right angle

root the inverse of a power

rotation a transformation in which a figure is turned a given angle and direction around a point

rounding approximating a number by analyzing a specific place value

sample a part of the total population; used in statistics to make predictions about the characteristics of the entire group

sample space a list of all possible outcomes of an activity

scale (1) an instrument used for weighing; (2) a system of marks at fixed intervals used in measurement or graphing

scale factor the ratio between the lengths of corresponding sides of two similar figures

scalene triangle a triangle with no sides the same length and no congruent angles

scatter plot a graph of the points representing a collection of data

scientific notation a form of writing a number expressed as a power of 10 and a decimal number greater than or equal to 1 and less than 10

secant a line that intersects a circle at exactly two points; a line that contains a chord of a circle

sector a region defined by a central angle and an arc

signed number a positive or negative number

similar figures figures that are the same shape but not necessarily the same size

sine in a right triangle, the ratio of the length of the leg opposite the given angle to the hypotenuse

skip counting counting by equal intervals (i.e., 2, 4, 6 . . . or 4, 8, 12 . . .)

slope of a line the measure of steepness of a line; the ratio of rise over run, or the change in y over the change in x

solid a three-dimensional figure

solution a value for a variable that makes an equation or inequality true

solution set a set consisting of all values that make an equation or inequality true

space the set of all possible points

sphere a three-dimensional figure made up of all points in space equidistant from a given point called the center

square a parallelogram with four congruent sides and four right angles

square root one of the two equal factors of a number

standard notation a number written with one digit for each place value in base 10; the most familiar way of representing whole numbers, integers, and decimals is standard notation (e.g., three hundred fifty six is 356)

statistics the collection, organization, description, and analysis of data; statistics are quantitative data

stem-and-leaf plot a display of data in which digits with larger place values (10's) are "stems" and digits with smaller place values (1's) are "leaves"

straight angle an angle whose measure is 180°; it is formed by two opposite rays

subscript a number written to the right of and slightly below a term; usually used for indexing

substitution property the property that allows equal values to replace each other

subtraction a mathematical operation that gives the difference between two numbers; subtraction also is used to compare two numbers or sets

sum the result of an addition

supplementary angles two angles the sum of whose measures is 180°

surface area the total area of the faces (including the bases) and curved surfaces of a three-dimensional figure

symbol a sign or token used to represent something, such as an operation, quantity, or relation

symmetric property the property that states for real numbers a and b, if $a = b$, then $b = a$

symmetry a correspondence in size, form, and arrangement of parts, related to a plane, line, or point; for example, a figure that has line symmetry has two halves that coincide if folded along a line of symmetry

system of equations a set of two or more equations with the same number of unknowns

tangent in a right triangle, the ratio of the length of the leg opposite an acute angle to the leg adjacent to the acute angle

tangent to a circle a line in the plane of a circle that touches a circle in exactly one point (tangent line)

t-chart a mathematical organizer to display and record data, patterns, or functions/rules in an organized way

term a product or quotient of numerals or variables or both; terms are separated by plus or minus signs in an expression

terminating decimal a decimal that contains a finite number of digits

tessellation a covering of a plane without overlaps or gaps, using combinations of congruent figures

theorem a mathematical statement or proposition derived from previously accepted results

theoretical probability the probability of an event without doing an experiment or analyzing data

transformation an operation that creates an image from an original figure or pre-image

transitive property the property that states for real numbers a, b, and c, if $a = b$ and $b = c$, then $a = c$, or if $a > b$ and $b > c$ then, $a > c$, or if $a < b$ and $b < c$, then $a < c$

translation a transformation that moves every point on a figure a given distance in a given direction

transversal a line that intersects two or more lines in a plane at different points

trapezoid a quadrilateral that has exactly one pair of parallel sides

tree diagram a branched diagram used to find all the possible permutations for a set of items or the prime factorization of a number of a number

trend the general drift, tendency, or direction of data

trend line a line that represents a general pattern for a set of data

triangle a polygon with three sides

trigonometric ratios   the ratios of the lengths of pairs of sides in a right triangle (i.e., sine, cosine, and tangent)

unit fraction   a fraction with a numerator of 1

unit price   the price of something for one unit of measure

valid argument   an argument that is correctly inferred or deduced from a premise

variability   numbers that describe how spread out a set of data is (i.e., range and quartile)

variable   a symbol that represents a quantity

Venn diagram   a representation that uses circles to show relationships between sets

vertex   the point at which the rays of an angle, two sides of a polygon, or the edges of a polyhedron meet (vertices)

vertex-edge graph   a structure consisting of vertices and edges, where the edges indicate a mapping among the vertices (e.g., the vertices may represent players in a tournament, and the edges indicate who plays whom)

vertical   at right angles to the horizon

vertical angles   the opposite angles formed when two lines intersect

volume   the measure of the capacity of a three-dimensional figure, measured in cubic units

whole   the entire object, collection of objects, or quantity being considered

whole numbers   the set of numbers consisting of the counting numbers and zero (i.e., 0, 1, 2, 3 . . .)

x-intercept   the coordinate at which the graph of a line intersects the x-axis

y-intercept   the coordinate at which the graph of a line intersects the y-axis

Arizona AIMS Mathematics Standard

Strand 1: Number Sense and Operations

Concept 1: Number Sense

PO 1   Classify real numbers as members of one or more of subsets of the real number system: natural, whole, integers, rational, or irrational numbers.

PO 2   Identify properties of the real number system: commutative, associative, distributive, identity, inverse, and closure.

PO 3   Distinguish between finite and infinite sets of numbers.

Concept 2: Numerical Operations

PO 1   Select the grade-level-appropriate operation to solve word problems.

PO 2   Solve word problems using grade-level-appropriate operations and numbers.

PO 3   Simplify numerical expressions including those containing signed numbers and absolute values.

PO 4   Apply subscripts to represent ordinal position.

PO 5   Use grade-level-appropriate mathematical terminology.

PO 6   Compute using scientific notation.

PO 7   Simplify numerical expressions using the order of operations.

Concept 3: Estimation

PO 1   Solve grade-level-appropriate problems using estimation.

PO 2   Determine if a solution to a problem is reasonable.

PO 3   Determine rational approximations of irrational numbers.

Strand 2: Data Analysis, Probability, and Discrete Math

Concept 1: Data Analysis (Statistics)

PO 1   Formulate questions to collect data in contextual situations.

PO 2   Organize collected data into an appropriate graphical representation.

PO 3   Display data as lists, tables, matrices, and plots.

PO 4   Construct equivalent displays of the same data.

PO 5   Identify graphic misrepresentations and distortions of sets of data.

PO 6   Identify which of the measures of central tendencies is most appropriate in a given situation.

PO 7   Make reasonable predictions based upon linear patterns in data sets or scatter plots.

PO 8   Make reasonable predictions for a set of data, based on patterns.

PO 9   Draw inferences from charts, tables, graphs, plots, or data sets.

PO 10 Apply the concepts of mean, median, mode, range, and quartiles to summarize data sets.

PO 11 Evaluate the reasonableness of conclusions drawn from data analysis.

PO 12 Recognize and explain the impact of interpreting data (making inferences or drawing conclusions) from a biased sample.

PO 13 Draw a line of best fit for a scatter plot.

PO 14   Determine whether displayed data has positive, negative, or no correlation.
PO 15   Identify a normal distribution.
PO 16   Identify differences between sampling and census.
PO 17   Identify differences between biased and unbiased samples.

Concept 2: Probability

PO 1   Find the probability that a specific event will occur, with or without replacement.
PO 2   Determine simple probabilities related to geometric figures, given necessary formulas.
PO 3   Predict the outcome of a grade-level-appropriate probability experiment.
PO 4   Record the data from performing a grade-level-appropriate probability experiment.
PO 5   Compare the outcome of an experiment to predictions made prior to performing the experiment.
PO 6   Distinguish between independent and dependent events.
PO 7   Compare the results of two repetitions of the same grade-level-appropriate probability experiment.

Concept 3: Discrete Mathematics: Systematic Listing & Counting

PO 1   Determine the number of possible outcomes for a contextual event using a chart, a tree diagram, or the counting principle.
PO 2   Determine when to use combinations versus permutations in counting objects.
PO 3   Use combinations and permutations to solve contextual problems.

Strand 3: Patterns, Algebra, and Functions

Concept 1: Patterns

PO 1   Communicate a grade-level-appropriate iterative or recursive pattern, using symbols or numbers.
PO 2   Find the nth term of an iterative or recursive pattern.
PO 3   Evaluate problems using simple or basic recursion formulas.

Concept 2: Functions and Relationships

PO 1   Determine if a relationship is a function, given a graph, table, or set of ordered pairs.
PO 2   Describe a contextual situation that is depicted by a given graph.
PO 3   Identify a graph that models a given real-world situation.
PO 4   Sketch a graph that models a given contextual situation.
PO 5   Determine domain and range for a function that represents a contextual situation.
PO 6   Determine the solution to a contextual maximum/minimum problem, given the graphical representation.
PO 7   Express the relationship between two variables using tables/matrices, equations, or graphs.
PO 8   Interpret the relationship between data suggested by tables/matrices, equations, or graphs.
PO 9   Determine from two linear equations whether the lines are parallel, perpendicular, coincident, or intersecting but not perpendicular.

Concept 3: Algebraic Representations

PO 1   Evaluate algebraic expressions, including absolute value and square roots.

PO 2   Simplify algebraic expressions.

PO 3   Multiply and divide monomial expressions with integral exponents.

PO 4   Translate a written expression or sentence into a mathematical expression or sentence.

PO 5   Translate a sentence written in context into an algebraic equation involving multiple operations.

PO 6   Write a linear equation for a table of values.

PO 7   Write a linear algebraic sentence that represents a data set that models a contextual situation.

PO 8   Solve linear equations in one variable (including absolute value).

PO 9   Solve linear inequalities in one variable.

PO 10   Write an equation of the line given: two points on the line, the slope and a point on the line, or the graph of the line.

PO 11   Solve an algebraic proportion.

PO 12   Solve systems of linear equations in two variables (integral coefficients and rational solutions).

PO 13   Add, subtract, and perform scalar multiplication with matrices.

PO 14   Calculate powers and roots of real numbers, both rational and irrational, using technology when appropriate.

PO 15   Simplify square roots and cube roots with monomial radicands (including those with variables) that are perfect squares or perfect cubes.

PO 16   Solve square root radical equations involving only one radical.

PO 17   Solve quadratic equations.

PO 18   Identify the sine, cosine, and tangent ratios of the acute angles of a right triangle.

Concept 4: Analysis of Change

PO 1   Determine slope and x- and y-intercepts of a linear equation.

PO 2   Solve formulas for specified variables.

Strand 4: Geometry and Measurement

Concept 1: Geometric Properties

PO 1   Identify the attributes of special triangles (isosceles, equilateral, right).

PO 2   Identify the hierarchy of quadrilaterals.

PO 3   Make a net to represent a three-dimensional object.

PO 4   Make a three-dimensional model from a net.

PO 5   Draw two-dimensional and three-dimensional figures with appropriate labels.

PO 6   Solve problems related to complementary, supplementary, or congruent angle concepts.

PO 7   Solve problems by applying the relationship between circles, angles, and intercepted arcs.

PO 8   Solve problems by applying the relationship between radii, diameters, chords, tangents, and secants.

PO 9   Solve problems using the triangle inequality property.

PO 10   Solve contextual problems using special-case right triangles.

PO 11   Determine when triangles are congruent by applying SSS, ASA, AAS, or SAS.

PO 12   Determine when triangles are similar by applying SAS, SSS, or AAA similarity postulates.

PO 13   Construct a triangle congruent to a given triangle.

PO 14   Solve contextual situations using angle and side length relationships.

Concept 2: Transformation of Shapes

PO 1   Sketch the planar figure that is the result of two or more transformations.

PO 2   Identify the properties of the planar figure that is the result of two or more transformations.

PO 3   Determine the new coordinates of a point when a single transformation is performed on a planar geometric figure.

PO 4   Determine whether a given pair of figures on a coordinate plane represents a translation, reflection, rotation, or dilation.

PO 5   Classify transformations based on whether they produce congruent or similar figures.

PO 6   Determine the effects of a single transformation on linear or area measurements of a planar geometric figure.

Concept 3: Coordinate Geometry

PO 1   Graph a quadratic equation with lead coefficient equal to one.

PO 2   Graph a linear equation in two variables.

PO 3   Graph a linear inequality in two variables.

PO 4   Determine the solution to a system of equations in two variables from a given graph.

PO 5   Determine the midpoint between two points in a coordinate system.

PO 6   Determine the changes in the graph of a linear function when constants and coefficients in its equation are varied.

PO 7   Determine the distance between two points in the coordinate system.

Concept 4: Measurement

- Units of Measure
- Geometric Objects

PO 1   Calculate the area of geometric shapes composed of two or more geometric figures.

PO 2   Calculate the volumes of three-dimensional geometric figures.

PO 3   Calculate the surface areas of three-dimensional geometric figures.

PO 4   Compare perimeter, area, and volume of two- and three-dimensional figures when dimensions are changed.

PO 5   Find the length of a circular arc.

PO 6   Find the area of a sector of a circle.

PO 7   Solve for missing measures in a pyramid (i.e., height and slant height).

PO 8   Find the sum of the interior and exterior angles of a polygon.

PO 9   Solve scale factor problems using ratios and proportions.

PO 10   Solve applied problems using similar triangles.

Strand 5: Structure and Logic

Concept 1: Algorithms and Algorithmic Thinking

PO 1   Determine whether a given procedure for simplifying an expression is valid.

PO 2   Determine whether a given procedure for solving an equation is valid.

PO 3   Determine whether a given procedure for solving a linear inequality is valid.

PO 4   Select an algorithm that explains a particular mathematical process.

PO 5   Determine the purpose of a simple mathematical algorithm.

PO 6   Determine whether given simple mathematical algorithms are equivalent.

Concept 2: Logic, Reasoning, Arguments, and Mathematical Proof

PO 1   Draw a simple valid conclusion from a given *if . . . then* statement and a minor premise.
PO 2   List related *if . . . then* statements in logical order.
PO 3   Write an appropriate conjecture given a certain set of circumstances.
PO 4   Analyze assertions related to a contextual situation by using principles of logic.
PO 5   Identify a valid conjecture using inductive reasoning.
PO 6   Distinguish valid arguments from invalid arguments.
PO 7   Create inductive and deductive arguments concerning geometric ideas and relationships, such as congruence, similarity, and the Pythagorean relationship.
PO 8   Critique inductive and deductive arguments concerning geometric ideas and relationships, such as congruence, similarity, and the Pythagorean relationship.
PO 9   Identify a counterexample for a given conjecture.
PO 10   Construct a counterexample to show that a given conjecture is false.
PO 11   State the inverse, converse, or contrapositive of a given statement.
PO 12   Determine if the inverse, converse, or contrapositive of a given statement is true or false.
PO 13   Construct a simple formal or informal deductive proof.
PO 14   Verify characteristics of a given geometric figure using coordinate formulas such as distance, midpoint, and slope to confirm parallelism, perpendicularity, and congruency.

NCTM Process Standard

Strand 1: Communication
Organize and consolidate thinking
Express thinking in a clear and coherent way
Analyze and evaluate others orally
Use mathematical vocabulary
Additional practices
    Use inquiry and questioning
    Use "argument" and justification
    Use discourse to deepen understanding of concepts
    Use prior knowledge
    Use information in context

Strand 2: Connections
Recognize the connections of mathematical ideas
Understand the interconnections
Build and synthesize the "whole"
Recognize and apply mathematics in various contexts
Spiral through and make connections within the five strands of mathematics

Strand 3: Problem Solving
Build new mathematical knowledge
Solve mathematical problems in many contexts
Use a variety of strategies
Utilize prior knowledge and prior use of strategies
Monitor and reflect on the processes

Strand 4: Reasoning and Proof
Use the fundamentals of mathematics
Make and investigate conjectures
Develop and evaluate arguments and proofs
Select various reasoning and methods of proof

Strand 5: Representation
Create and use data displays, models, tables, graphs, pictures, or charts to organize and record
Select, apply, and translate to solve problems
Model and interpret physical, social, and mathematical phenomena

AIMS Mathematics Blueprint for High School

	Concept 1	Concept 2	Concept 3	Concept 4	Total by Strand
Strand 1	5%	5%	4%	NA	14%
Strand 2	9%	5%	5%	NA	19%
Strand 3	5%	7%	14%	5%	31%
Strand 4	9%	5%	7%	6%	27%
Strand 5	5%	4%	NA	NA	9%

NA means there are no Concept numbers for that Strand.

Total by strand refers to the number of questions out of 100 on the AIMS test that will be related to that Strand.

Index

387

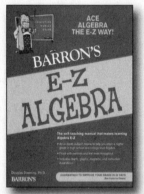
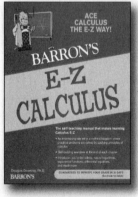

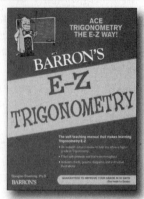
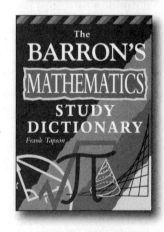